Cannabinoid Receptors

NEUROSCIENCE PERSPECTIVES

Series Editor
Peter Jenner
Pharmacology Group
Biomedical Sciences Division
King's College London
Manresa Road
London SW3 6LX

Titles in this series
Roger Horton and Cornelius Katona (eds), Biological Aspects of Affective Disorders
Trevor Stone (ed), Adenosine in the Nervous System
Judith Pratt (ed), The Biological Bases of Drug Tolerance and Dependence
Michel Hamon (ed), Central and Peripheral 5-HT_3 Receptors
John Waddington (ed), $D_1 : D_2$ Dopamine Receptor Interactions
Thomas Barnes (ed), Antipsychotic Drugs and Their Side-Effects
Eva Giesen-Crouse (ed), Peripheral Benzodiazepine Receptors
David Nicholson (ed), Anti-dementia Agents
Alan Harvey (ed), Natural and Synthetic Neurotoxins
John Wood (ed), Capsaicin in the Study of Pain
Yossef Itzhak (ed), Sigma Receptors
John Mayer and Ian Brown (eds), Heat Shock Proteins in the Nervous System
Chris Carter (ed) The Neuropharmacology of Polyamines
Steven Vincent (ed) Nitric Oxide in the Nervous System

Cannabinoid Receptors

edited by

R. G. Pertwee

Department of Biomedical Sciences
Marischal College, University of Aberdeen
Aberdeen, Scotland

ACADEMIC PRESS
Harcourt Brace & Company, Publishers
London San Diego New York
Boston Sydney Tokyo Toronto

12049018.

ACADEMIC PRESS LIMITED
24/28 Oval Road,
London NW1 7DX

United States Edition published by
ACADEMIC PRESS INC.
San Diego, CA 92101

This book is printed on acid free paper

A catalogue record for this book
is available from the British Library

ISBN 0-12-551460-3

Typeset by Servis Filmsetting Ltd, Manchester, UK
Printed and bound in Great Britain
by T J Press, Padstow, Cornwall

Contents

Contributors

Tom I. Bonner Laboratory of Cell Biology, National Institute of Mental Health, Bethesda, MD 20892, USA.

Steven R. Childers Center for the Neurobiological Investigation of Drug Abuse, Department of Physiology and Pharmacology, Bowman Gray School of Medicine, Wake Forest University, Winston, Salem, NC 27157, USA

Sam A. Deadwyler Center for the Neurobiological Investigation of Drug Abuse, Department of Physiology and Pharmacology, Bowman Gray School of Medicine, Wake Forest University, Winston, Salem, NC 27157, USA

Ester Fride Department of Natural Products, Hebrew University, Medical Faculty, Jerusalem 91120, Israel

Robert E. Hampson Center for the Neurobiological Investigation of Drug Abuse, Department of Physiology and Pharmacology, Bowman Gray School of Medicine, Wake Forest University, Winston, Salem, NC 27157, USA

Miles Herkenham Section on Functional Neuroanatomy, Bldg 36, Rm 2D15, NIMH, Bethesda, MD 20892, USA

Allyn C. Howlett Department of Pharmacological and Physiological Science, Saint Louis University School of Medicine, 1402 South Grand Boulevard, St. Louis, MO 63104 USA

Alexandros Makriyannis School of Pharmacy, Institute of Materials Science, University of Connecticut, CT 06268, USA

Billy R. Martin Department of Pharmacology and Toxicology, Virginia Commonwealth University, Richmond, VA 23298, USA

Lisa A. Matsuda Department of Psychiatry and Behavioral Sciences, Medical University of South Carolina, 171 Ashley Avenue, Charleston, SC 29425, USA.

Raphael Mechoulam Department of Natural Products, Hebrew University, Medical Faculty, Jerusalem 91120, Israel

Roger G. Pertwee Department of Biomedical Sciences, Marischal College, University of Aberdeen, Aberdeen AB9 1AS, Scotland

Raj K. Razdan Organix, Inc., Cambridge, MA 01801, USA

Brian F. Thomas Research Triangle Institute, Research Triangle Park, NC 27709, USA

Series Preface

The neurosciences are one of the most diverse and rapidly changing areas in the biological sphere. The need to understand the workings of the nervous system pervades a vast array of different research areas. By definition research in the neurosciences encompasses anatomy, pathology, biochemistry, physiology, pharmacology, molecular biology, genetics and therapeutics. Ultimately, we are striving to determine how the human brain functions under normal circumstances and perhaps more importantly how function changes in organic disease and in altered states of mind. The key to many of these illnesses will unlock one of the major therapeutic challenges remaining in this era.

The difficulty lies in the vastness of the subject matter. However I try, I find it almost impossible to keep abreast of the changes occurring in my immediate sphere of interest, let alone those exciting advances being made in other areas. The array of journals dealing with neurosciences is enormous and the flow of reprints needed to keep me updated is daunting. Inevitably piles of papers accumulate on my desk and in my briefcase. Many lie there unread until sufficient time has passed for their content to be overtaken by yet more of the ever rising tide of publications.

There are various approaches that one can take to deal with this problem. There is the blinkered approach in which you pretend that literature outside your area does not exist. There is the ignore it totally option. Indeed, one colleague of mine has ceased to read the literature in the belief that, if there is a publication of critical importance to his research, someone will tell him about it. I am not that brave and instead I arrived at what I thought was the ideal solution. I started to read critical reviews of areas of current interest. But I soon came unstuck as I realized that, for many subjects of importance to the neurosciences, such authoritative works did not exist.

Instead of simply moaning that the literature was incomplete, I conceived the idea of *Neuroscience Perspectives*. For purely selfish reasons I wanted to have available a series of individually edited monographs dealing in depth with issues of current interest to those working in the neuroscience area. Already a number of volumes have been published which have been well received and the series is thriving with books on a range of topics in preparation or in production. Each volume is designed to bring a multidisciplinary approach to the subject matter by pursuing the topic from the laboratory to the clinic. The editors of the individual volumes are producing balanced critiques of each topic to provide the reader with an up-to-date, clear and comprehensive view of the state of the art.

As with all ventures of this kind, I am simply the individual who initiates a chain of events leading to the production of the series. In reality, it is key individuals at Academic Press who are really responsible for the success of *Neuroscience Perspectives*. In particular, Dr Carey Chapman and Léona Daw have the unenviable task of recruiting editors and authors and keeping the ship on an even keel.

Finally, I hope that *Neuroscience Perspectives* will continue to be enjoyed by my colleagues in the neurosciences. Already the series is being read, understood and enjoyed by a wide audience and it is fast becoming a reference series in the field.

Peter Jenner

Preface

The discovery of cannabinoid receptors and of endogenous cannabinoid receptor ligands heralds the beginning of a truly exciting new phase of cannabinoid research which must now extend well beyond the realms of pharmacology and molecular biology into those of physiology and pathophysiology. Such research will hopefully lead on to a better understanding of how the body functions in both health and disease and also to the development of new medicines. These advances will require input from a wide range of disciplines and it is my objective to encourage this by encapsulating current knowledge and ideas about cannabinoid receptors in a single volume.

The first chapter in this book is a general overview in which I have tried to identify the main pharmacological, physiological and clinical implications of the discovery of cannabinoid receptors and their endogenous ligands. This chapter also introduces the reader to some of the major themes of the other chapters and to the chemical structures of those cannabinoids that have most often been used for research purposes in recent times. Chapter 1 ends with a brief description of an exciting recent advance in cannabinoid research that occurred during the proof stage of this book: the development of potent and selective cannabinoid receptor antagonists. Chapter 2 discusses the relationship between cannabinoid structure and pharmacological activity and then proceeds to deduce a cannabinoid pharmacophore. There are also descriptions of the main cannabinoid bioassays and an outline of strategies that have been and still are being used in the search for cannabinoid receptor antagonists. Chapter 3 focuses on two properties that are common to all known psychotropic cannabinoids: marked lipophilicity and amphipathicity. Evidence is presented that, in combination, these properties facilitate the interaction of cannabinoid molecules with their receptors. The extent to which these same two properties give rise to non-receptor mediated changes in cellular membranes is also considered, together with the pharmacological significance of these changes. Chapter 4 contains an account of the first cloning of the central cannabinoid receptor and includes descriptions of the main structural features, both of the receptor protein and of the genetic material responsible for expressing this protein. Chapter 5 describes the central and peripheral distribution patterns of cannabinoid receptors and cannabinoid receptor mRNA and discusses the extent to which these patterns can account for cannabinoid pharmacology. Chapter 6 reviews the various signal transduction pathways with which cannabinoids are known to interact. As well as identifying likely second messenger systems for cannabinoid receptors, consideration is given to the significance of non-receptor mediated effects of cannabinoids on the production of known second messengers. Chapter 7 continues the theme of cannabinoid receptor signal transduction mechanisms, concentrating in particular on the involvement of presynaptic N-type calcium channels and A-type potassium channels. This chapter also puts forward evidence that presynaptic cannabinoid receptors share effector systems with other presynaptic receptor types and considers how present ideas about the mode of action of cannabinoids can be used to explain some of their behavioural effects. The final chapter contains a fascinating first-hand account of the scientific strategies and experiments that led to the extraction of arachidonyl ethanolamide from the brain, to the demonstration that this compound (anandamide) is a cannabinoid receptor agonist and to the discovery in brain tissue of other cannabimimetic anandamide-like compounds.

Finally, several schemes of nomenclature for cannabinoid receptors have made their way into the literature. The receptor nomenclature chosen for use in the present volume is described in Section 1.1.3 of Chapter 1.

CHAPTER 1

PHARMACOLOGICAL, PHYSIOLOGICAL AND CLINICAL IMPLICATIONS OF THE DISCOVERY OF CANNABINOID RECEPTORS: AN OVERVIEW

R. G. Pertwee

Department of Biomedical Sciences, Marischal College, University of Aberdeen, Aberdeen AB9 1AS, Scotland

Table of Contents

Cannabinoid Receptors
ISBN 0-12-551460-3

1.1 Introduction

1.1.1 The discovery of cannabinoid receptors

Two findings in particular have led to the discovery of cannabinoid receptors: first, that the brain contains specific, high-affinity cannabinoid binding sites that are appropriately distributed (Table 1) and second, that the brain contains genetic material capable of expressing functional cannabinoid receptors (Table 2). The existence of cannabinoid receptors is also supported, albeit less directly, by several other findings. Among these are the observations that many cannabinoids are no less potent than agents already known to act through specific receptors, that the pharmacological activity of psychotropic cannabinoids is governed by chemical structure and that enantiomeric cannabinoids can show a marked degree of stereoselectivity (see Pertwee, 1993 and Chapter 2). It is also known that psychotropic cannabinoids can interact with certain classical second messenger systems. While the evidence that cannabinoid receptors are negatively coupled to adenylate cyclase through G_I proteins is especially strong (Table 3), there is also evidence that some cannabinoid receptors are negatively coupled to voltage gated N-type calcium channels through G_I or G_0 protein and that some are positively coupled to voltage-gated A-type potassium channels, possibly through G protein mediated inhibition of adenylate cyclase (Chapters 6 and 7).

The chemical structures of just a few of the many cannabimimetic agents that have helped to establish cannabinoid structure–activity relationships are depicted in Figures 1 to 3. The structures shown are those of the compounds that have been used most widely in cannabinoid research in recent times.

Table 1 Properties of cannabinoid binding sites that indicate them to be the recognition sites of functional cannabinoid receptors.

Properties of cannabinoid binding sites	References
Saturated by low concentrations of cannabimimetic agents	1, 2, 3, 4, 5, 6
High affinity for cannabimimetic agents	1, 2, 3, 4, 5, 6, 7
Selectively bind cannabimimetic agents	6, 8, 9
Present in the brain in high concentrations	1, 2, 4, 5, 6, 7, 10, 11, 12
The structure–affinity relationships of cannabinoids for binding correspond to their structure–activity relationships in cannabinoid bioassays	1, 2, 3, 4, 5, 6, 7
The stereoselectivity for cannabinoids shown by the binding sites corresponds to that shown by cannabinoids in functional bioassays	1, 2, 4, 6, 7, 13
The distribution of cannabinoid binding sites in the brain is largely consistent with the pharmacological properties of psychotropic cannabinoids	2, 4, 6, 7, 8, 10, 12, 14
The distribution of cannabinoid binding sites in mouse cerebellum is similar to that of adenylate cyclase which is thought to form part of the signal transduction mechanism for at least some cannabinoid receptors	15

References: 1, Devane *et al.* (1988); 2, Herkenham *et al.* (1990); 3, Martin *et al.* (1991); 4, Thomas *et al.* (1992); 5, Felder *et al.* (1992); 6, Kuster *et al.* (1993); 7, Herkenham *et al.* (1991b); 8, Bidaut-Russell *et al.* (1990); 9, Howlett *et al.* (1992); 10, Mailleux and Vanderhaeghen (1992a); 11, Mailleux *et al.* (1992a); 12, Jansen *et al.* (1992); 13, Howlett *et al.* (1990); 14, Mailleux and Vanderhaeghen (1992b); 15, Herkenham *et al.* (1991a).

Table 2 Properties of genetic material responsible for expressing cannabinoid receptors in the brain.

Properties of the genetic material	References
The material encodes a novel (cannabinoid) receptor	1, 2, 3
The material has been detected in the brains of man, rat, guinea-pig, dog and mouse*	1, 2, 3
The predicted receptor protein product contains 472 or 473 amino acids arranged in a sequence different from that of any other known receptor	1, 3
The genomic location of the human cannabinoid receptor gene is the q14–q15 region of chromosome 6	4, 5
The predicted amino acid sequences of human and rat cannabinoid receptor protein are 98% identical	2, 3
The structure of the cloned cannabinoid receptor indicates it to be G protein linked	1, 3
Insertion of the genetic material into cells triggers the expression of specific high-affinity cannabinoid binding sites	3, 6, 7, 8
Insertion of the genetic material into cells confers an ability to respond to psychotropic cannabinoids (suppression of cyclic AMP production)	1, 3, 6, 7, 8
There are similarities in the brain between the distribution pattern of cannabinoid receptor mRNA and the distribution pattern of cannabinoid binding sites (the distribution patterns do not match perfectly)†	1, 9, 10, 11, 12
Cannabinoid receptor mRNA is detectable in brain tissue from human and rat neonates	11, 13

*See Chapter 4
†See Chapter 5
References: 1, Matsuda *et al.* (1990); 2, Gérard *et al.* (1990); 3, Gérard *et al.* (1991); 4, Caenazzo *et al.* (1991); 5, Hoehe *et al.* (1991); 6, Felder *et al.* (1992); 7, Felder *et al.* (1993); 8, Vogel *et al.* (1993); 9, Mailleux and Vanderhaeghen (1992a); 10, Mailleux *et al.* (1992b); 11, Mailleux and Vanderhaeghen (1992b); 12, Matsuda *et al.* (1993); 13, McLaughlin and Abood (1993).

Table 3 Evidence that cannabinoid receptors are negatively coupled to adenylate cyclase through G_i proteins.

	References
The cannabinoid receptor has a structure typical of a G protein-linked receptor	1, 2
Cannabinoid-binding site affinity is reduced by stable GTP analogues	3, 4, 5, 6
Cannabinoids attenuate forskolin-induced or hormone-induced stimulation of cyclic AMP production in preparations containing cannabinoid receptors	1, 2, 7, 8, 9, 10, 11, 12, 13, 14
There is a correlation between the ability of cannabinoids to inhibit cyclic AMP production and psychotropic potency	1, 2, 11, 15, 16, 17
The presence of cannabinoid receptors is a prerequisite for inhibition of cyclic AMP production by psychotropic cannabinoids	1, 2, 18, 19
The inhibitory effect of psychotropic cannabinoids on cyclic AMP production is not mediated by certain non-cannabinoid receptors known to be negatively coupled to adenylate cyclase	1, 9, 13
The distribution pattern of cannabinoid-binding sites in brain tissue resembles that of adenylate cyclase	5, 20
The inhibitory effect of psychotropic cannabinoids on cyclic AMP production can be blocked by pertussis toxin	1, 10, 12, 14, 18, 19, 21
There is evidence that psychotropic cannabinoids do not interact directly with adenylate cyclase or hasten the catabolism of cyclic AMP	8, 22
There is evidence that psychotropic cannabinoids only inhibit cyclic AMP production in preparations in which adenylate cyclase is membrane-bound	10

References: 1, Matsuda *et al.* (1990); 2, Gérard *et al.* (1991); 3, Devane *et al.* (1988); 4, Herkenham *et al.* (1990); 5, Herkenham *et al.* (1991b); 6, Kuster *et al.* (1993); 7, Pacheco *et al.* (1991); 8, Howlett (1984); 9, Howlett and Fleming (1984); 10, Howlett *et al.* (1986); 11, Felder *et al.* (1992); 12, Bidaut-Russell *et al.* (1990); 13, Bidaut-Russell and Howlett (1991); 14, Pacheco *et al.* (1993); 15, Howlett (1987); 16, Howlett *et al.* (1988); 17, Howlett *et al.* (1990); 18, Vogel *et al.* (1993); 19, Felder *et al.* (1993); 20, Herkenham *et al.* (1991a); 21, Howlett *et al.* (1992); 22, Howlett (1985).

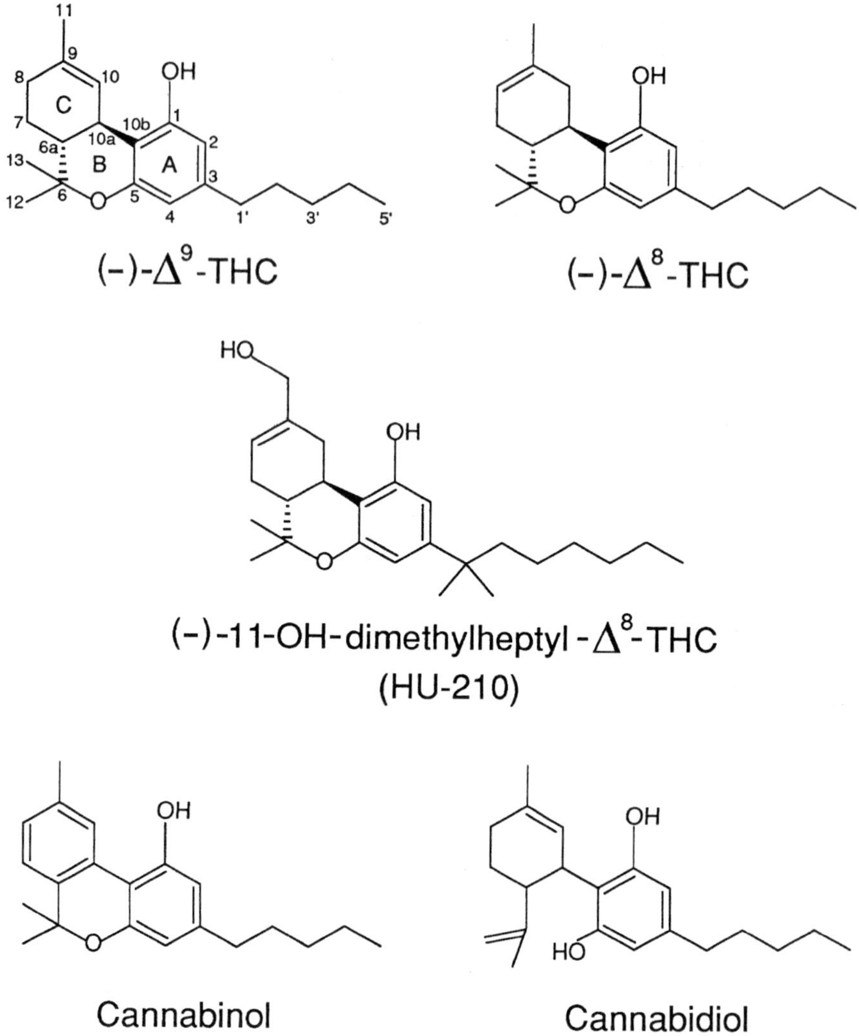

Figure 1 Structures of the psychoactive enantiomers of Δ^9-THC, Δ^8-THC and 11-hydroxy-dimethylheptyl-Δ^8-THC (HU-210). The structures of cannabinol (marginally psychoactive) and cannabidiol (non-psychoactive) are also shown.

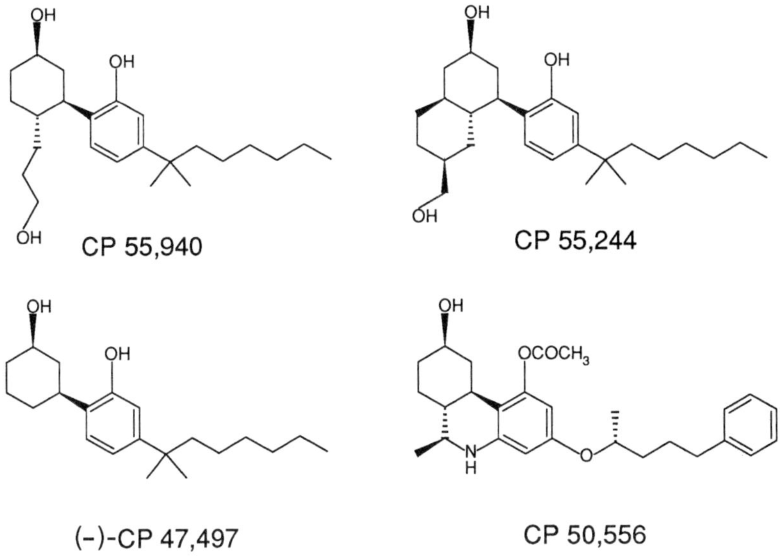

Figure 2 Structures of the (−)-enantiomers of CP 55,940, CP 55,244, C 47,497 and CP 50,556 (levonantradol). The corresponding less active (+)-enantiomers are CP 56,667, CP 55,243, (+)-CP 47,497 and CP 53,870 (dextronantradol) respectively.

1.1.2 The discovery of endogenous cannabinoid receptor ligands

The realization that cannabinoids act through receptors sparked off a search for an endogenous ligand. This led to the discovery, in porcine brain, of the cannabinoid receptor agonist, anandamide (arachidonyl ethanolamide) (Figure 4, Table 4 and Chapter 8). More recently, two other endogenous unsaturated fatty acid ethanolamides, also with significant affinity for cannabinoid binding sites, have been extracted from the brain. These are homo-γ-linolenylethanolamide and 7, 10, 13, 16-docosatetraenylethanolamide (Figure 4; see also Hanuš *et al.*, 1993 and Chapters 2 and 8). Both compounds share the ability of anandamide and other cannabimimetic agents to inhibit the electrically evoked twitch response of the mouse isolated vas deferens at sub-micromolar concentrations (IC_{50}=99.3 and 95.5 nM respectively) indicating that they not only bind to cannabinoid receptors but also possess pharmacological activity (efficacy) (Pertwee *et al.*, 1994a). The idea that homo-γ-linolenylethanolamide and 7, 10, 13, 16-docosatetraenylethanolamide are both cannabinoid receptor agonists is supported by the finding that Δ^9-tetrahydrocannabinol (Δ^9-THC) can induce a sixfold tolerance to the inhibitory effects of these fatty acid ethanolamides in isolated vasa deferentia (Pertwee *et al.*, 1994a).

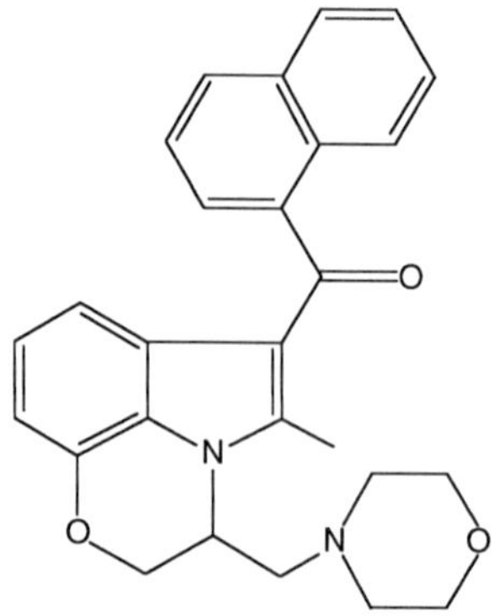

WIN 55,212-2

Figure 3 Structure of WIN 55,212-2.

1.1.3 The discovery of a second type of cannabinoid receptor

Until recently, there was little evidence to support the existence of more than one type of cannabinoid receptor (see Pertwee, 1993). Indeed, this situation still holds true for the central nervous system, which has so far been found to contain just a single class of cannabinoid binding site and only one type of cannabinoid receptor mRNA (see Pertwee, 1993). However, genetic material capable of expressing a second type of cannabinoid receptor has now been discovered outside the central nervous system: in a human leukaemic cell line and in rat spleen macrophages (Munro *et al.*, 1993). The newly discovered cannabinoid receptor shows 44% homology with the receptor first cloned by Matsuda *et al.* (1990).

Various names for the same cannabinoid receptors are in current use: CB_1 (or CB1) and CBR1 for the receptor cloned by Matsuda *et al.* (1990) and CB_2 (or CB2) and CBR2 for the peripheral receptor. The names selected for use in this book are CB_1 and CB_2 as it is this nomenclature which is being proposed to the Committee on Receptor Nomenclature and Drug Classification of the International Union of Pharmacology by its subcommittee on Cannabinoid Receptor Nomenclature and Classification (see also Watson and Girdlestone, 1994).

The structure of the CB_2 receptor indicates that, like the CB_1 receptor, it is a member of the superfamily of G protein-linked receptors. One difference between these two receptor types is that the CB_2 receptor appears only to be expressed outside the brain while the CB_1 receptor occurs both centrally and peripherally (Chapters 4 and 5). A second difference concerns the recognition sites of CB_1 and CB_2 receptors, the rank order of the affinities of certain cannabinoids for CB_2 binding sites transfected into cultured cells (cannabinol>Δ^9-THC>cannabidiol) differing from their rank order for CB_1 binding sites (Δ^9-THC>cannabinol>

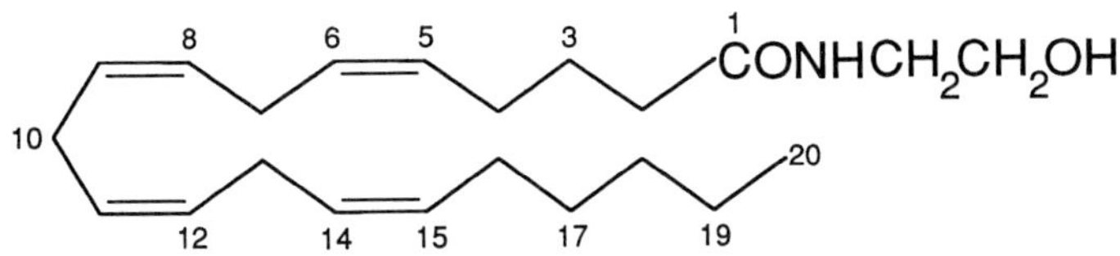

Arachidonyl ethanolamide
Anandamide (20:4, n-6)

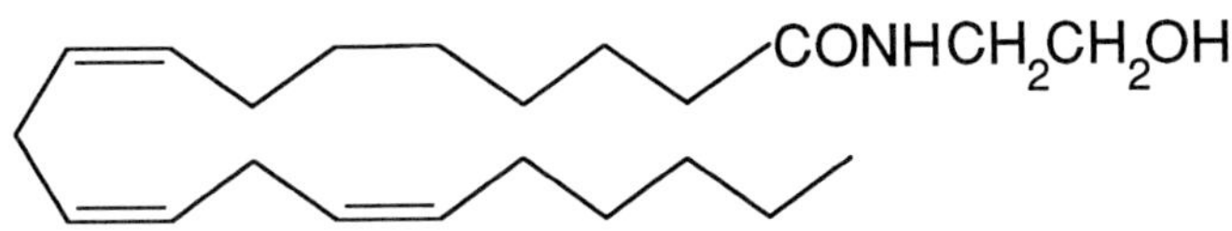

Homo-γ-linolenylethanolamide
Anandamide (20:3, n-6)

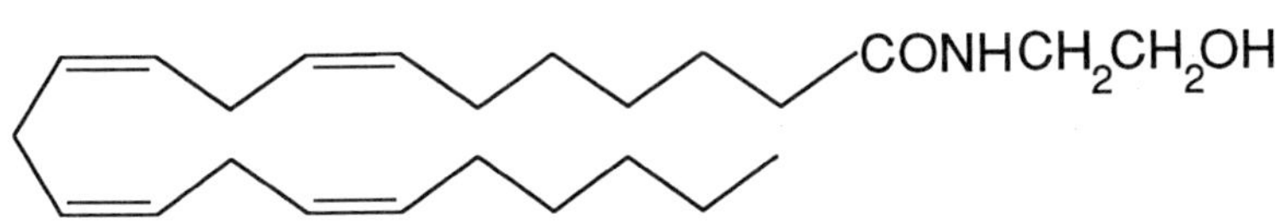

7, 10, 13, 16-Docosatetraenylethanolamide
Anandamide (22:4, n-6)

Figure 4 Structures of anandamide (20:4, *n*-6), anandamide (20:3, *n*-6) and anandamide (22:4, *n*-6).

cannabidiol) (Devane *et al.*, 1988; Herkenham *et al.*, 1990; Munro *et al.*, 1993). Anandamide binds to both types of receptor, although like Δ^9-THC, its affinity for the CB_2 site is lower than that for the CB_1 site (Devane *et al.*, 1992; Munro *et al.*, 1993). While it is known that CB_2 receptors can be occupied by cannabinoids, it has not been established whether such occupancy can trigger a response.

1.1.4 The term 'anandamide system'

For convenience, the system that is made up of cannabinoid receptors and endogenous anandamide-related compounds is referred to in this chapter as the anandamide system. It is important to bear in mind, however, that endogenous ligands for cannabinoid receptors may exist with chemical properties that are quite different from those of the anandamide group of compounds (see Chapter 7).

Table 4 The evidence that anandamide is an endogenous cannabinoid receptor agonist.

	References
Anandamide has been extracted from porcine brain	1
Anandamide binds avidly to rat brain cannabinoid receptors*	1
Anandamide shows selectivity for cannabinoid binding sites	2
Anandamide interacts with cannabinoid receptors expressed by cells transfected with cannabinoid receptor DNA*	3, 4
Anandamide has cannabimimetic properties *in vitro*	1, 3, 4, 5, 6, 7
Anandamide has cannabimimetic properties *in vivo*	5, 8, 9, 10, 11
Anandamide shows cross-tolerance with THC in the mouse isolated vas deferens	5

*Indicates anandamide to be a cannabinoid receptor ligand as cannabinoid receptors/binding sites are highly selective (see Table 1).
References: 1, Devane *et al.* (1992); 2, A.M. Snowman and S.H. Snyder unpublished (see Section 8.6.4); 3, Vogel *et al.* (1993); 4, Felder *et al.* (1993); 5, Pertwee *et al.* (1993); 6, Mackie *et al.* (1993); 7, Childers *et al.* (1994); 8, Fride and Mechoulam (1993); 9, Pertwee and Stevenson (1994); 10, Wickens and Pertwee (1993); 11, Crawley *et al.* (1993).

1.2 Pharmacological implications of the discovery of the anandamide system

1.2.1 Identification of the sites of action of cannabinoids

When current information about the distribution pattern of cannabinoid binding sites and of cannabinoid receptor mRNA is placed alongside existing knowledge about the functional anatomy of the body it becomes possible to predict likely locations of the central and peripheral sites at which cannabinoids produce at least some of their many effects.

1.2.1.1 Central sites of action

As detailed in Chapters 5 and 7, at least some of the sites at which cannabis produces its effects on cognition and memory may prove to be located within the cerebral cortex and hippocampus, both regions of the brain that are particularly rich in

CB_1 cannabinoid binding sites. These regions may also mediate the well-documented effects of cannabis on the perception of time, sound, colour and taste (Section 1.3.2). The sites at which cannabis produces its effects on motor activity and postural control probably reside in the basal ganglia and cerebellum, as these brain areas are also densely populated with cannabinoid binding sites (Chapters 5 and 7).

Some areas of the central nervous system that are known to be responsive to cannabinoids contain far fewer cannabinoid binding sites than others. Even so, effects produced by cannabinoids in such areas, may well be cannabinoid receptor mediated, as the absolute number of cannabinoid binding sites present in these areas is quite high (Chapter 5). Sites that are relatively sparsely populated with cannabinoid binding sites include parts of the central nervous system in which cannabinoids can act to produce hypothermia (hypothalamus) or antinociception (spinal cord). The brainstem, where cannabinoids may act to suppress nausea and vomiting, also contains relatively few cannabinoid binding sites.

1.2.1.2 Sites of action outside the central nervous system

The discovery that cells of the immune system contain both cannabinoid binding sites and cannabinoid receptor mRNA indicates that the immunosuppressive activity of psychotropic cannabinoids may be cannabinoid receptor mediated. To date, cannabinoid binding sites have been observed to occur in mouse spleen cells and in membranes obtained from a human monocytic cell line (U937) while cannabinoid receptor mRNA has been detected in human, rat and mouse spleen cells, in human thymus and tonsil cells and in human peripheral blood mononuclear cells, albeit in significantly lower concentrations than in brain tissue (Kaminski *et al.*, 1992; Bouaboula *et al.*, 1993; Munro *et al.*, 1993). Of the various types of human leukocytes investigated by Bouaboula *et al.* (1993), B-lymphocytes were found to contain the greatest amount of cannabinoid receptor mRNA. In rat spleen, expression of the cannabinoid receptor is concentrated in the marginal zone (Munro *et al.*, 1993). This is the site of entry into the spleen of blood-borne cells and antigens. It contains macrophages whose functions are to digest and process bacterial antigens and to direct lymphocyte recirculation.

Other peripheral tissues that have been found to contain cannabinoid receptor mRNA or cannabinoid binding sites are dog testis and guinea-pig small intestine (Chapter 5). Cannabinoid receptors may also be present in the vasa deferentia of mice, rats and guinea-pigs and in the mouse small intestine, as cannabinoids show significant stereoselectivity in all these tissues (see Pertwee, 1993). Evidence for the presence of cannabinoid receptors in the mouse vas deferens is particularly strong as cannabinoids added to this tissue also show remarkably high potency and a strong correlation is known to exist between the ability of cannabinoids to inhibit electrically evoked contractions of this tissue and their psychotropic activity (Kuster *et al.*, 1993; Pertwee, 1993). No cannabinoid receptor mRNA has yet been detected in heart, lung, liver, kidney or nasal membranes, at least in rats (Matsuda *et al.*, 1990; Munro *et al.*, 1993).

1.2.2 The potential for developing new drugs

1.2.2.1 Cannabinoid receptor agonists and antagonists

The existence of cannabinoid receptors opens up the possibility of designing selective cannabinoid antagonists. To date, progress in this area of cannabinoid research has been disappointing (Chapter 2)[1]. Thus although one compound, the aminoalkylindole analogue, WIN 56,098 (Figure 5), has been reported to antagonize Δ^9-THC and levonantradol in the mouse isolated vas deferens and to show some degree of selectivity, the degree of antagonism it produced was modest, even when quite high concentrations were used (10 μM) (Pacheco *et al.*, 1991).

As there appear to be significant differences between the recognition sites of cannabinoid CB_1 and CB_2 receptors (Section 1.1.3), it may be possible to develop drugs that can interact selectively with the CB_2 receptor. These drugs, which would be expected to lack psychoactivity, might well have clinical applications either as immunosuppressants (CB_2 agonists), for example in the treatment of autoimmune diseases such as rheumatoid arthritis, or as enhancers of immune function (CB_2 antagonists). However, it has first to be established whether CB_2 receptors can trigger responses when occupied and, if so, whether they mediate cannabinoid-induced changes in immune function.

Anandamide is structurally different from all other known cannabinoid receptor agonists. Its discovery has sparked off the syntheses of a rapidly growing number of new analogues (e.g. see Chapter 2). These will help to establish the structure–activity relationships of anandamide. One new analogue, *R*-(+)-arachidonyl-1′-hydroxyl-2′-propanolamide, appears to be significantly more resistant to metabolic degradation than anandamide and there is little doubt that it is largely because of this that the analogue is somewhat more potent than anandamide as a cannabimimetic agent (Abadji *et al.*, 1994).

1.2.2.2 Drugs which alter the synthesis, storage, release or fate of endogenous cannabinoid receptor agonists

It may eventually become possible to develop drugs that can selectively modulate the activity of the anandamide system by interfering with the processes that govern the concentrations of endogenous cannabinoid receptor agonists at their sites of action. Progress in this area of research must await a detailed elucidation of the mechanisms responsible for the synthesis, storage (if any), release and fate of endogenous cannabinoids. Characterization of the enzymes that catalyse the synthesis and/or metabolism of endogenous cannabinoids may also allow the distribution pattern of these compounds to be established, for example by the application of immunohistochemical techniques. For anandamide, experiments to date have revealed arachidonic acid to be both an anandamide precursor (in rat brain) and

[1]See end of this chapter for a recent development.

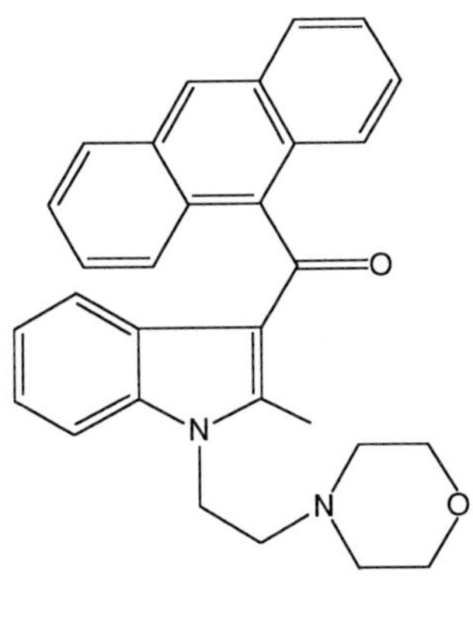

WIN 56,098

Figure 5 Structure of WIN 56,098.

a product of its degradation (Deutsch and Chin, 1993). Metabolism of anandamide has been detected in rat brain, liver, kidney and lung and in neuroblastoma and glioma cells but not in rat serum, heart or skeletal muscle or in HeLa, Hep2 or HepG2 cells (Deutsch and Chin, 1993). It is also now known that the protease inhibitor, phenylmethylsulfonyl fluoride (PMSF), can potentiate anandamide in certain tissue preparations. This it presumably does by inhibiting anandamide metabolism. In particular, PMSF (50 μM) has been found to enhance the ability of anandamide to displace a radioactively labelled probe from cannabinoid binding sites (Abadji *et al.*, 1994; Childers *et al.*, 1994). As shown in Figure 6, PMSF also enhances the ability of anandamide to inhibit electrically evoked contractions of the guinea-pig myenteric plexus preparation, a preparation which can serve as a quantitative bioassay for psychotropic cannabinoids (Pertwee *et al.*, 1992).

1.2.3 Cannabinoid tolerance

Tolerance can develop rapidly to many of the effects of psychotropic cannabinoids (see Pertwee, 1991). Little is known about the mechanisms responsible for cannabinoid tolerance other than that they are largely pharmacodynamic in nature. The discovery of the anandamide system raises the possibility that cannabinoids induce tolerance by provoking compensatory decreases in the sensitivity or density of cannabinoid receptors, perhaps by inducing changes in the genetic expression of the receptor protein. Reductions in the sensitivity of cannabinoid receptors could be caused by cannabinoid-induced changes either in their signal transduction mechanisms or in the affinity of their recognition sites for cannabinoids.

Experiments with rats pretreated with Δ^9-THC or CP 55,940 have yielded results which support the hypothesis that tolerance to cannabinoids is brought about by drug-induced decreases in brain cannabinoid receptor density (Table 5). Conversely,

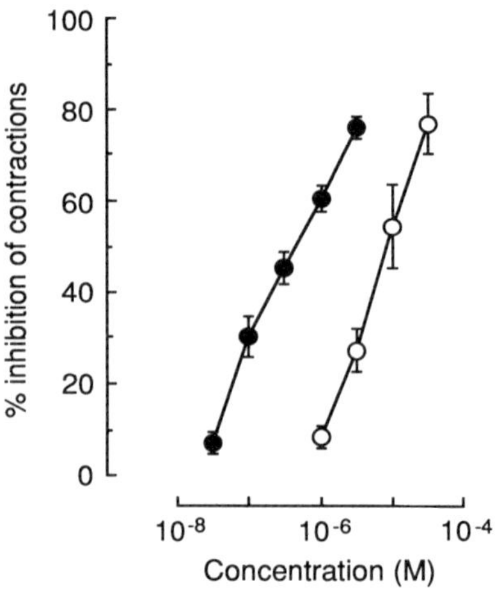

Figure 6 Mean cumulative concentration–response curves for anandamide in the myenteric plexus-longitudinal muscle preparation of guinea-pig small intestine constructed in the presence (closed symbols) or absence of 50 μM phenylmethylsulfonyl fluoride (PMSF) (Pertwee *et al.*, 1995). Each symbol represents the mean value ± SE of inhibition of electrically evoked contractions expressed as a percentage of the amplitude of the twitch response measured immediately before the first addition of anandamide to the organ bath (n=6 different segments of tissue). Each preparation was mounted in a 4 ml organ bath at an initial tension of 0.5 g. Baths contained Krebs solution maintained at 37°C and bubbled with 95% O_2 and 5% CO_2. Isometric contractions were elicited by field stimulation through platinum electrodes attached to the upper and lower ends of each bath (single bipolar rectangular pulses of 110% maximal voltage, 0.5 ms duration and 0.1 Hz frequency). Drug additions were made in volumes of 10 μl after the tissues had equilibrated. Anandamide and PMSF were mixed with two parts of Tween 80 by weight and then dispersed in water (PMSF) or saline. PMSF was added to the organ bath 15 min before the first addition of anandamide. Thereafter, tissues were incubated for 135 min without replacing the fluid in the bath. Additions of anandamide were made at 30 min intervals. Courtesy of Elsevier Science B.V., publishers of *European Journal of Pharmacology*.

results from mouse experiments suggest that tolerance to Δ^9-THC can develop without any concomitant changes in the density of cannabinoid receptors (Table 5) or, indeed, without any changes in cannabinoid receptor affinity or in cannabinoid receptor mRNA levels (Abood *et al.*, 1993). This discrepancy could have arisen because the rat cannabinoid binding data were obtained using tissue from a number of different brain areas, whereas the corresponding mouse data were obtained using tissue from whole brain. Thus, it may well be that cannabinoid pretreatment can bring about cannabinoid receptor losses that are sufficient to induce tolerance to certain effects and yet insufficient to produce detectable decreases in the receptor population of the whole brain because the losses occur only in a limited number of brain areas. This idea, that cannabinoid receptors in different brain areas are not equally susceptible to down-regulation, is supported by the fact that cannabinoid tolerance develops less readily to some effects than others (Pertwee, 1991). It is also possible that the decreases in binding of the radiolabelled cannabinoid receptor probe observed in the rat experiments resulted not from cannabinoid receptor down-regulation but rather from competition between the probe and (unlabelled) cannabinoid molecules that had accumulated in the brain during the pretreatment phase. Any unlabelled cannabinoid molecules that may have accumulated in this way were given far less time to disperse from the vicinity of cannabinoid receptors in the rat experiments (less than 1 h) than in the mouse experiments (24 h) (Table 5).

1.2.4 Selectivity of bioassays for cannabimimetic agents

In the absence of selective cannabinoid receptor antagonists it has, in the past, been impossible to distinguish convincingly between cannabimimetic and non-cannabimimetic agents by conventional bioassay (cannabinoid binding assays are specific but provide no measure of efficacy). The discovery of cannabinoid receptors, and particularly the cloning of these receptors, has opened up an alternative strategy by making it possible to compare responses elicited by drugs in cell lines lacking cannabinoid receptors with those elicited in identical cell lines after their transfection with genetic material encoding cannabinoid receptor protein (Chapter 2). Another strategy for distinguishing between cannabimimetic and non-cannabimimetic agents that has always been open to the pharmacologist is to desensitize a bioassay system to cannabimimetics selectively. The problem, of course, has been to find a system in which selective tolerance of this sort can be induced. The discovery of cannabinoid receptors has raised the possibility that tolerance is brought about by some change in the recognition site of these receptors (Section 1.2.3). In this case it should indeed be possible to achieve selective cannabinoid tolerance, at least in some bioassay systems. One such system may be a neuroblastoma cell line investigated by Dill and Howlett (1988). Another may be the mouse isolated vas deferens as pretreatment with Δ^9-THC (20 mg/kg/day i.p for 2 days) can render vasa deferentia tolerant to cannabinoids without also making them tolerant to a range of non-cannabimimetic inhibitors of the twitch response: the α_2-adrenoceptor agonist, clonidine, the $GABA_B$ receptor agonist, baclofen, the opioid receptor agonist, morphine, and selective agonists for μ (D-Ala^2, *N*-Me-Phe^4, Gly-ol^5- enkephalin), δ (D-Ala^2, D-Leu^5-enkephalin) and κ opioid receptors (U-69,593) (Pertwee *et al.*, 1993, 1994a; Griffin and Pertwee, 1994; Pertwee and Griffin, 1995).

1.2.5 Actions of cannabinoids that are not receptor-mediated

Among these are actions that occur in response to quite low cannabinoid concentrations, no higher than those required to activate cannabinoid receptors (concentrations of 1 μM or less; see Pertwee, 1990). They include the facilitation or inhibition of the neuronal uptake of certain neurotransmitters, the inhibition of some membrane-bound enzymes and the perturbation of membrane phospholipids (see Chapters 3 and 6; see also Pertwee, 1990). Further research is required to

Table 5 Effects of cannabinoid pretreatment on the binding of [^{3}H]CP 55,940 to central cannabinoid receptors and on cannabinoid-induced hypokinesia.

Cannabinoid pretreatment	[^{3}H]CP 55,940 binding experiments			Occurrence of tolerance to the hypokinetic effect of the pretreatment cannabinoid	Reference
	Tissue	Decrease in affinity of binding sites	Decrease in density of binding sites		
CP 55,940 given i.p. once daily for 7 or 14 days to rats killed 30 min after the final injection:					
1 mg/kg	Striatal brain sections	No	Yes	No	1
3 mg/kg		No	Yes	Yes	
10 mg/kg		Slight	Yes	Yes	
Cannabidiol, 10 mg/kg i.p., given once daily for 7 or 14 days to rats killed 30 min after the final injection	Striatal brain sections	No	No	†	1
Δ^{9}-THC, 10 mg/kg i.p., given once daily for 7 or 14 days to rats killed 30 min after the final injection	Striatal brain sections	No	Yes	Yes	1

Table 5 *(contd)*

Cannabinoid pretreatment	[^{3}H]CP 55,940 binding experiments			Occurrence of tolerance to the hypokinetic effect of the pretreatment cannabinoid	Reference
	Tissue	Decrease in affinity of binding sites	Decrease in density of binding sites		
Δ^9-THC, 6.4 mg/kg i.p., given once daily for 7 days to rats killed 50 min after the final injection	Striatum	No	Yes	Yes	2
	LFb	No	Yes	Yes	
	VMe	No	Insignificant	Yes	
Δ^9-THC, 10 mg/kg i.p., given twice daily for 6.5 days to mice killed 24 h after the final injection	WB	No	No	Yes	3

LFb=limbic forebrain, VMe=ventral mesencephalon and WB=whole brain.
†Cannabidiol was inactive as a hypokinetic agent.
References: 1, Oviedo *et al.* (1993); 2, de Fonseca *et al.* (1994a); 3, Abood *et al.* (1993).

determine the extent to which actions of this sort contribute to the pharmacological effects produced by cannabinoids in the whole organism and to establish the underlying mechanisms. It will also be important to find out which, if any, of the non-receptor-mediated actions of cannabinoids extend to anandamide or other endogenous cannabinoid receptor agonists, and which do not.

1.3 Physiological implications of the discovery of the anandamide system

1.3.1 Anandamide as an endogenous mediator

Although it is known that anandamide is present in the brain and can activate cannabinoid receptors, it has not been established whether this compound ever comes into contact with cannabinoid receptors under physiological and/or pathophysiological conditions and if it does, what type of chemical mediator it is. Anandamide may, for example, be a hormone, capable of activating cannabinoid receptors that are remote from its site of release and, indeed, there is already evidence that this compound is not degraded while in the circulation (Section 1.2.2.2). Another possibility is that anandamide is a neurotransmitter, in which case there should be neurons that are capable both of synthesizing anandamide and of rapidly placing the compound in contact with cannabinoid receptors. Whether the latter process involves the release of anandamide into the extracellular space or into membranes (Chapter 5) has not been established. Unless anandamide is produced on demand (like prostaglandins), anandamide-releasing neurons would, in addition, be expected to have a means both of storing the compound and of regulating its free concentration within the cytosol. There are also likely to be enzymic and/or transport mechanisms that can rapidly remove anandamide from the vicinity of cannabinoid receptors after its release. One location for cannabinoid receptors is the nerve terminal and it is probable that cannabinoids can act presynaptically to modulate the neuronal release of a range of different transmitters (see Chapters 5 and 7), in which case anandamide may function as a neuromodulator.

1.3.2 Physiological processes in which endogenous cannabinoid receptor agonists may serve as chemical mediators

These processes have still to be identified. Likely candidates are processes that are known to be susceptible to the actions of cannabis and a selection of these is listed below together with a brief description of how each process is affected by cannabis.

(1) Cognition (see Paton and Pertwee, 1973a; Dewey 1986). Cannabis induces mental confusion and loss of ability to concentrate.
(2) Memory (Miller and Branconnier, 1983 and Chapters 5 and 7). There is

evidence that accessibility of information acquired while experiencing the effects of cannabis is disrupted.

(3) Mood (Paton and Pertwee, 1973a). Although cannabis usually induces euphoria it can also induce dysphoria.

(4) Perception (Paton and Pertwee, 1973a). Cannabis can heighten the perception of sound, colour and taste to produce an enhanced vividness and freshness of sensory experience. Cannabis can also induce feelings of increased insight, significance and meaning, alter the perception of time such that felt time passes more slowly than clock time, and reduce the perception of pain.

(5) Movement, coordination, posture and skeletal muscle tone (Paton and Pertwee, 1973a, b; Dewey, 1986; Razdan, 1986). Cannabis usually causes a reduction in physical activity which may be preceded by a period of increased activity. It has subtle effects on spinal reflexes that may be either facilitatory or inhibitory, and can impair postural control. In animals, other prominent signs of altered extrapyramidal function include ataxia (in dogs) and catalepsy (in rats and mice).

(6) Sleep: cannabis induces sleepiness (Paton and Pertwee, 1973a).

(7) Thermoregulation (Pertwee, 1985). Cannabis can produce both rises and falls in deep body temperature. Some experiments, mainly with rats, have indicated the direction of cannabinoid-induced changes in body temperature to be dose-dependent; doses at the lower end of the effective dose range produce hyperthermia and higher doses, hypothermia.

(8) Appetite. Cannabis can stimulate hunger in man, particularly for solid foods that are sweet tasting (Greenberg *et al.*, 1976; Foltin *et al.*, 1988). The human data are supported by results from animal experiments, performed mainly with rats. These have shown that both cannabis and Δ^9-THC can stimulate food intake (Glick and Milloy, 1972; Gluck and Ferraro, 1974; Abel, 1975; Brown *et al.*, 1977; Anderson-Baker *et al.*, 1979). The dose–response relationship seems to be biphasic, with the stimulatory effect on food consumption that occurs in response to relatively low doses being replaced by an inhibitory effect when higher doses are given. In addition to increasing spontaneous food intake, Δ^9-THC can facilitate feeding induced by electrical stimulation of the lateral hypothalamus of unfasted Lewis rats (Trojniar and Wise, 1991). In rats, as in humans, cannabis seems to induce an increased preference for 'sweet calories' (Milano *et al.*, 1988). Although most animal experiments have been carried out with rats, cannabinoid-induced stimulatory effects on food intake have also been observed in sheep (McLaughlin *et al.*, 1979).

(9) Immune response (Munson and Fehr, 1983; Hollister, 1986; Martin, 1986). Cannabis has an immunosuppressant effect.

It is, of course, tempting to use information of this sort as a basis for speculating about the physiological roles of the anandamide system. Consider, for example, the well-known ability of psychotropic cannabinoids to induce hypothermia. This, cannabinoids are thought to do by decreasing the thermoregulatory set point – there being evidence that they can interact with hypothalamic thermoregulatory centres to produce coordinated changes in autonomic and behavioural heat gain and heat loss mechanisms directed at decreasing deep body temperature (Pertwee, 1985). As anandamide can also induce hypothermia (Crawley *et al.*, 1993; Fride and Mechoulam, 1993; Pertwee *et al.*, 1993; Pertwee and Stevenson, 1994) we are faced with the intriguing possibility that this compound is a physiological cryogen that can, for example, contribute to the production of circadian changes in body temperature or trigger hypothermia in hibernating animals. The much-documented ability of cannabimimetic agents, including anandamide (Fride and Mechoulam, 1993), to induce catalepsy is also worth considering in this context. Thus, it is always possible that anandamide is an endogenous cataleptogen, responsible perhaps for the cataleptic state that can be induced in some species without the use of drugs, for example by firmly gripping folds of skin at the back of the neck. Finally, cannabis is known to impair the functioning of certain physiological processes, for example, cognition, memory and the immune response. It has not been established whether or not disruptive effects of this sort are ever produced by endogenous cannabinoid receptor agonists under physiological or pathophysiological conditions.

Any attempt to base predictions about the physiological roles of endogenous cannabinoids on current knowledge about the pharmacological properties of cannabis must allow for the likelihood that there will be differences as well as similarities between the effects produced by the exogenous administration of cannabis and those that are elicited by an endogenous cannabinoid receptor agonist following its release. This is because the pharmacological properties of cannabis probably depend not only on its ability to mimic the endogenous compound by interacting with cannabinoid receptors, but also on its ability to act through other mechanisms (Section 1.2.5) which it may not share with the endogenous ligand. In addition, the pattern of activation of cannabinoid receptors is likely to be quite different after the administration of cannabis than after the physiological release of an endogenous ligand. After cannabis administration, it is probable that all cannabinoid receptors will be activated, the extent and duration of activation being determined largely by dosage and by the pharmacokinetics of cannabis. Conversely, the activation of cannabinoid receptors following the release of a cannabinoid receptor agonist (particularly one that functions as a transmitter) is expected to be far less extensive as it is likely that all tissues capable of releasing such an agonist will normally not do so simultaneously. A further difficulty in correctly predicting physiological roles of endogenous cannabinoid receptor agonists from what is known about the pharmacology of cannabis stems from the shape of cannabinoid dose–response relationships, which are sometimes biphasic in nature so that the measured response may increase when one set of doses is used but decrease when another set of doses is used (see above and Pertwee, 1990). The question then arises whether it is the increase or the decrease that is mimicking the physiological response to endogenous cannabinoid receptor agonists.

1.3.3 Future research

Clearly, further experiments are required to characterize the nature and establish the precise locations of the pathways and mechanisms that are responsible for the biosynthesis, storage or metabolism of anandamide and other endogenous cannabinoid receptor agonists. It will be particularly important to determine the extent to which the distribution pattern of these processes, both within and without the central nervous system, matches that of cannabinoid receptors. Another important goal for future research must be to elucidate the physiological roles of the anandamide system. One obvious strategy is to investigate the consequences of inactivating the system, for example by administering cannabinoid antagonists or drugs that selectively inhibit anandamide synthesis, storage or release (when these become available), by rendering animals tolerant to the effects of anandamide or by breeding strains of animals that lack cannabinoid receptors (Chapter 4) or are unable to produce ligands for the receptor. Another strategy would be to monitor the release of anandamide and other endogenous cannabinoid receptor agonists in specific areas of the brain under various conditions. Future research should also be directed at characterizing the mechanisms that regulate the anandamide system. There are already reports that the number or affinity of cannabinoid receptors in rat brain fluctuates during the oestrous cycle and is affected by sex steroids (de Fonseca *et al.*, 1994b) and that cannabinoid receptor gene expression in rat striatum can be regulated by dopamine, glutamic acid and glucocorticoids (Mailleux and Vanderhaeghen, 1993a,b 1994).

1.4 Clinical implications of the discovery of the anandamide system

1.4.1 Introduction

The history of the use of cannabis as a medicine stretches back thousands of years to a time when the Chinese Emperor, Shen Nung, produced the first known written account of the medicinal properties of cannabis (see Mechoulam, 1986). Previous therapeutic uses of cannabis have included the treatment of tetanus, convulsive disorders, neuralgia, migraine, dysmenorrhea, postpartum psychoses, senile insomnia, depression, opium and chloral hydrate dependence and gonorrhea (see Hollister, 1986). As more selective and effective drugs became available, the use of cannabis as a medicine declined. In Britain, the prescription of cannabis by doctors ceased completely in the early 1970s when the Misuse of Drugs Act came into force.

The identification in the mid 1960s of (−)-$\Delta^{6a,\,10a}$-*trans*-THC as the principal psychoactive constituent of cannabis and its subsequent synthesis (Gaoni and Mechoulam, 1964; Mechoulam and Gaoni, 1965) was followed by a massive upsurge of interest in the pharmacology, toxicology and potential uses in modern medicine of cannabis and of individual cannabinoids, both natural and synthetic. By the mid 1980s, shortly before the discovery of cannabinoid receptors, cannabinoids were being used in the clinic for the suppression of nausea and vomiting associated with cancer chemotherapy. The possible use of cannabinoids in the treatment of glaucoma also looked promising (see Hollister, 1986). However, experiments conducted in the same period and directed at investigating other possible clinical uses of cannabinoids, yielded results that were somewhat less encouraging (see Hollister, 1986). These results made it seem unlikely that cannabinoids would find a place in modern medicine to relieve pain or to treat bronchitis, asthma, hypertension, insomnia or opiate or alcohol abstinence syndromes. Moreover, insufficient progress had been made in the 1980s to allow any firm conclusions to be drawn about the use of cannabinoids in the treatment of epilepsy. One advance during this period that did prove to be of particular importance was the development by Pfizer of CP 55,940, a cannabinoid which was thought to have therapeutic potential as an analgesic (see Figure 2). Although the analgesic properties of CP 55,940 have never been exploited clinically, it was the availability of this compound that first made possible the detection of specific cannabinoid binding sites in the brain – a vital step in the discovery of the anandamide system (Devane *et al.*, 1988 and Section 1.1.1).

1.4.2 Clinical importance of the anandamide system

A prominent feature of the pharmacology of cannabinoids is that they elicit a very wide range of responses. In the past, this was attributed to the rather non-specific mechanisms by which cannabinoids were then considered to produce their characteristic effects and there was thought to be little scope for improving cannabinoid selectivity through conventional pharmacological strategies. However, the discovery of the anandamide system has changed this situation dramatically by opening up possible new ways of capitalizing more effectively on the beneficial effects of cannabinoids and of minimizing the incidence of some of their less desirable adverse effects. There is now the intriguing prospect that cannabinoids can be designed that are capable of producing therapeutically useful effects but that lack psychoactivity because they interact exclusively with the peripheral CB_2 receptor (Section 1.2.2.1).

The discovery of the anandamide system has also opened up the possibility that malfunctions of this system can disrupt normal physiological processes and/or that the signs and symptoms of certain disorders are mediated by pathways that release anandamide or other cannabinoid receptor agonists. Even if such pathways are not directly involved in the production of unwanted pathological effects, it is possible that they modulate the pathways that do. It follows that there may be some disorders that can be treated with drugs that selectively activate or inactivate the anandamide system, for example, by interacting with cannabinoid receptors or with processes responsible for the biosynthesis, storage, release or fate of anandamide (or other endogenous cannabinoid receptor agonists).

On examining the potential physiological roles of the anandamide system (Section 1.3.2), it becomes clear that there are a number of disorders that it might cause or mediate. These include dementia, unipolar or bipolar depression and disorders of memory, of motor function, of appetite and/or of the immune system. The memory changes induced by cannabis resemble those seen in patients with herpes simplex encephalitis, Korsakoff syndrome or Alzheimer's disease (see Miller and Branconnier, 1983). Consequently, it will be of particular interest to establish whether inhibitors of the anandamide system (when developed), improve cognition and memory both in patients suffering from dementia and in normal subjects. It is also possible that cannabinoids have therapeutic potential as modulators of immune function (Section 1.2.2.1) and that they may be effective in the treatment of chronic pain, migraine, epilepsy and certain movement disorders, including dystonias, Tourette's syndrome and Parkinson's disease (Consroe and Snider, 1986; Hollister, 1986; Consroe and Sandyk, 1992). Finally, there is growing evidence to suggest that cannabinoids would be effective in relieving the muscle spasms, tremor, ataxia and bladder dysfunctions that are often associated with multiple sclerosis or spinal cord injury. This evidence is summarized below.

(1) Preclinical data. The data indicate that cannabinoids can exert inhibitory effects on motor function and attenuate polysynaptic reflexes and that cannabinoid receptors are present in areas of the central nervous system that control or mediate motor function (see Consroe and Sandyk, 1992, and Sections 1.2.1 and 1.3.2). These include the cerebral cortex, cerebellum, basal ganglia, brainstem and spinal cord. There is also evidence that Δ^9-THC can counter autoimmune encephalomyelitis in Lewis rats and in strain 13 guinea-pigs (Lyman *et al.*, 1989). However, the validity of autoimmune encephalomyelitis as a model for multiple sclerosis is equivocal.

(2) Clinical trials. Although only a few investigations have been carried out, and these with rather small numbers of patients, the results obtained indicate that cannabinoids (or cannabis) can relieve at least some of the signs and symptoms that are provoked by multiple sclerosis or spinal injury (Table 6). Interestingly, there is a suggestion that patients with multiple sclerosis prefer cannabis to Δ^9-THC (Petro and Ellenberger, 1981). This may be because Δ^9-THC has been taken orally at an inappropriate dose: usually, cannabis is smoked, a procedure which allows the recipient to control the intensity of the effects experienced by regulating the amount of drug absorbed. It is also possible that other constituents of cannabis contribute directly to its putative beneficial effects or act indirectly to augment the sought-after effects of Δ^9-THC or to attenuate its adverse effects. The psychically inactive constituent of cannabis, cannabidiol, could be one such compound as it may have antidystonic properties (see Consroe and Sandyk, 1992) and as it seems to be capable of enhancing some effects of Δ^9-THC and of attenuating others (see Dewey, 1986).

(3) Anecdotal data. There is an ever-growing number of claims from patients with multiple sclerosis that cannabis reduces spasticity, relieves pain associated with skeletal muscle spasms and improves bladder function. The assertion is often made that cannabis is more effective than conventional skeletal muscle relaxants and that its side-effects are more tolerable. Some anecdotal reports also make the claim that cannabis can be effective in suppressing signs and symptoms of multiple sclerosis at smaller doses than those that are required to produce a definite 'high'. Several accounts of cannabis use for multiple sclerosis

Table 6 The treatment of patients with multiple sclerosis or spinal injury: clinical trials with Δ^9-THC or cannabis.

Signs and symptoms	Study design	Treatment	Measured effects	Reference
Spasticity related to multiple sclerosis in nine patients naive to cannabis	db, P	Δ^9-THC p.o. 5 or 10 mg	'Spasticity score' improved (four patients); feel 'better able to walk' (three patients); feel 'high' (one patient)	1
		Placebo	'Spasticity score' improved (one patient); feel 'high' (one patient)	1
Disabling tremor and mild ataxia related to multiple sclerosis in one cannabis-experienced male patient (aged 30)*	P	Δ^9-THC p.o. 5 mg	Performance in handwriting test improved; decrease in head and neck tremor; little change in mild hand ataxia; mild 'high'	2
		Placebo	No improvement; 'high' sensation	2
Disabling tremor and other signs related to multiple sclerosis in one female patient (aged 30)	P	Δ^9-THC p.o. 15 mg	Performance in handwriting test improved (5 and 10 mg ineffective)	2
		Placebo	No improvement	2

Table 6 *(contd)*

Signs and symptoms	Study design	Treatment	Measured effects	Reference
Spasticity, weakness, hyperactive reflexes, and impaired coordination and gait related to multiple sclerosis in 13 patients with a history of intolerable side-effects from conventional drug treatments*	db, P, c	Δ^9-THC p.o. 7.5 mg	Subjective improvement in spasticity; performance in objective function tests not improved; tolerable side-effects; (10 mg was intolerable to some patients) (2.5 and 5 mg ineffective)	3
		Placebo	No improvement Δ^9-THC-like subjective effects in five patients	3
Spasticity, intention tremor, ataxia and other signs related to multiple sclerosis in one cannabis-experienced male patient (aged 30)	ot	Cannabis (inhaled)	Marked reduction in spasticity; intention tremor almost abolished; ataxia reduced	4
Spasticity and pain due to spinal cord injury in one male patient (aged 28)	db, P	Δ^9-THC p.o. 5 mg†	Marked reductions in spasticity and pain; improvements in bladder control, quality of sleep, mood and ability to concentrate	5

*Standard antispasticity drugs unsuccessful or induced intolerable side-effects.
†Δ^9-THC and placebo were taken in conjunction with baclofen (40 mg) and clonazepam (1 mg).
c=crossover, db=double blind, ot=open trial, P=placebo.
References: 1, Petro and Ellenberger (1981); 2, Clifford (1983); 3, Ungerleider *et al.* (1987); 4, Meinck *et al.* (1989); 5, Maurer *et al.* (1990).

have recently appeared in British newspapers (Davies, 1992; Doyle, 1992; Hodges, 1992; 1993; Ferriman, 1993; Handscomb, 1993; James, 1993). In her 1993 article, Clare Hodges, who herself suffers from multiple sclerosis, wrote:

> 'I was being prescribed a whole range of medicines. There were pills to stop me feeling sick. These made me clumsy and drowsy. There were pills to relieve bladder spasms but they made me feel sick and gave me blurred vision. There were pills to help me sleep but they made me anxious and were habit-forming' and then
>
> 'For about a year now, I have been regularly taking a small amount of cannabis resin – less than the size of half a pea – late at night. I used to smoke it . . . but I was worried that my children might see me smoking so now I eat it. After a short time, my body completely relaxes, which relieves my tension and spasms. During the day I have to use a catheter whenever I want to empty my bladder and, most notably, cannabis relieves the discomfort and difficulty I have controlling it. It has also stopped the nausea that kept me awake at night' and
>
> 'I don't often take enough to "get high". When I do, I'm sure the feeling of calm and euphoria does my spirits a lot of good, too.'

1.4.3 Adverse effects of cannabinoids

After having considered the main adverse effects of cannabis, Hollister (1986) stated that:

> '. . . one is forced to conclude that cannabis is a relatively safe drug as social drugs go. To date, it compares favorably with tobacco and alcohol, if not with caffeine. One should bear in mind, however, the very long time that it took to determine the ill effects on health of these accepted social drugs'.

This still seems to be the state of play. Of course, cannabinoids are no different from other drugs in that they do give rise to adverse effects (see Hollister, 1986). Certain groups of patients are more at risk from some of these effects than others. In particular, the ability of cannabis to aggravate existing psychoses, induce tachycardia and suppress circulating levels of reproductive hormones means that it would be imprudent to give psychotropic cannabinoids to patients with schizophrenia (overt or latent), coronary arteriosclerosis or congestive heart failure, or to adolescent patients. In addition, there is evidence that when smoked, cannabis can decrease airway conductance, an effect which would put at risk patients with impaired pulmonary function. Cannabis smoke may also contribute to the development of premalignant and malignant lesions of lung cells and to the induction of chronic bronchitis. The clinical significance of the ability of cannabinoids to suppress immune function or to retard foetal development or induce foetal resorption in animals has still to be established (see Hollister, 1986). In the meantime it would be unwise to administer cannabinoids during pregnancy. The question of whether cannabinoids can induce permanent brain damage in man also still needs to be resolved. However, it is noteworthy that daily inhalation of cannabis smoke for a year did not produce any detectable morphological or neurochemical changes in the brains of rhesus monkeys when these were examined 7 months after drug withdrawal (Slikker *et al.*, 1992). Cannabinoid dependence is unlikely to be of major clinical significance. Thus, although dependence on psychotropic cannabinoids can be induced, the withdrawal syndrome is mild (Hollister, 1986; Pertwee, 1991). Finally, psychotropic cannabinoids have some properties that probably preclude their use in the treatment of certain disorders. For example, the ability of cannabinoids to aggravate existing psychoses probably rules out their use in the treatment of tardive dyskinesia where this has been provoked by major tranquillizers in schizophrenics. Similarly, because of their effects on cognitive function, memory, perception and mood it may prove unwise to give psychotropic cannabinoids to patients with Huntington's disease.

In conclusion, there is no evidence to suggest that psychotropic cannabinoids (or cannabis) are particularly unsafe or that their adverse effects are any more severe or unacceptable than those of many drugs now used clinically, for example L-DOPA, lithium, the major tranquillizers and some antidepressants. Indeed, it is already permissible to prescribe certain cannabinoids as antiemetics in Britain (nabilone) and North America (nabilone and Δ^9-THC) (Figures 1 and 7). Some patients may tolerate the side-effects of cannabinoids less well than others. However, a similar state of affairs exists for many drugs that are already used in the clinic.

1.4.4 Future research

The challenge now is for pharmacologists, both preclinical and clinical, to establish objectively whether drugs that interact with the anandamide system can be exploited to clinical advantage and, if so, in what ways. In particular, it will be important to identify any conditions for which such drugs might provide treatments that are more effective than existing treatments, or that are as effective as existing treatments but provoke adverse effects that are less serious or unpleasant. One strategy for future research is to investigate the pathophysiology of the anandamide system, for example by searching for abnormalities in the distribution pattern, density or properties of cannabinoid receptors in diseased tissues. Indeed, there is already evidence from a recent post mortem investigation, that patients with Huntington's disease have far fewer cannabinoid receptors in the substantia nigra pars reticulata than neurologically normal patients (Glass *et al.*, 1993). The question of whether changes of clinical significance can occur in genetic material responsible for expressing cannabinoid receptors or for expressing proteins that have a role in the synthesis, storage, transport or metabolism of anandamide should also be addressed. One possibility, that the human cannabinoid receptor gene might be linked to manic-depressive illness, has already been investigated (Hoehe *et al.*, 1993): no such linkage was detected. However, cannabinoid receptor gene expression is affected by age, experiments with rat striatal tissue having demonstrated age-related decreases both in the concentration of cannabinoid receptor mRNA and in the number of specific cannabinoid binding sites (Mailleux and Vanderhaeghen, 1992c).

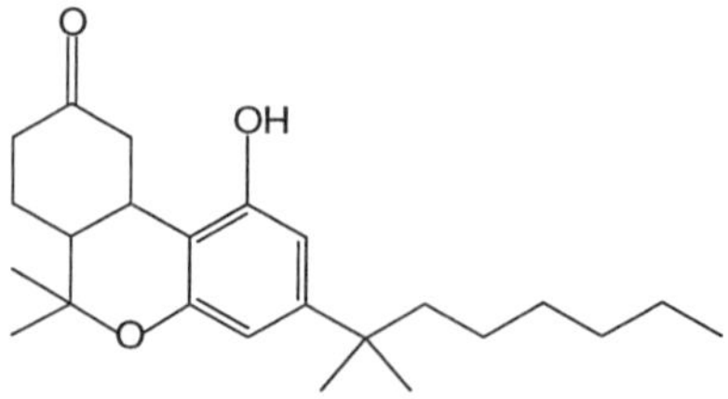

Nabilone

Figure 7 Structure of nabilone.

The question of whether similar changes occur in the human brain during ageing or in age-related disorders such as Alzheimer's disease or Parkinsons's disease, has yet to be investigated. Above all, it is important that the ongoing search for additional types of cannabinoid receptors and for selective cannabinoid antagonists continues. The existence of the former would increase the chance of developing cannabinoid receptor ligands with improved selectivity. The latter, as well as forming a potential new class of therapeutic agents, would be powerful experimental tools that would greatly facilitate the classification of cannabinoid receptors and speed up the elucidation of the physiological and pathophysiological roles of the anandamide system.

Further Developments

A major development that occurred during the proof stage of this book was the discovery of two potent cannabinoid receptor antagonists, SR141716A and AM 630 (Barth *et al.*, 1994; Pertwee *et al.*, 1994b; Rinaldi-Carmona *et al.*, 1994).

SR141716A, readily displaces tritiated CP 55,940 from cannabinoid binding sites in the brain (Ki=2 nM) and antagonizes the inhibitory effects of cannabinoid receptor agonists (CP 55,940 or WIN 55,212-2) both on the electrically-evoked twitch response of the mouse isolated vas deferens (dissociation constant=10 nM) and on adenylate cyclase activity, measured in rat brain membranes (IC_{50}=48 nM). SR141716A has also been shown to antagonize several effects of a cannabinoid receptor agonist *in vivo* (WIN 55,212-2). Effects of WIN 55,212-2 antagonized by doses of SR141716A below 2 mg/kg i.p. or p.o. include hypothermia, ring immobility (catalepsy) and antinociception. SR141716A seems to be a selective CB_1 receptor antagonist as concentrations in the micromolar range have been reported to interact neither with CB_2 receptors nor with a number of different non-cannabinoid receptors (Barth *et al.*, 1994; Rinaldi-Carmona *et al.*, 1994).

Research with AM 630 is still at an early stage (Pertwee *et al.*, 1994b). The results obtained so far indicate that the ability of this compound to antagonize the inhibitory effects of cannabinoid receptor agonists (CP 55,940, WIN 55,212-2 and Δ^9-THC) on the electrically-evoked twitch response of the mouse isolated vas deferens differs little from that of SR141716A (dissociation constant=20 nM). AM 630 has also been found to antagonize anandamide-induced inhibition of the twitch response. Concentrations of AM 630 that antagonize the effects of cannabinoids on the vas deferens have been found not to block the effects of two non-cannabinoid twitch inhibitors, morphine and clonidine.

SR141716A and AM 630 both behave like competitive, surmountable antagonists. Thus, when added to mouse isolated vasa deferentia, they can produce large rightward shifts in the log concentration response curves of cannabinoid receptor agonists (well over 100 fold) that are parallel and that do not involve any decrease in the size of the maximum responses to these agonists (Barth *et al.*, 1994; Pertwee et al., 1994b; Rinaldi-Carmona *et al.*, 1994).

References

Abadji, V., Lin, S., Taha, G., Griffin, G., Stevenson, L.A., Pertwee, R.G. & Makriyannis, A. (1994) *R*-methanandamide: a chiral novel anandamide possessing higher potency and metabolic stability. *J. Med. Chem.* **37**, 1889–1893.

Abel, E.L. (1975) Cannabis: effects on hunger and thirst. *Behav. Biol.* **15**, 255–281.

Abood, M.E., Sauss, C., Fan, F., Tilton, C.L. & Martin, B.R. (1993) Development of behavioral tolerance to Δ^9-THC without alteration of cannabinoid receptor binding or mRNA levels in whole brain. *Pharmacol. Biochem. Behav.* **46**, 575–579.

Anderson-Baker, W.C., McLaughlin, C.L. & Baile, C.A. (1979) Oral and hypothalamic injections of barbiturates, benzodiazepines and cannabinoids and food intake in rats. *Pharmacol. Biochem. Behav.* **11**, 487–491.

Barth, F., Héaulme, M., Shire, D., Calandra, B., Congy, C., Martinez, S., Maruani, J., Néliat, G., Caput, D., Ferrara, P., Soubrié, P., Brelière, J.-C., Le Fur, G. & Rinaldi-Carmona, M. (1994) SR141716A, a potent and selective antagonist of the brain cannabinoid receptor. *Int. Cann. Res. Soc. Abst.*

Bidaut-Russell, M. & Howlett, A.C. (1991) Cannabinoid receptor-regulated cyclic AMP accumulation in the rat striatum. *J. Neurochem.* **57**, 1769–1773.

Bidaut-Russell, M., Devane, W.A. & Howlett, A.C. (1990) Cannabinoid receptors and modulation of cyclic AMP accumulation in the rat brain. *J. Neurochem.* **55**, 21–26.

Bouaboula, M., Rinaldi, M., Carayon, P., Carillon, C., Delpech, B., Shire, D., Le Fur, G. & Casellas, P. (1993) Cannabinoid-receptor expression in human leukocytes. *Eur. J. Biochem.* **214**, 173–180.

Brown, J.E., Kassouny, M. & Cross, J.K. (1977) Kinetic studies of food intake and sucrose solution preference by rats treated with low doses of Δ^9-tetrahydrocannabinol. *Behav. Biol.* **20**, 104–110.

Caenazzo, L., Hoehe, M.R., Hsieh, W.-T., Berrettini, W.H., Bonner, T.I. & Gershon, E.S. (1991) HindIII identifies a two allele DNA polymorphism of the human cannabinoid receptor gene (CNR). *Nucl. Acids Res.* **19**, 4798.

Childers, S.R., Sexton, T. & Roy, M.B. (1994) Effects of anandamide on cannabinoid receptors in rat brain membranes. *Biochem. Pharmacol.* **47**, 711–715.

Clifford, D.B. (1983) Tetrahydrocannabinol for tremor in multiple sclerosis. *Ann. Neurol.* **13**, 669–671.

Consroe, P. & Snider, S.R. (1986) Therapeutic potential of cannabinoids in neurological disorders. In *Cannabinoids as Therapeutic Agents* (ed. Mechoulam, R.), pp. 21–49. CRC Press Inc., Boca Raton, FL.

Consroe, P. & Sandyk, R. (1992) Potential role of cannabinoids for therapy of neurological disorders. In *Marijuana/Cannabinoids. Neurobiology and Neurophysiology* (eds Murphy, L. & Bartke, A.), pp. 459–524. CRC Press Inc., Boca Raton, FL.

Crawley, J.N., Corwin, R.L., Robinson, J.K., Felder, C.C., Devane, W.A. & Axelrod, J. (1993) Anandamide, an endogenous ligand of the cannabinoid receptor, induces hypomotility and hypothermia *in vivo* in rodents. *Pharmacol. Biochem. Behav.* **46**, 967–972.

Davies, C. (1992) Drug dealers saved my wife from her MS hell. In *The Mail on Sunday*, 15 November.

de Fonseca, F.R., Gorriti, M.A., Fernández-Ruiz, J.J., Palomo, T. & Ramos, J.A. (1994a) Downregulation of rat brain cannabinoid binding sites after chronic Δ^9-tetrahydrocannabinol treatment. *Pharmacol. Biochem. Behav.* **47**, 33–40.

de Fonseca, F.R., Cebeira, M., Ramos, J.A., Martín, M. & Fernández-Ruiz, J.J. (1994b) Cannabinoid receptors in rat brain areas: sexual differences, fluctuations during estrous cycle and changes after gonadectomy and sex steroid replacement. *Life Sci.* **54**, 159–170.

Deutsch, D.G. & Chin, S.A. (1993) Enzymatic synthesis and degradation of anandamide, a cannabinoid receptor agonist. *Biochem. Pharmacol.* **46**, 791–796.

Devane, W.A., Dysarz, F.A., Johnson, M.R., Melvin, L.S. & Howlett, A.C. (1988) Determination and characterization of a cannabinoid receptor in rat brain. *Mol. Pharmacol.* **34**, 605–613.

Devane, W.A., Hanuš, L., Breuer, A., Pertwee, R.G., Stevenson, L.A., Griffin, G., Gibson, D., Mandelbaum, A., Etinger, A. & Mechoulam, R. (1992) Isolation and structure of a brain constituent that binds to the cannabinoid receptor. *Science* **258**, 1946–1949.

Dewey, W.L. (1986) Cannabinoid Pharmacology. *Pharmacol. Rev.* **38**, 151–178.

Dill, J.A. & Howlett, A.C. (1988) Regulation of adenylate cyclase by chronic exposure to cannabimimetic drugs. *J. Pharmacol. Exp. Ther.* **244**, 1157–1163.

Doyle, C. (1992) High, dry and happier. In *The Daily Telegraph*, 24 November.

Felder, C.C., Veluz, J.S., Williams, H.L., Briley, E.M. & Matsuda, L.A. (1992) Cannabinoid agonists stimulate both receptor- and non-receptor-mediated signal transduction pathways in cells transfected with and expressing cannabinoid receptor clones. *Mol. Pharmacol.* **42**, 838–845.

Felder, C.C., Briley, E.M., Axelrod, J., Simpson, J.T., Mackie, K. & Devane, W.A. (1993) Anandamide, an endogenous cannabimimetic eicosanoid, binds to the cloned human cannabinoid receptor and stimulates receptor-mediated signal transduction. *Proc. Natl Acad. Sci. USA* **90**, 7656–7660.

Ferriman, A. (1993) Marijuana: the best medicine? In *The Times*, 4 May.

Foltin, R.W., Fischman, M.W. & Byrne, M.F. (1988) Effects of smoked marijuana on food intake and body weight of humans living in a residential laboratory. *Appetite* **11**, 1–14.

Fride, E. & Mechoulam, R. (1993) Pharmacological activity of the cannabinoid receptor agonist, anandamide, a brain constituent. *Eur. J. Pharmacol.* **231**, 313–314.

Gaoni, Y. & Mechoulam, R. (1964) Isolation, structure and partial synthesis of an active constituent of hashish. *J. Amer. Chem. Soc.* **86**, 1646–1647.

Gérard, C., Mollereau, C., Vassart, G. & Parmentier, M. (1990) Nucleotide sequence of a human cannabinoid receptor cDNA. *Nucl. Acids Res.* **18**, 7142.

Gérard, C.M., Mollereau, C., Vassart, G. & Parmentier, M. (1991) Molecular cloning of a human cannabinoid receptor which is also expressed in testis. *Biochem. J.* **279**, 129–134.

Glass, M., Faull, R.L.M. & Dragunow, M. (1993) Loss of cannabinoid receptors in the substantia nigra in Huntington's disease. *Neuroscience* **56**, 523–527.

Glick, S.D. & Milloy, S. (1972) Increased and decreased eating following THC administration. *Psychon. Sci.* **29**, 6.

Gluck, J.P. & Ferraro, D.P. (1974) Effects of Δ^9-THC on food and water intake of deprivation experienced rats. *Behav. Biol.* **11**, 395–401.
Greenberg, I., Kuehnle, J., Mendelson, J.H. & Bernstein, J.G. (1976) Effects of marihuana use on body weight and caloric intake in humans. *Psychopharmacology* **49**, 79–84.
Griffin, G. & Pertwee, R.G. (1994) The mouse vas deferens as a bioassay for cannabimimetic agents: the question of selectivity. *Int. Cannabis Res. Soc. Abstr.*
Handscomb, M. (1993) Cannabis: why doctors want it to be legal. *The Independent*, 23 February.
Hanuš, L., Gopher, A., Almog, S. & Mechoulam, R. (1993) Two new unsaturated fatty acid ethanolamides in brain that bind to the cannabinoid receptor. *J. Med. Chem.* **36**, 3032–3034.
Herkenham, M., Lynn, A.B., Little, M.D., Johnson, M.R., Melvin, L.S., de Costa, B.R. & Rice, K.C. (1990) Cannabinoid receptor localization in brain. *Proc. Natl Acad. Sci. USA* **87**, 1932–1936.
Herkenham, M., Groen, B.G.S., Lynn, A.B., de Costa, B.R. & Richfield, E.K. (1991a) Neuronal localization of cannabinoid receptors and second messengers in mutant mouse cerebellum. *Brain Res.* **552**, 301–310.
Herkenham, M., Lynn, A.B., Johnson, M.R., Melvin, L.S., de Costa, B.R. & Rice, K.C. (1991b) Characterization and localization of cannabinoid receptors in rat brain: a quantitative *in vitro* autoradiographic study. *J. Neurosci.* **11**, 563–583.
Hodges, C. (1992) Very alternative medicine. In *The Spectator*, 1 August.
Hodges, C. (1993) I wish I could get it at the chemist's. In *The Independent*, 23 February.
Hoehe, M.R., Caenazzo, L., Martinez, M.M., Hsieh, W.-T., Modi, W.S., Gershon, E.S. & Bonner, T.I. (1991) Genetic and physical mapping of the human cannabinoid receptor gene to chromosome 6q14–q15. *New Biologist* **3**, 880–885.
Hoehe, M.R., Goldin, L.R., Martinez, M.M., Caenazzo, L., Hsieh, W.-T., Gershon, E.S., Bonner, T.I. & Church, G.M. (1993) Molecular genetic studies of the human cannabinoid receptor gene. *Int. Cannabis Res. Soc. Abstr.*
Hollister, L.E. (1986) Health aspects of cannabis. *Pharmacol. Rev.* **38**, 1–20.
Howlett, A.C. (1984) Inhibition of neuroblastoma adenylate cyclase by cannabinoid and nantradol compounds. *Life Sci.* **35**, 1803–1810.
Howlett, A.C. (1985) Cannabinoid inhibition of adenylate cyclase. Biochemistry of the response in neuroblastoma cell membranes. *Mol. Pharmacol.* **27**, 429–436.
Howlett, A.C. (1987) Cannabinoid inhibition of adenylate cyclase: relative activity of constituents and metabolites of marihuana. *Neuropharmacology* **26**, 507–512.
Howlett, A.C. & Fleming, R.M. (1984) Cannabinoid inhibition of adenylate cyclase. Pharmacology of the response in neuroblastoma cell membranes. *Mol. Pharmacol.* **26**, 532–538.
Howlett, A.C., Qualy, J.M. & Khachatrian, L.L. (1986) Involvement of G_i in the inhibition of adenylate cyclase by cannabimimetic drugs. *Mol. Pharmacol.* **29**, 307–313.
Howlett, A.C., Johnson, M.R., Melvin, L.S. & Milne, G.M. (1988) Nonclassical cannabinoid analgetics inhibit adenylate cyclase: development of a cannabinoid receptor model. *Mol. Pharmacol.* **33**, 297–302.
Howlett, A.C., Champion, T.M., Wilken, G.H. & Mechoulam, R. (1990) Stereochemical effects of 11-OH-Δ^8-tetrahydrocannabinol-dimethylheptyl to inhibit adenylate cyclase and bind to the cannabinoid receptor. *Neuropharmacology* **29**, 161–165.
Howlett, A.C., Evans, D.M. & Houston, D.B. (1992) The cannabinoid receptor. In *Marijuana/Cannabinoids. Neurobiology and Neurophysiology* (eds Murphy, L. & Bartke, A.), pp 35–72. CRC Press Inc., Boca Raton, FL.
James, T. (1993) Breaking the law to beat MS. In *The Yorkshire Post*, 27 September.
Jansen, E.M., Haycock, D.A., Ward, S.J. & Seybold, V.S. (1992) Distribution of cannabinoid receptors in rat brain determined with aminoalkylindoles. *Brain Res.* **575**, 93–102.

Kaminski, N.E., Abood, M.E., Kessler, F.K., Martin, B.R. & Schatz, A.R. (1992) Identification of a functionally relevant cannabinoid receptor on mouse spleen cells that is involved in cannabinoid-mediated immune modulation. *Mol. Pharmacol.* **42**, 736–742.
Kuster, J.E., Stevenson, J.I., Ward, S.J., D'Ambra, T.E. & Haycock, D.A. (1993) Aminoalkylindole binding in rat cerebellum: selective displacement by natural and synthetic cannabinoids. *J. Pharmacol. Exp. Ther.* **264**, 1352–1363.
Lyman, W.D., Sonett, J.R., Brosnan, C.F., Elkin, R. & Bornstein, M.B. (1989) Δ^9-tetrahydrocannabinol: a novel treatment for experimental autoimmune encephalomyelitis. *J. Neuroimmunol.* **23**, 73–81.
Mackie, K., Devane, W.A. & Hille, B. (1993) Anandamide, an endogenous cannabinoid, inhibits calcium currents as a partial agonist in N18 neuroblastoma cells. *Mol. Pharmacol.* **44**, 498–503.
Mailleux, P. & Vanderhaeghen, J.-J. (1992a) Distribution of neuronal cannabinoid receptor in the adult rat brain: a comparative receptor binding radioautography and *in situ* hybridization histochemistry. *Neuroscience* **48**, 655–668.
Mailleux, P. & Vanderhaeghen, J.-J. (1992b) Localization of cannabinoid receptor in the human developing and adult basal ganglia. Higher levels in the striatonigral neurons. *Neurosci. Lett.* **148**, 173–176.
Mailleux, P. & Vanderhaeghen, J.-J. (1992c) Age-related loss of cannabinoid receptor binding sites and mRNA in the rat striatum. *Neurosci. Lett.* **147**, 179–181.
Mailleux, P. & Vanderhaeghen, J.-J. (1993a) Glucocorticoid regulation of cannabinoid receptor messenger RNA levels in the rat caudate-putamen. An *in situ* hybridization study. *Neurosci. Lett.* **156**, 51–53.
Mailleux, P. & Vanderhaeghen, J.-J. (1993b) Dopaminergic regulation of cannabinoid receptor mRNA levels in the rat caudate-putamen: an *in situ* hybridization study. *J. Neurochem.* **61**, 1705–1712.
Mailleux, P. & Vanderhaeghen, J.-J. (1994) Glutamatergic regulation of cannabinoid receptor gene expression in the caudate-putamen. *Eur. J. Pharmacol. – Mol. Pharmacol. Section* **266**, 193–196.
Mailleux, P., Verslijpe, M. & Vanderhaeghen, J.-J. (1992a) Initial observations on the distribution of cannabinoid receptor binding sites in the human adult basal ganglia using autoradiography. *Neurosci. Lett.* **139**, 7–9.
Mailleux, P., Parmentier, M. & Vanderhaeghen, J.-J. (1992b) Distribution of cannabinoid receptor messenger RNA in the human brain: an *in situ* hybridization histochemistry with oligonucleotides. *Neurosci. Lett.* **143**, 200–204.
Martin, B.R. (1986) Cellular effects of cannabinoids. *Pharmacol. Rev.* **38**, 45–74.
Martin, B.R., Compton, D.R., Thomas, B.F., Prescott, W.R., Little, P.J., Razdan, R.K., Johnson, M.R., Melvin, L.S., Mechoulam, R. & Ward, S.J. (1991) Behavioral, biochemical, and molecular modeling evaluations of cannabinoid analogs. *Pharmacol. Biochem. Behav.* **40**, 471–478.
Matsuda, L.A., Bonner, T.I. & Lolait, S.J. (1993) Localization of cannabinoid receptor mRNA in rat brain. *J. Comp. Neurol.* **327**, 535–550.
Matsuda, L.A., Lolait, S.J., Brownstein, M.J., Young, A.C. & Bonner, T.I. (1990) Structure of a cannabinoid receptor and functional expression of the cloned cDNA. *Nature* **346**, 561–564.
Maurer, M., Henn, V., Dittrich, A. & Hofmann, A. (1990) Delta-9-tetrahydrocannabinol shows antispastic and analgesic effects in a single case double-blind trial. *Eur. Arch. Psychiatry Clin. Neurosci.* **240**, 1–4.
McLaughlin, C.R. & Abood, M.E. (1993) Developmental expression of cannabinoid receptor mRNA. *Develop. Brain Res.* **76**, 75–78.
McLaughlin, C.L., Baile, C.A. & Bender, P.E. (1979) Cannabinols and feeding in sheep. *Psychopharmacology* **64**, 321–323.

Mechoulam, R. (1986) The pharmacohistory of Cannabis sativa. In *Cannabinoids as Therapeutic Agents* (ed. Mechoulam, R.), pp. 1–19. CRC Press Inc., Boca Raton, FL.
Mechoulam, R. & Gaoni, Y. (1965) A total synthesis of dl-Δ^1-tetrahydrocannabinol, the active constituent of hashish. *J. Amer. Chem. Soc.* **87**, 3273–3275.
Meinck, H.-M., Schönle, P.W. & Conrad, B. (1989) Effect of cannabinoids on spasticity and ataxia in multiple sclerosis. *J. Neurol.* **236**, 120–122.
Milano, W.C., Wild, K.D., Hui, Y., Hubbell, C.L. & Reid, L.D. (1988) PCP, THC, ethanol, and morphine and consumption of palatable solutions. *Pharmacol. Biochem. Behav.* **31**, 893–897.
Miller, L.L. & Branconnier, R.J. (1983) Cannabis: effects on memory and the cholinergic limbic system. *Psychol. Bull.* **93**, 441–456.
Munro, S., Thomas, K.L. & Abu-Shaar, M. (1993) Molecular characterization of a peripheral receptor for cannabinoids. *Nature* **365**, 61–65.
Munson, A.E. & Fehr, K.O. (1983) Immunological effects of cannabis. In *Adverse Health and Behavioral Consequences of Cannabis Use* (eds Fehr, K.O. & Kalant, H.), pp. 257–353. Addiction Research Foundation, Toronto.
Oviedo, A., Glowa, J. & Herkenham, M. (1993) Chronic cannabinoid administration alters cannabinoid receptor binding in rat brain: a quantitative autoradiographic study. *Brain Res.* **616**, 293–302.
Pacheco, M., Childers, S.R., Arnold, R., Casiano, F. & Ward, S.J. (1991) Aminoalkylindoles: actions on specific G-protein-linked receptors. *J. Pharmacol. Exp. Ther.* **257**, 170–183.
Pacheco, M.A., Ward, S.J. & Childers, S.R. (1993) Identification of cannabinoid receptors in cultures of rat cerebellar granule cells. *Brain Res.* **603**, 102–110.
Paton, W.D.M. & Pertwee, R.G. (1973a) The actions of cannabis in man. In *Marijuana. Chemistry, Pharmacology, Metabolism and Clinical Effects* (ed. Mechoulam, R.), pp. 287–333. Academic Press, New York.
Paton, W.D.M. & Pertwee, R.G. (1973b) The pharmacology of cannabis in animals. In *Marijuana. Chemistry, Pharmacology, Metabolism and Clinical Effects* (ed. Mechoulam, R.), pp. 191–285. Academic Press, New York.
Pertwee, R.G. (1985) Effects of cannabinoids on thermoregulation: a brief review. In *Marihuana '84* (ed. Harvey, D.J.), pp. 263–277. IRL Press, Oxford.
Pertwee, R.G. (1990) The central neuropharmacology of psychotropic cannabinoids. In *Psychotropic Drugs of Abuse* (ed. Balfour, D.J.K.), pp. 355–429. Pergamon Press, New York.
Pertwee, R.G. (1991) Tolerance to and dependence on psychotropic cannabinoids. In *The Biological Bases of Drug Tolerance and Dependence* (ed. Pratt, J.A.), pp. 231–263. Academic Press, London.
Pertwee, R. (1993) The evidence for the existence of cannabinoid receptors. *Gen. Pharmacol.* **24**, 811–824.
Pertwee, R.G. & Griffin, G. (1995) A preliminary investigation of the mechanisms underlying cannabinoid tolerance in the mouse vas deferens. *Eur. J. Pharmacol.* **272**, 67–72.
Pertwee, R.G. & Stevenson, L.A. (1994) Effect of pretreatment with delta-9-tetrahydrocannabinol on the ability of certain cannabimimetic agents to induce hypothermia in mice. In *Temperature Regulation. Recent Physiological and Pharmacological Advances* (ed. A.S. Milton), pp. 177–182. Birkhäuser, Basel.
Pertwee, R.G., Stevenson, L.A., Elrick, D.B., Mechoulam, R. & Corbett, A.D. (1992) Inhibitory effects of certain enantiomeric cannabinoids in the mouse vas deferens and the myenteric plexus preparation of guinea-pig small intestine. *Br. J. Pharmacol.* **105**, 980–984.
Pertwee, R.G., Stevenson, L.A. & Griffin, G. (1993) Cross-tolerance between delta-9-tetrahydrocannabinol and the cannabimimetic agents, CP 55,940, WIN 55,212-2 and anandamide. *Br. J. Pharmacol.* **110**, 1483–1490.
Pertwee, R.G., Griffin, G., Hanuš, L. & Mechoulam, R. (1994a) Effects of two endogenous fatty acid ethanolamides on mouse vasa deferentia. *Eur. J. Pharmacol.* **259**, 115–120.

Pertwee, R.G., Griffin, G., Fernando, S.R., Li, X., Abadji, V. & Makriyannis, A. (1994b) AM630, a competitive surmountable cannabinoid receptor antagonist? *Int. Cann. Res. Soc. Abst.*

Pertwee, R.G., Fernando, S.R., Griffin, G., Abadji, V. & Makriyannis, A. (1995) Effect of phenylmethylsulfonyl fluoride on the potency of anandamide as an inhibitor of electrically evoked contractions in two isolated tissue preparations. *Eur. J. Pharmacol.* **272**, 73–78.

Petro, D.J. & Ellenberger, C. (1981) Treatment of human spasticity with Δ^9-tetrahydrocannabinol. *J. Clin Pharmacol* **21** (Suppl.), 413S-416S.

Razdan, R.K. (1986) Structure-activity relationships in cannabinoids. *Pharmacol. Rev.* **38**, 75–149.

Rinaldi-Carmona, M., Barth, F., Héaulme, M., Shire, D., Calandra, B., Congy, C., Martinez, S., Maruani, J., Néliat, G., Caput, D., Ferrara, P., Soubrié, P., Brelière, J.-C. & Le Fur, G. (1994) SR141716A, a potent and selective antagonist of the brain cannabinoid receptor. *FEBS Letts.* **350**, 240–244.

Slikker, W., Paule, M.G., Ali, S.F., Scallet, A.C. & Bailey, J.R. (1992) Behavioural, neurochemical, and neurohistological effects of chronic marijuana smoke exposure in the non-human primate. In *Marijuana/Cannabinoids. Neurobiology and Neurophysiology* (eds Murphy, L. & Bartke, A.), pp. 219–273. CRC Press Inc, Boca Raton, FL.

Thomas, B.F., Wei, X. & Martin, B.R. (1992) Characterization and autoradiographic localization of the cannabinoid binding site in rat brain using [^{3}H]11-OH-Δ^9-THC-DMH. *J. Pharmacol. Exp. Ther.* **263**, 1383–1390.

Trojniar, W. & Wise, R.A. (1991) Facilitory effect of Δ^9-tetrahydrocannabinol on hypothalamically induced feeding. *Psychopharmacology* **103**, 172–176.

Ungerleider, J.T., Andyrsiak, T., Fairbanks, L., Ellison, G.W. & Myers, L.W. (1987) Delta-9-THC in the treatment of spasticity associated with multiple sclerosis. *Adv. Alcohol Subst. Abuse* **7**, 39–50.

Vogel, Z., Barg, J., Levy, R., Saya, D., Heldman, E. & Mechoulam, R. (1993) Anandamide, a brain endogenous compound, interacts specifically with cannabinoid receptors and inhibits adenylate cyclase. *J. Neurochem*, **61**, 352–355.

Watson, S. & Girdlestone, D. (1994) 1994 receptor and ion channel nomenclature supplement. *Trends in Pharmacological Sciences (Suppl.)*, 5th edn. Elsevier, Cambridge.

Wickens, A.P. & Pertwee, R.G. (1993) Δ^9-tetrahydrocannabinol and anandamide enhance the ability of muscimol to induce catalepsy in the globus pallidus of rats. *Eur. J. Pharmacol.* **250**, 205–208.

CHAPTER 2

STRUCTURAL REQUIREMENTS FOR CANNABINOID RECEPTOR PROBES

Billy R. Martin[1], Brian F. Thomas[2] and Raj K. Razdan[3]

[1]*Department of Pharmacology and Toxicology, Virginia Commonwealth University, Richmond, VA 23298, USA*
[2]*Research Triangle Institute, Research Triangle Park, NC 27709, USA*
[3]*Organix, Inc., Cambridge, MA 01801, USA*

Table of Contents

Cannabinoid Receptors
ISBN 0-12-551460-3

2.1 Introduction

The objectives of this chapter are to provide a critical description of the models which are used to establish cannabinoid activity, discuss the relevance of findings in these models, describe structural determinants for psychotropic activity, discuss structure–activity anomalies and the possibility of receptor subtypes, review possible strategies for developing antagonists and discuss promising leads.

2.2 Critical description of *in vivo* and *in vitro* assay systems

Like most centrally acting agents, cannabinoids produce a complex array of pharmacological effects. These diverse effects provide a plethora of measures which can be assessed, while at the same time complicating interpretation of the findings. Numerous models for evaluating cannabinoids have arisen because of the complex pharmacological nature of cannabinoids and as a consequence of the different research questions which have been asked. In general, most of the research questions are directed toward determining structural prerequisites for psychotomimetic activity, identifying biochemical correlates for specific pharmacological actions, and determining whether therapeutically useful agents can be developed from the cannabinoids. The models which have been developed range from a single measure in a simple system to multiple measures in a complex system. Obviously, in order to establish structural requirements, the system should be quantitative (i.e. potency can be established with a reasonable degree of precision) and the measure must be specific for the class of compounds in question. Precision becomes especially important with mechanistic studies which rely on correlation of *in vivo* and *in vitro* effects. For example, to determine whether a particular structural modification is involved in the production of a specific pharmacological effect, a semi-quantitative analysis of the pharmacological effect is usually sufficient. Conversely, determination of whether a particular pharmacological effect is mediated through a receptor requires quantitative measures in both assays. The objective of this section is to discuss some of the more prominent assay systems with regard to their strengths and limitations.

2.2.1 In *vivo* models in laboratory animals

The strengths and weaknesses of *in vivo* models are summarized in Table 1.

2.2.1.1 Dog static-ataxia

One of the earliest models was introduced by Walton more than 50 years ago (Walton *et al.*, 1937). His description of the effects of cannabinoids in dogs effectively documented the uniqueness of this syndrome for this class of compounds. The effects include sedation, catalepsy, motor incoordination and hyperexcitability. However, it is the combination of these effects which causes dogs to weave to and fro while remaining fixed in one spot, hence the anomalous term static-ataxia. The primary advantage of this model is that these behaviours describe a highly specific profile for cannabinoids which is not confused with that produced by other behaviourally active compounds. In addition, Walton developed a semi-quantitative scale which allows for the reliable estimation of potency. This model has been used successfully during the past 20 years by our laboratory. However, the subjective nature of the assessment and the semi-quantitative scale are less than ideal. These shortcomings can be largely overcome by requiring the experimenter be 'blinded' with regard to treatment and by proper experimental design. Our typical experimental design involves five dogs that receive an i.v. administration of either vehicle, a positive control (Δ^9-tetrahydrocannabinol (Δ^9-THC), 0.5 mg/kg), or one of three doses of the test compound. Attempts are made to have at least three replicates for each treatment group. The animals can be used more than once if 1 week is allowed to elapse before the animal is retested. We have observed no tolerance using this practice. The strength of this model is that the results obtained correlate well with psychoactivity. Needless to say, one of the greatest disadvantages, which is associated with all studies conducted in large animals, is the effort and cost which is required to maintain the animals. Because of these considerations, it is difficult to determine dose–response relationships with a sufficient number of animals at each dose.

2.2.1.2 Overt behaviour in monkeys

In the late 1960s and early 1970s, R. Mechoulam's laboratory synthesized a large number of cannabinoid analogues which were evaluated primarily in the monkey (Edery *et al.*, 1971). Δ^9-THC produced sedation, ptosis, body sag, etc., which was reasonably selective for cannabinoids. These investigators also developed a semi-quantitative rating scale which allowed them to make potency determinations. However, most of the disadvantages discussed with the dog model also apply to the monkey.

Table 1 *In vivo* models in laboratory animals and humans for assessing cannabinoid effects

Model	Species	Route	Dose (mg/kg)	Advantages	Disadvantages	Reference
Static-ataxia	Dog	i.v.	0.1–0.5	High specificity, good correlation with psychotomimetic activity.	Semi-quantitative, time-consuming, experimental difficulties associated with large animals. Possible tolerance with repeated administration.	Walton *et al.* (1937)
Overt behavior	Monkeys	i.m.	0.05–0.5	Reasonable specificity, good correlation with psychotomimetic activity.	Semi-quantitative, time-consuming, experimental difficulties associated with large animals. Possible tolerance with repeated administration.	Edery *et al.* (1971)
Drug discrimination	Monkeys	i.m.	0.05–0.20	High specificity, quantitative, excellent correlation with psychotomimetic activity.	Time-consuming, extensive training required, expensive.	Gold *et al.* (1992)
Drug discrimination	Rats	i.p.	3.0	High specificity, quantitative, excellent correlation with psychotomimetic activity.	Time-consuming, extensive training required.	Järbe and McMillan (1980); Balster and Prescott (1992)

Table 1 *(Contd)*

Model	Species	Route	Dose (mg/kg)	Advantages	Disadvantages	Reference
Catalepsy	Rats	i.p.	0.3–10.0	Quantitative, minimal training required.	Low specificity.	Prescott *et al.* (1992)
Tetrad model	Mice	i.v.	1.0–3.0	Quantitative, minimal training required, not very labour intensive, provides a profile of action, ideal for establishing dose–response and time course.	High specificity.	Compton *et al.* (1993)
Psychotomimetic	Humans	i.v.	0.05	Direct measure of psychotomimetic effects. No training required.	Time-consuming and expensive. Limited to compounds with known pharmacological and toxicological profiles. Semi-quantitative.	Perez-Reyes *et al.* (1972); Hollister (1974)
Tachycardia	Humans	i.v.	0.05	Quantifiable, reliable.	Same as above. Not selective.	Perez-Reyes *et al.* (1972)

2.2.1.3 Drug discrimination in rats

Drug discrimination is considered to be one of the most reliable means of predicting whether test drugs produce behavioural effects similar to those of a known drug. Initially, an animal is trained to press a lever for food reward and then subsequently trained to press a specific lever for this reward when under the influence of Δ^9-THC and another lever when any other drug is administered. Therefore, on test days the animal's lever choice tells the experimenter whether the test compound is perceived as THC-like or not. Much of the rat drug discrimination literature for the cannabinoids has been generated by Järbe's laboratory (Järbe and Ohlin, 1977; Järbe and McMillan, 1979, 1980; Järbe *et al.*, 1989; Järbe and Mathis, 1992) and has been reviewed recently (Balster and Prescott, 1992). In our laboratory, three to six months are required to train animals to discriminate between vehicle and Δ^9-THC (3.0 mg/kg, i.p.). At least 80% of the animal's responses must be on the correct drug lever for a compound to be considered THC-like. Once trained, the animals can be used for up to 1 year.

The discriminative stimulus cue is reliable and dose-related with a relatively short onset (testing 15 min after an i.p. injection). We have also trained animals to discriminate between CP 55,940 and vehicle (Gold *et al.*, 1992). Almost all centrally acting agents are depressants at some dose, which produces a decrease in the rate at which the animals press the lever for food. Measurement of rate suppression provides some indication of the depressant properties of the drug. There are several drawbacks to drug discrimination. Extensive training is required and this training must be maintained throughout. It is therefore rather labour intensive to maintain a large colony of trained animals. Another drawback is that weak agonists frequently produce severe rate suppression before complete generalization occurs. It is also difficult to characterize partial agonists because the criterion for generalization requires 80% correct drug lever response. However, the high selectivity of this assay far outweighs these drawbacks. In addition, the drug discrimination is easily quantified and potency ratios readily determined.

2.2.1.4 Drug discrimination in monkeys

The above description of drug discrimination in rats applies to monkeys. One concern with drug discrimination is that, despite an excellent correlation between rat drug discrimination and human behavioural effects, the experimenter can never be absolutely sure that a unique behavioural effect of the drug is serving as the cue. The rationale for using primates rather than rats is that the results may provide a more accurate reflection of cannabinoid behavioural effects in humans. Fortunately, monkeys are highly sensitive to cannabinoids so that large quantities of the test drugs are not required. Needless to say, considerable time and effort are required to maintain this behavioural model. In addition, monkeys exhibit a wide variability in sensitivity to cannabinoids thereby making presentation of group data difficult. Despite these shortcomings, this model has provided reassuring data that novel cannabinoids, such as CP 55,940 (Gold *et al.*, 1992) and WIN 55,212 (Compton *et al.*, 1992a) are likely to produce cannabinoid behavioural effects in humans. Establishing this fact is particularly crucial since these compounds are being used widely as cannabinoid probes.

2.2.1.5 Rat catalepsy

Measurement of catalepsy using the elevate-ring procedure has proven to be a very useful model for studying numerous classes of centrally acting drugs because the neurochemical substrates have been well characterized (Pertwee and Greentree, 1988). Cannabinoids readily produce catalepsy in rats 30 min after i.p. administration (Prescott *et al.*, 1992). However, the doses required for catalepsy are somewhat higher than those necessary for drug discrimination. Fortunately, catalepsy has a long duration which has allowed us to measure successfully the discriminative cue and catalepsy in the same animal (Prescott *et al.*, 1992). One of the drawbacks is that manually determining periods of immobility can be quite tedious and time consuming. It is for these reasons that we developed a computer system for quantitative analysis of movement (Martin *et al.*, 1992). The quantitative nature of catalepsy makes it a valuable tool for SAR studies. However, numerous classes of compounds produce catalepsy so that this assay when used alone does not provide cannabinoid specificity.

2.2.1.6 Mouse tetrad model

As mentioned above, cannabinoids are known to produce a wide range of pharmacological effects which include hyperstimulation, sedation, catalepsy and several other depressant properties. Individually, none of these effects can be considered unique for cannabinoids, since all of these properties are shared by numerous classes of centrally active agents. Several years ago we initiated efforts to develop a cannabinoid behavioural model that would allow us to establish the pharmacological profile of a large number of analogues. Ideally, such a model should be relatively simple and reasonably selective. We discovered that, following i.v. administration in mice, cannabinoids produce sedation, hypothermia, antinociception and catalepsy in the same dosage range and within the same time frame so that all four behaviours could be determined in the same animal for each injection. Our experience with this composite model indicates that compounds which produce all four behaviours always exhibit cannabinoid effects in models which we traditionally consider to be highly predictive, such as drug discrimination. The advantages of this model are that relatively small quantities of material are required for a comprehensive evaluation, no animal training is required, and all of the measures can be quantified (Compton *et al.*, 1993).

2.2.1.7 Summary

The choice of assay depends upon numerous factors, not least of which is the scientific objective of the study. The mouse behavioural model is ideal for an initial evaluation and for establishing a general pharmacological profile. It is also well suited for determining whether an analogue exhibits any separation in pharmacological effects, such as producing antinociception without sedation. If the goal of the research is to establish, beyond doubt, that a small number of compounds produce THC-like effects, then drug discrimination becomes a priority. Of course, the evaluation of a substance such as anandamide, the putative endogenous cannabinoid, should be undertaken in as many models as possible.

2.2.2 Human assays

Several cannabinoids have been evaluated in humans. These include the active constituents in marijuana, their metabolites and some agents with therapeutic potential (Razdan, 1986). Some of the earlier studies demonstrated that structure–activity relationship (SAR) experiments could be conducted in humans (Perez-Reyes *et al.*, 1972; Hollister, 1974). Perez-Reyes infused subjects with a test compound and determined the titration dose at which the subjects became 'high'. These evaluations in humans have provided the basis for correlating psychotomimetic potency to potency in animal models. The disadvantages of a human model are obvious. Studies conducted with human subjects have to be done with extreme caution and are confined to drugs with known pharmacological and toxicological properties. Therefore, novel compounds with little background information cannot be tested in humans. In addition, numerous variables can have impact on the findings, such as prior drug histories, the research setting, etc.

It is worth mentioning in passing that one of the most consistent findings with the cannabinoids in humans is tachycardia which provides a reliable measure of potency. Obviously, measurement of tachycardia alone does not provide any specificity.

2.2.3 *In vitro* assays

Cannabinoids exert numerous biochemical effects which could be used for establishing potency and ultimately SAR. However, it appears that most of the effects of the cannabinoids can be attributed to interactions with the cannabinoid receptor. There is also considerable evidence linking the G-protein coupled cannabinoid receptor to adenylate cyclase and, to a lesser extent, to calcium channels. Therefore, determination of receptor affinity and the ability of cannabinoids to inhibit adenylate cyclase and modulate calcium channels are obvious choices for exploring cannabinoid mechanism of action. In addition, smooth muscle preparations provide an excellent means for assessing *in vitro* potency and function. The *in vitro* assays are summarized in Table 2.

2.2.3.1 Receptor binding

Clearly, the most useful *in vitro* assay is receptor binding which has been conducted primarily with [^{3}H]CP 55,940 (Devane *et al.*, 1988) and to a lesser extent with [^{3}H]11-OH-Δ^9-THC-dimethylheptyl (Thomas *et al.*, 1992), [^{3}H]WIN 55,212-2 (Pacheco *et al.*, 1993) and [^{3}H]HU-243 (Devane *et al.*, 1992a). All of these ligands appear to label the same site and would appear to be interchangeable. These receptor assays have been found to be highly selective for cannabinoids (Howlett *et al.*, 1992). Of course, receptor binding assays have both advantages (cheaper and less time-consuming than *in vivo* models) and disadvantages (fail to distinguish agonists and antagonists; do not assess function). One of the greatest benefits of receptor assays is delineation among receptor subtypes by the use of highly selective ligands, as has been accomplished with the opioids. To date no selective cannabinoid ligands have emerged.

2.2.3.2 Adenylate cyclase

The demonstration that the cannabinoid receptor was coupled to adenylate cyclase made this enzyme a logical target for a functional assay (Howlett, 1984). While manipulation of adenylate cyclase represents an important means for elucidating cannabinoid actions, there are numerous limitations which restrict its usefulness for SAR studies. First, Δ^9-THC inhibition of adenylate cyclase generates a very small signal so that reliable quantitative changes usually occur only with potent agonists. In addition, selected brain areas should be used as many are not susceptible to cannabinoid-induced adenylate cyclase inhibition (Bidaut-Russell *et al.*, 1990). Cell culture, either transformed cells (Howlett, 1984) or primary cerebellar granule cells (Pacheco *et al.*, 1993), have proven to provide a greater signal than brain tissue, but they too are not very suitable for agonists less potent than Δ^9-THC. The original observation that adenylate cyclase is involved in cannabinoid action was based on a limited SAR study in neuroblastoma cells (Howlett, 1984), while attempts to establish SAR using whole-brain homogenates were largely futile (Little and Martin, 1991). Notwithstanding these limitations, measurement of adenylate cyclase can be valuable for SAR studies under selected circumstances.

2.2.3.3 Smooth muscle preparations

It has long been known that cannabinoids will inhibit electrically stimulated contractions of smooth muscle preparations. Although it had been demonstrated much earlier that cannabinoids inhibited contractions of the guinea-pig ileum (Rosell *et al.*, 1976), only recently has this preparation been revived for SAR studies (Pertwee *et al.*, 1992). In addition, it has been demonstrated that the mouse vas deferens is much more sensitive to cannabinoids than the guinea-pig ileum (Pacheco *et al.*, 1991; Pertwee *et al.*, 1992). The advantages of this preparation are that function can be assessed in a quantitative manner and that potency can be determined

Table 2 Advantages and disadvantages of *in vitro* assays for cannabinoids

Assay	Substrate or tissue	Advantages	Disadvantages	Reference
Receptor binding	[^{3}H]CP 55,940	Highly selective, quick to perform, and quantitative.	Does not assess function, fails to delineate between agonists and antagonists.	Devane *et al.* (1988)
	[^{3}H]11-OH-Δ^9-THC-dimethylheptyl	Highly selective, quick to perform, and quantitative.	Does not assess function, fails to delineate between agonists and antagonists.	Thomas *et al.* (1992)
	[^{3}H)-WIN 55,212	Highly selective, quick to perform, and quantiative.	Does not assess function, fails to delineate between agonists and antagonists.	Pacheco *et al.* (1993)
	[^{3}H]-HU-243	Highly selective, quick to perform, and quantitative.	Does not assess function, fails to delineate between agonists and antagonists.	Devane *et al.* (1992a)
Adenylate cyclase	Neuroblastoma cells (N18TG2)	Provides some indication of function. Ideal for potent agonists. Quantitative.	Not selective. Difficult to evaluate weak agonists. Cell culture requires considerable research commitment.	Howlett (1984)
	Cerebellar granule cells	Provides some indication of function. Ideal for potent agonists. Quantitative.	Not selective. Difficult to evaluate weak agonists. Cell culture requires considerable research commitment.	Pacheco *et al.* (1993)
	Rodent brain	Provides some indication of function. Ideal for potent agonists. Quantitative.	Not selective. Difficult to evaluate weak agonists. Need to utilize brain areas.	Bidaut-Russell *et al.* (1990); Little and Martin (1991)

Table 2 *(contd)*

Assay	Substrate or tissue	Advantages	Disadvantages	Reference
Smooth muscle	Mouse vas deferens	Quantitative, provides functional measure.	Not selective	Pacheco *et al.* (1991); Pertwee *et al.*, (1992)
	Guinea-pig ileum	Quantitative, provides functional measure.	Not selective	Pertwee *et al.* (1992)
Calcium channels	Neuroblastoma cells	Quantitative, provides functional measure.	Not selective	Caulfield and Brown (1992); Mackie and Hille (1992); Felder *et al.* (1993)
Transfected cells	COS cells	Highly specific because the receptor is transfected into a cell which is normally unresponsive to cannabinoids.	Sometimes difficult to obtain sufficient number of receptors to conduct binding studies. The sample preparation is more difficult than brain tissue.	Munro *et al.* (1993)
	CHO cells			Felder *et al.* (1993)

with very small quantities of material. Unfortunately, use of this assay alone does not provide much assurance that the agent being tested is a cannabinoid as many other substances will inhibit contractions of smooth muscle preparations.

2.2.3.4 Calcium channels

The recent discovery that cannabinoids inhibit N-type calcium channels has provided a new method for assessing SAR and, more importantly, a potential means for distinguishing between actions of the cannabinoids (Caulfield and Brown, 1992; Mackie and Hille, 1992; Felder *et al.*, 1993; Mackie *et al.*, 1993). Cannabinoid potency can be determined accurately; however, inhibition of N-type channels is not unique to cannabinoids.

2.2.3.5 Transfected cells

Transfection of receptors into cells that do not normally contain them offers a unique method of evaluating SAR. The ability of a cell to either respond or bind to a ligand once the receptor has been incorporated into it offers compelling evidence for a specific receptor–ligand interaction. Although this approach is particularly useful for establishing receptor selectivity, only a few SAR studies have been carried out with cannabinoids. A peripheral cannabinoid receptor was transfected into COS cells, and five cannabinoids were evaluated for binding properties (Munro *et al.*, 1993). The emphasis of this study was to establish that cannabinoids bind to this receptor rather than establish an SAR. In another study (Felder *et al.*, 1993), CHO cells were transfected with the brain cannabinoid receptor for assessment of both functional and binding properties. These transfected cell preparations provide numerous advantages over tissue preparations which may contain multiple receptor and second messenger systems. However, the findings should be interpreted with caution because it is an artificial system. For example, Felder *et al.* (1993) reported that anandamide's affinity for the cannabinoid receptor in transfected CHO cells is an order of magnitude lower than that in brain tissue. This discrepancy could merely be the result of methodological differences between cell and brain preparations, or it could represent distinctive receptor-binding characteristics in different environments.

2.2.4 Correlation between *in vivo* and *in vitro* data

The fundamental principle of SAR resides in a correlation between structural alterations in the molecule and changes in either pharmacological or biochemical potency. As alluded to earlier, the research question dictates the type of correlation which is sought. It is rather straightforward to characterize the structural requirements for a particular pharmacological effect or to determine whether a particular pharmacological effect may be mediated through a specific neurochemical system. The latter simply involves correlating the pharmacological and biochemical effects produced by a series of analogues. However, considerable attention has been directed towards elucidating the mechanism by which cannabinoids produce their psychotomimetic effects. Since human studies are not easily carried out, it is usually necessary to choose an animal model which is highly correlated with psychotomimetic activity. To further complicate matters, cannabinoids produce a wide range of pharmacological effects in laboratory animals (described in Table 1) so that it is not easy to arrive at a single choice for an animal model. In addition, it seems highly probable that cannabinoids produce their central effects by interacting with several biochemical systems. Therefore, it is often necessary to rely on several pharmacological measures, particularly when the question is whether a single receptor is responsible for multiple pharmacological effects. Indeed, a major goal of our laboratory has been to determine whether multiple mechanisms are responsible for all of the diverse centrally mediated actions of the cannabinoids.

Questions arise as to what extent SARs have been established for all of the models presented in Tables 1 and 2 and as to how well these *in vivo* and *in vitro* models correspond to each other. Tables 3–9 summarize some examples of analogues evaluated in many of the assays. Unfortunately, a large number of analogues have not been evaluated in each model, which makes comparison of some of the models difficult. For example, the monkey overt behaviour and dog static-ataxia were used extensively in the 1970s and early 1980s, and they have largely given way to drug discrimination, catalepsy and the mouse tetrad model. *In vitro* assays either did not come into existence or were not used for SAR until the mid to late 1980s. Therefore, these summaries are necessarily confined to analogues developed in the recent past.

In general, there is relatively good agreement among all of the assays for the analogues summarized in Tables 3–9. Most of these analogues have been evaluated in the mouse tetrad model and in receptor binding. To date, we have found that, with few exceptions, all analogues are uniformly active in all four mouse behavioural tests, which in turn correlates with receptor binding. In a comparison using more than 60 analogues, an excellent correlation was obtained with all four behaviours and receptor binding (Compton *et al.*, 1993). As can be seen in Table 3, Δ^8-THC-DMH and 5′-bromo-Δ^8-THC-DMH are very potent in all four mouse behaviours and exhibit high receptor affinity. In contrast, all of the nitro analogues are inactive in the behavioural assays and have very low receptor affinity.

The novel bicyclic and tricyclic CP analogues were developed almost exclusively on the basis of antinociceptive activity (Johnson and Melvin, 1986; Melvin and Johnson, 1987). As the results in Table 4 show, this series of analogues also exhibited a pharmacological profile identical to that of the traditional cannabinoids. In addition to the results obtained in the mouse behavioural assays, these compounds also engendered THC-like responses in drug discrimination. One benefit of the CP analogues is that they represented the first analogues which were recognized as being much more potent than the traditional cannabinoids. The summary in Table 5 describes the potency of several of these very potent analogues. Again, there is a good correlation among all of the models which is best exemplified by the

Table 3 Pharmacological activity of cannabinoids with substitutions at C2, C4 and in the side-chain[a]

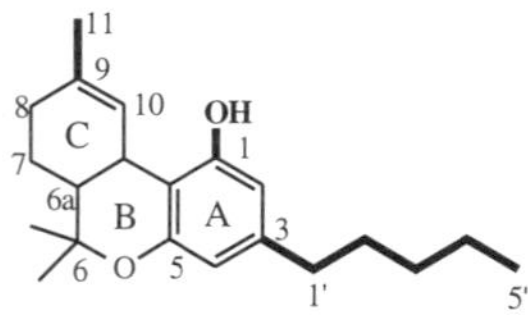

Δ^9-THC

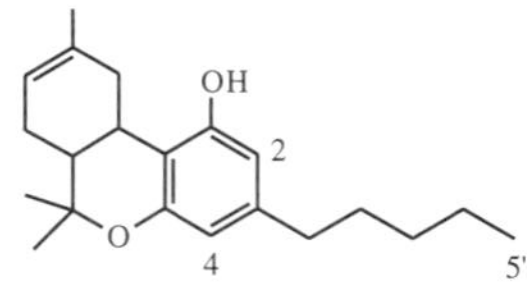

Δ^8-THC

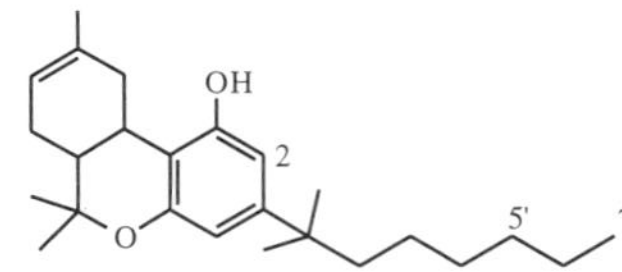

Δ^8-THC-dimethylheptyl

	Spontaneous activity	Tail flick	Temperature	Immobility	K_I
	(μmol/kg)				(nM)
Standards					
Δ^9-THC	3.2	4.5	4.5	4.8	41±2
Δ^8-THC	2.9	4.8	4.5	4.8	44±12
Δ^8-THC-dimethylheptyl	0.27	0.14	0.15	0.17	0.77±0.11
Substitutions at C2 and C4					
2-Iodo-Δ^8-THC	8.0	0.68	20	4.3	89±15
4-Bromo-Δ^8-THC	>250	>250	>250	>250	5250±2090
2,4-Diiodo-Δ^8-THC	>175	>175	>175	>175	10 600±2090
2-Nitro-Δ^8-THC	28.7	278	101	>300	>10 000
4-Nitro-Δ^8-THC	>275	>275	40.1	52.7	1630±360
2,4-Dinitro-Δ^8-THC	>75	>75	>75	>75	>10 000
2-Iodo-Δ^8-THC-dimethylheptyl	4.4	5.0	10	1.7	61±9
Substitutions at C5′					
5′-F_3-Δ^8-THC	0.82	0.46	2.9	2.6	25±6

Table 3 *(contd)*

	Spontaneous activity	Tail flick	Temperature	Immobility	K_i
	(μmol/kg)				(nM)
5′-F_3-Δ^9-THC	2.4	1.9	7.6	1.1	38±3
5′-Bromo-Δ^8-THC-dimethylpentyl	0.08	0.05	0.09	0.10	0.43±0.09
Substitutions at C7′of dimethylheptyl					
7′-Bromo-Δ^8-THC-dimethylheptyl	0.13	0.13	0.19	0.21	ND

[a]Reported previously (Charalambous *et al.*, 1991; Martin *et al.*, 1993).
ND, not determined.

Table 4 Pharmacological potency of bicyclic analogues

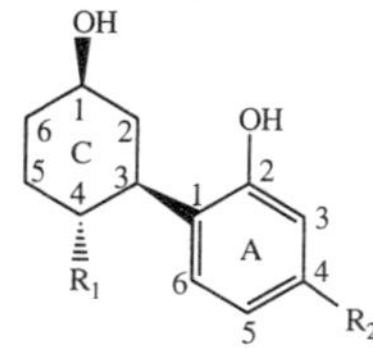

1 to **14** and **16**

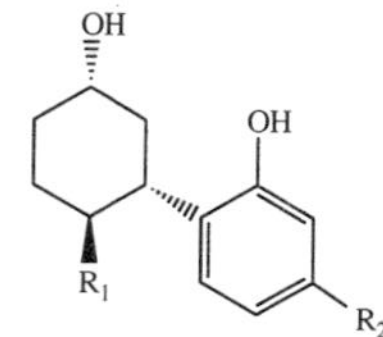

15

Analogue	R_1	R_2	Spontaneous[a] Activity (μmol/kg)	Tail[a] Flick (μmol/kg)	Temp[a] (μmol/kg)	Imm[a] (μmol/kg)	K_I[b] (nM)	Drug Disc.[c] (μmol/kg)
1	H	$C(CH_3)_2CH_3$	>100	>100	>100	>100	11,300	>10
2	H	$C(CH_3)_2CH_2CH_3$	>100	>100	>100	>100	8100	
3	H	$C(CH_3)_2(CH_2)_2CH_3$	>100	>100	>100	>100	3760	
4	H	$C(CH_3)_2(CH_2)_3CH_3$	>100	34.0	>100	>100	73.0	
5	H	$C(CH_3)_2(CH_2)_4CH_3$	1.97	4.60	12.5	9.52	163	≈1
6	H	$C(CH_3)_2(CH_2)_5CH_3$	0.31	0.94	4.08	3.45	9.45	≈1
7	H	$C(CH_3)_2(CH_2)_6CH_3$	0.30	0.30	0.90	0.60	4.73	≈1
8	H	$C(CH_3)_2(CH_2)_7CH_3$	4.04	6.35	4.04	2.31	28.5	>30
9	H	$C(CH_3)_2(CH_2)_8CH_3$	64.3	13.0	114	76.8	163	>10
10	H	$C(CH_3)_2(CH_2)_9CH_3$	>80	18.42	>80	>80	381	>30
11	CH_3	$C(CH_3)_2(CH_2)_5CH_3$	0.60	1.80	2.41	3.31	6.16	
12	CH_3	$C(CH_3)_2(CH_2)_5CH_3$	0.30	0.60	1.50	0.90	7.70	
13	$(CH_2)_3OH$	H	>60	>60	>60	>60	>10 000	

Table 4 *(contd)*

Analogue	R_1	R_2	Spontaneous[a] Activity (μmol/kg)	Tail[a] Flick (μmol/kg)	Temp[a] (μmol/kg)	Imm[a] (μmol/kg)	K_I[b] (nM)	Drug Disc.[c] (μmol/kg)
14[d]	$(CH_2)_3OH$	$C(CH_3)_2(CH_2)_5CH_3$	0.11	0.23	0.93	0.92	0.92	
15	$(CH_2)_3OH$	$C(CH_3)_2(CH_2)_5CH_3$	7.98	16.0	>25	>25	61.7	
16	$(CH_2)_4OH$	$C(CH_3)_2(CH_2)_5CH_3$	0.77	0.13	1.54	2.05	2.20	

[a]Spontaneous activity, tail-flick (antinociception), temp. (rectal temperature), imm. (immobility) were determined as described before (Compton *et al.*, 1992b). Data are presented as ED_{50}s. > indicates the highest dose tested which failed to produce 50% effect.
[b]Affinity for the cannabinoid receptor as determined by [^{3}H]CP 55,940 binding (Compton *et al.*, 1993). The results are presented as K_I(nM). > indicates this concentration produced less than 50% displacement of binding.
[c]W. R. Prescott and B. R. Martin, unpublished drug discrimination (Drug Disc.) data. > indicates the highest dose tested which failed to produce 80% generalization.
[d]CP 55,940.

Table 5 Structural features resulting in potent cannabinoid analogues

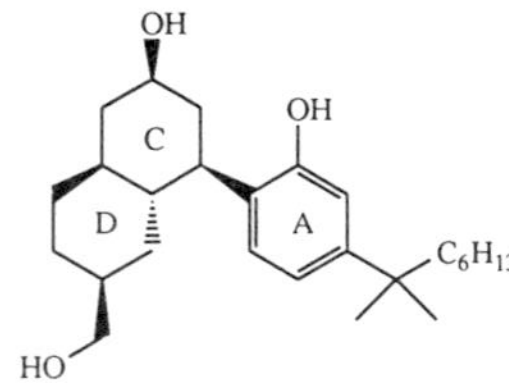

ACD

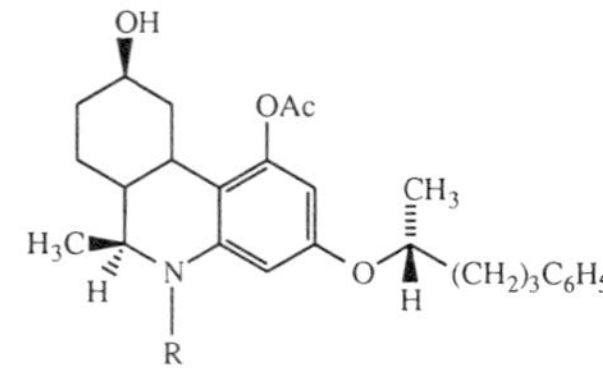

R = H, levonantradol
R = CH_3, *N*-methyllevonantradol

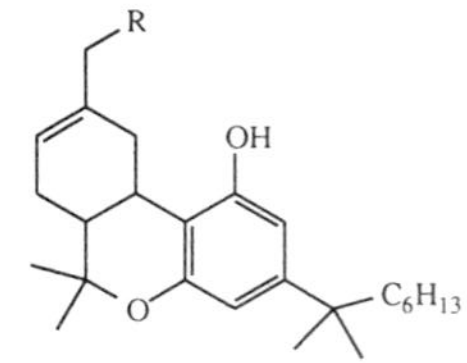

R = CH_2OH; 11-OH-Δ^8-THC-DMH

	Spontaneous[a] Activity (μmol/kg)	Tail[a] Flick (μmol/kg)	Temp[a] (μmol/kg)	Imm[a] (μmol/kg)	K_I[b] (nM)	Drug Disc.[c] (μmol/kg)	A.C.[d] nM	MVD[e] (nM)
Δ^9-THC	3.2[a]	4.5[a]	4.5[a]	4.8[a]	41[b]	2.2[c]	430[d]	6.3[e]
(−)-ACD	0.01[a]	0.02[a]	0.12[a]	0.21[a]	ND		5[d]	
(+)-ACD	19.9[a]	>25[a]	>25[a]	>25[a]	ND			
Levonantradol	0.26[a]	0.61[a]	1.26[a]	3.16[a]	1.06[b]	0.04[g]	100[d]	
Dextronantradol	>20[a]	>20[a]	>20[a]	>20[a]	3100[b]	>3.0[g]		
N-Methyllevonantradol	0.02[a]	0.01[a]	0.02[a]	0.03[a]	6.70[b]		126[d]	
N-Methyldextronantradol	15.6[a]	25.4[a]	10.0[a]	16.4[a]	607[b]			
CP 55,940	0.11[a]	0.23[a]	0.93[a]	0.92[a]	0.92[b]	0.08[c]	25[d]	
(−)-11-OH-Δ^8-THC-DMH	0.01[f]	0.02[f]	0.05[f]	0.05[f]	0.73[b]	0.01[h]	1.8[d]	0.15[e]
(+)-11-OH-Δ^8-THC-DMH	>80[f]	>80[f]	>80[f]	80[f]	1990[b]	>5.0[h]	>1,000[d]	>30[e]

[a] Spontaneous activity, tail-flick (antinociception), temp. (rectal temperature), imm. (immobility) were determined as described earlier (Compton *et al.*, 1992b). The results are presented as ED_{50} (μmol/kg).
[b] Affinity for the cannabinoid receptor as determined by [^{3}H]CP 55,940 binding (Compton *et al.*, 1993). The results are presented as K_I (nM).
[c] Drug discrimination in rats as described previously (Gold *et al.*, 1992). Results presented as ED_{50} (μmol/kg).

Table 5 *(contd)*

[d]Inhibition of adenylate cyclase activity in neuroblastoma cells as reported (Howlett *et al.*, 1988, 1990). Results presented as K_{inh} (nM).

[e]Inhibition of the mouse vas deferens (MVD) twitch as reported (Pertwee *et al.*, 1992). The concentrations that decreased twitch height by 50% are provided.

[f]Little *et al.* (1989).

[g]Browne and Weissman (1981).

[h]Järbe *et al.* (1989).

enantioselectivity. Adenylate cyclase was determined for some of these analogues and a reasonable correlation was obtained. It would appear that levonantradol and *N*-methyl-levonantradol were somewhat less potent in the adenylate cyclase assay than expected. The synthesis of (−)-11-OH-Δ^8-THC-DMH resulted in one of the most potent cannabinoids produced to date (Mechoulam *et al.*, 1987). It is clear, in Table 5, that this enantiomer is extremely potent in all assays, including mouse vas deferens, and that there is enantioselectivity.

The value of having numerous cannabinoid models became evident when the Sterling research group began to suspect that pravadoline analogues might also be cannabinoids as discussed later (Section 2.3.3). The structural dissimilarity between these compounds, termed aminoalkylindoles (AAI), (see structures in Table 6) and cannabinoids demanded ample evidence that these two classes of compounds were pharmacologically equivalent. Again, the data in Table 7 provide strong evidence that compounds such as (+)-WIN 55,212-2 are cannabinoids.

Of course, the recent identification of an arachidonic acid derivative, anandamide, as an endogenous cannabinoid ligand raises questions similar to those provoked by the CP analogues and AAIs. The structural requirements which had been developed for cannabinoids did not appear to accommodate eicosanoids. However, the evaluation of anandamide and its analogues in receptor binding and adenylate cyclase assays support the contention that these compounds are indeed cannabinoids (Table 9). Unfortunately, these analogues have not yet been evaluated in behavioural assays.

To summarize briefly, these behavioural and *in vitro* cannabinoid assays fulfil the criteria of models used in establishing SAR and have been used effectively to identify several classes of compounds which would not have been identified otherwise.

2.3 Pharmacological activity in laboratory animals

With regard to cannabinoid SAR analysis, the lack of a clearly defined mechanism of action hindered the formulation of structural requirements for specific behavioural effects (Martin, 1986). Indeed, prior to the discovery of a specific, high-affinity binding site for cannabinoids, membrane perturbation, a relatively non-specific mechanism of action was also considered to be a possible mechanism of action (Paton *et al.*, 1972). However, the SAR indicated that there were strict structural requirements for behavioural activity (Edery *et al.*, 1971) – a three-point requirement for receptor interaction was subsequently postulated (Binder *et al.*, 1984; Howlett *et al.*, 1988). The three structural requirements for cannabinoid activity were proposed to consist of: (1) an appropriate substituent at the C9 position to allow the compound to fit into a receptor cavity; (2) a free phenolic hydroxyl group for hydrogen bonding to the receptor; and (3) a lipophilic side-chain permitting interaction with a hydrophobic region of the receptor to occur (the three features indicated in bold in the Δ^9-THC structure in Table 3). Although these requirements

Table 6 Development of WIN 55,212[a]

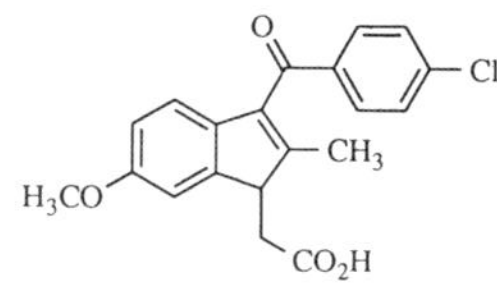

Indomethacin

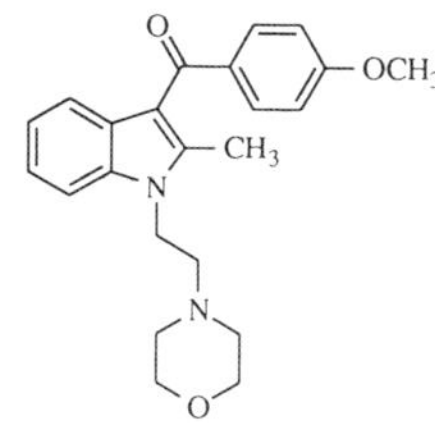

Pravadoline

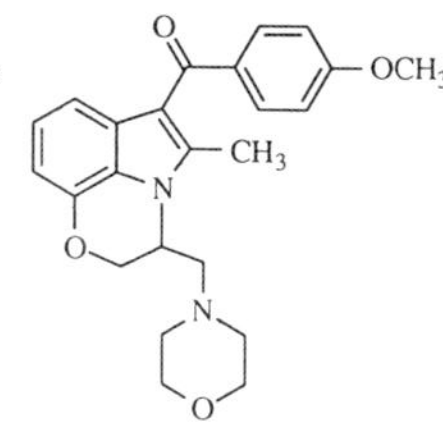

17

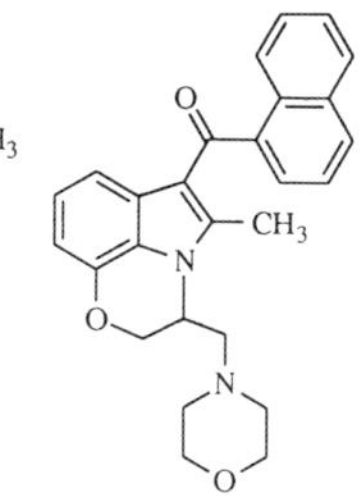

WIN55,212

Compound	MVD activity[b] (μM)	PG synthesis[c] (μM)	Antinociception[d] (μmol/kg)	Receptor binding[e] (nM)
Pravadoline	0.45	5	69	3155.0
17[f]	0.12	72	30% at 250	152.0
WIN 55,212 (R)	0.00043	9	62	2.8

[a] Reported previously (D'Ambra *et al.*, 1992).
[b] Inhibition of electrically stimulated mouse vas deferens expressed as IC_{50}.
[c] Inhibition of prostaglandin synthesis (or cyclooxygenase activity) expressed as IC_{50}.
[d] Inhibition of acetic acid stretching, results presented as ED_{50}.
[e] Competition for $[^3H]$WIN 55,212 binding expressed as IC_{50}.
[f] Analogue **17** is presented as analogue 20 in the original publication (D'Ambra *et al.*, 1992).

Table 7 Cannabinoid effects of pravadoline analogues

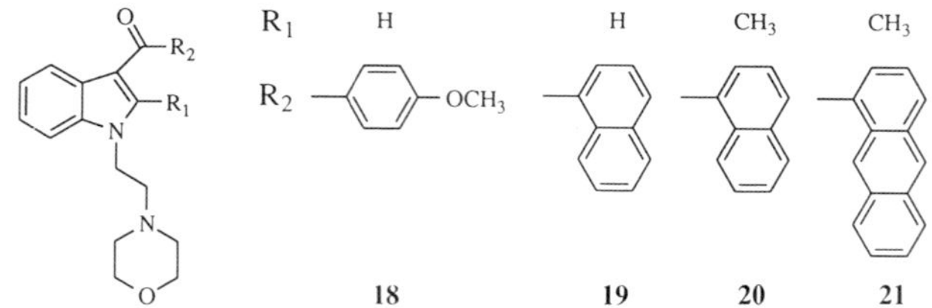

18 19 20 21

	Spontaneous[a] Activity	Tail[a] Flick	Temp.[a]	Imm.[a]	K_I[b] (nM)	Drug Disc.[a] µmol/kg	MVD[c] (nM)
	(µmol/kg)						
Δ^9-THC	0.9	0.8	2.3	0.9	41	0.6	
18	>10	>10	>10	>10	140	>20	
19	0.3	0.3	1.2	2.7	21	0.54	6[d]
20	2.0	1.9	7.2	3.9	39	1.7	15[d]
21	>10	>10	>10	>10	3230	>10	
WIN 55,212(+)	0.1	0.4	12[c]	1.1	6.4	0.17	0.43[e]
WIN 55,212(–)	>30	>30	>30	>30	>10,000	>10	>10,000[e]

[a] Data reported previously (Compton *et al.*, 1992a). Analogues **18–21** were presented as analogues I–IV in that publication.
[b] Competition for [^{3}H]CP 55,940 binding (unpublished data). IC_{50} values of 2.8 and 8000 nM were obtained for (+)- and (–)-WIN 55,212, respectively, in the [^{3}H]WIN 55,212 binding assay (D'Ambra *et al.*, 1992).
[c] Inhibition of electrically stimulated mouse vas deferens.
[d] Bell *et al.* (1991).
[e] D'Ambra *et al.* (1992).

do not appear to narrowly define structural specificity, they have provided a theoretical framework with which to test certain assumptions regarding cannabinoid SAR and have allowed some predictions regarding the potencies (active versus inactive) of compounds not yet synthesized.

2.3.1 Dibenzopyran derivatives

The three-point receptor attachment described above was postulated based upon the notion that cannabinoids existed as dibenzopyrans, depicted by Δ^9-THC. Although we now know that this three-ringed structure is not required for cannabinoid activity, structural modifications in this series are still crucial for defining the pharmacophore, for possible development of antagonists and for the fact that this series is most closely related to Δ^9-THC. The structural requirements for the C9 position, phenolic hydroxyl, and side-chain have been critically reviewed (Razdan, 1986). However, confining the description of these interactions to the basic structural requirements at these three sites is too simplistic. One portion of the THC molecule that has not been investigated thoroughly is the aromatic ring; this is most likely due to the fact that substitutions at the C2 and C4 positions have rendered the cannabinoid inactive (Mechoulam and Edery, 1973). A series of halogenated derivatives were prepared (Table 3) which revealed that iodination of Δ^8-THC at the C2 position and bromination and iodination in the side-chain all enhanced potency (Charalambous *et al.*, 1991; Martin *et al.*, 1993). Substitution of fluorines for the hydrogens on C5′ of the straight pentyl side-chain had relatively little effect on the pharmacological activity of the THC, which is likely due to the fact that both fluorine and hydrogen atoms occupy comparable volumes. Both branching and extending the side chain markedly increased potency, while the addition of a bromine to the terminus of the dimethylheptyl (DMH) side chain had no influence on potency. These observations are consistent with earlier findings which demonstrated that potency is retained even when the side chain is extended up to eight carbon atoms (Mechoulam and Edery, 1973).

While these researchers showed that substitutions of strongly electronegative groups at positions C2 and C4 rendered the cannabinoid inactive, substitution in position C2 by either a methyl or ethyl group was tolerated (Edery *et al.*, 1972). Substitution at position C4 by any group was not tolerated. Despite these observations, the importance of positions C2 and C4 have not been evaluated critically for their role in ligand–receptor interaction. Alterations in the electron density map of Δ^8-THC by addition of a nitro group at C2, but not by iodination, suggests that electron densities of the phenol and the oxygen in the B ring are essential for activity. This observation was reinforced by the lack of activity of 4-nitro- and 2,4-dinitro-Δ^8-THC.

Despite the fact that Δ^9-THC produces potent psychotomimetic effects in humans, it has proven to be marginally effective in many *in vitro* biochemical systems (Martin, 1986). The question arose as to whether these effects were related to specific actions of Δ^9-THC. Indeed, it was not possible to characterize receptor binding using radiolabelled Δ^8-THC (Harris *et al.*, 1978) and Δ^9-THC (Roth and Williams, 1979). Earlier, it had been demonstrated that substitution of a dimethylheptyl side-chain for the traditional pentyl side-chain dramatically enhanced potency (Hardman *et al.*, 1971), a finding that has contributed to the recent development of several highly potent cannabinoids. Numerous studies had also documented enhanced potency upon 11 hydroxylation (Razdan, 1986). Therefore, 11-OH-Δ^8-THC-DMH (Table 3) was prepared (Mechoulam *et al.*, 1988), which proved to be several-hundred times more potent than Δ^8-THC in several behavioural tests (Little *et al.*, 1989). Razdan synthesized the corresponding 11-OH-Δ^9-THC-DMH which exhibited similar high potency (Martin *et al.*, 1991). More recently, the hexahydro analogue of 11-OH-THC-DMH was synthesized which has provided another potent analogue (Devane *et al.*, 1992a).

Enantioselectivity is an important criterion for drug–receptor interactions, because enantiomers share the same physicochemical characteristics. All of the initial studies with Δ^9-THC failed to demonstrate complete enantioselectivity. However, almost complete enantioselectivity can be achieved when very pure enantiomers are obtained (Table 5) as with the case of 11-OH-Δ^8-THC-dimethylheptyl (Mechoulam *et al.*, 1988; Järbe *et al.*, 1989; Little *et al.*, 1989; Howlett *et al.*, 1990).

2.3.2 Novel bicyclic and tricyclic analogues

It had generally been assumed that an intact B ring was essential for activity because cannabidiol lacks psychoactivity (Razdan, 1986). However, the Pfizer group (Melvin *et al.*, 1984) synthesized a derivative of 9-*nor*-9β-hydroxyhexahydrocannabinol that lacked the B ring entirely and had a dimethylheptyl side-chain rather than the traditional pentyl group at the C3 position. It proved to have a pharmacological profile similar to that of Δ^9-THC, although it was more potent. This synthetic strategy led to the synthesis of CP 55,940 (compound **14** in Table 4) which proved to be four to 25 times more potent than Δ^9-THC, depending upon the pharmacological measure (Compton *et al.*, 1992b). This analogue was also radiolabelled for the characterization of the cannabinoid receptor, as discussed later (Devane *et al.*, 1988).

The Pfizer group synthesized an impressive array of analogues in the bicyclic and tricyclic series which helped redefine many of the structural determinants of cannabinoid action (Johnson and Melvin, 1986). An example of one series of compounds which has been evaluated in several cannabinoid models is presented in Table 4. These results demonstrate that alkyl substitution at C4 in the C ring is important for activity and most likely mimics the B ring of Δ^9-THC. Evidently, the side chain is the dominant factor for activity. Analogue **6** is completely devoid of substitution at C4 in the C ring, yet it is only slightly less potent than CP 55,940 (**14**) because they both have a dimethylheptyl side-chain (R_2) at C4 in the A ring. Analogues **8** and **9** demonstrate the diminution in activity when the side-chain is extended beyond eight carbons.

The systematic approach taken by the Pfizer researchers (Johnson and Melvin, 1986) also resulted in extremely potent agonists. Some of these novel bi- and tri-cyclic analogues are as much as 700 times more potent than Δ^9-THC (Little *et al.*, 1988). One of the most notable examples is (–)-ACD, which is shown in Table 5. By incorporating the hydroxy-alkyl group in a ring system, Little *et al.* were able to position this constituent into its most favourable state and thereby achieve very high potency. As can be seen in Table 5, almost complete enantioselectivity was achieved with this compound.

2.3.3 Aminoalkylindoles

A rather fortuitous finding by the Sterling research group led to the discovery of a class of compounds, the AAIs, which appear to be structurally distinct from the cannabinoids yet produce cannabinoid pharmacological effects. Pravadoline was designed as a non-acidic analogue of the non-steroidal anti-inflammatory drug (NSAID) indomethacin. However, it exhibited greater analgesic efficacy than most NSAIDs which prompted a search for its mechanism of action (Table 6). While pravadoline inhibited prostaglandin synthesis (cyclooxygenase activity) like other NSAIDs, examination of a series of analogues revealed that analgesia was independent of cyclooxygenase inhibition. Analgesic potency correlated better with adenylate cyclase inhibition. Pravadoline was also found to inhibit electrically stimulated contractions of guinea-pig ileum and mouse vas deferens. Efforts were then undertaken to examine the SAR of these properties of pravadoline in order to gain insight into the mechanisms responsible for analgesia.

Although a large number of analogues were synthesized, the most noteworthy analogues resulted from restraining the amide side chain. For example, analogue **17** (analogue 20 in D'Ambra *et al.*, 1992) exhibited decreased cyclooxygenase inhibitory activity and attenuated analgesic activity, but increased activity in inhibiting contractions of the mouse vas deferens. Subsequently, WIN 55,212-2 [*R* or (+)-enantiomer] was synthesized. Radiolabelling WIN 55,212-2 led to the discovery that these AAIs were acting at the cannabinoid receptor.

WIN-55,212-2 (Table 6) is a prototypic AAI whose antinociceptive properties do not appear to be related to inhibition of cyclooxygenase or to an interaction with opioid mechanisms. Although AAIs are structurally distinct from Δ^9-THC, a comparison of the results obtained from a cannabinoid ligand-binding assay with those obtained from an assay utilizing radiolabelled (+)-WIN-55,212-2 as the ligand indicates similar rank order potencies and suggests identical binding sites (Pacheco *et al.*, 1991). Studies undertaken to confirm that the AAIs share a common pharmacological profile with Δ^9-THC indicated that the (+)-isomer exhibited ED_{50}s in the general range of those of Δ^9-THC in producing hypoactivity, antinociception, hypothermia and ring-immobility in mice (Compton *et al.*, 1992a). The (–)-isomer was inactive up to the highest dose tested. The (+)-isomer engendered THC-like responding in the drug discrimination paradigm in rats even though it produced considerable response rate suppression (Compton *et al.*, 1992a).

Some of the most convincing evidence that the AAIs are cannabinoids is summarized in Table 7. A series of pravadoline analogues has been evaluated in several cannabinoid models (Bell *et al.*, 1991; Compton *et al.*, 1992a; D'Ambra *et al.*, 1992), and all of the results indicated cannabinoid activity. Analogue **19** and **20** were active in all of the behavioural assays, competed for cannabinoid receptor binding and inhibited contractions of the mouse vas deferens. These analogues contain a naphthoyl substitution at R_2 which seems to be optimal for activity.

Models have been proposed which describe the potential structural commonalties between AAIs and Δ^9-THC (Estep *et al.*, 1991). In an attempt to determine the structural similarities between AAIs and Δ^9-THC, the overlap depicted in Table 8 was proposed (Huffman *et al.*, 1994). The indole is flipped to the left in order to superimpose the A, B and C rings. In addition, the indole nitrogen corresponds to C1′ in the side-chain of Δ^9-THC. In order to test this alignment, analogues **22** and **23** (Table 8) were synthesized (Huffman *et al.*, 1994). Analogues **23a** and **23b** lacked the naphthoyl substituent which corresponds to the B–C rings of Δ^9-THC and therefore lacked activity, as would be predicted. As for mimicking the side-chain in Δ^9-THC, analogues of **22** with side-chains of four to six carbons showed both *in vivo* and *in vitro* activity. Further studies will be necessary to firmly establish this proposed alignment.

2.3.4 Arachidonic acid derivatives

Despite the introduction of structural diversity by the CP compounds and AAIs, we were not prepared for cannabinoid activity residing in a fatty acid. The discovery that arachidonic acid ethanolamide (anandamide) is an endogenous cannabinoid ligand was not predicted (see Table 9 for structure), although structural commonalities between prostanoids and cannabinoids had been proposed (Milne and Johnson, 1981). The original publication reported the identification of one active compound only (Devane *et al.*, 1992b). Two additional constituents that are structurally related to anandamide have now been reported (Hanuš *et al.*, 1993). These compounds, homo-γ-linolenylethanolamide and docosatetraenylethanolamide also exhibit cannabinoid activity (**24** and **25**, respectively, in Table 9).

It has been suggested (Hanuš *et al.*, 1993) that the entire group of active compounds be named 'anandamides' with each individual member identified with the parent fatty acid (indicated in parenthesis) following the generally accepted fatty acid shorthand designation. Therefore, the ethanolamide derived from arachidonic acid is anandamide (20:4, *n*-6); that from homo-γ-linolenic acid is anandamide (20:3, *n*-6) (analogue **24**); and that from docosatetraenoic acid is anandamide (22:4, *n*-6) (analogue **25**).

It is anticipated that extensive SAR studies will be conducted in this series because of their relative ease of synthesis and the availability of a large number of congeners. To date, both the fatty acid moiety and the ethanol amide chain have been modified, and the new compounds have been tested for cannabinoid activity. The three naturally occurring anandamides: anandamide (20:4, *n*-6); anandamide (20:3, *n*-6) and anandamide (22:4, *n*-6) have been found to bind to the cannabinoid receptor using rat brain membranes (Devane *et al.*, 1992b; Hanuš *et al.*, 1993) and

Table 8 Aminoalkylindole analogues[a]

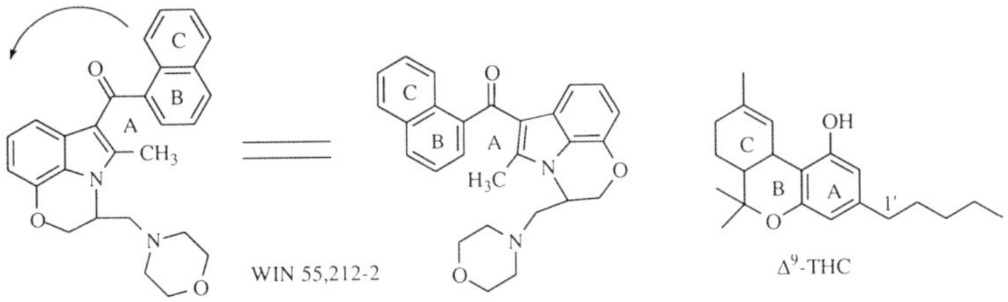

WIN 55,212-2 Δ[9]-THC

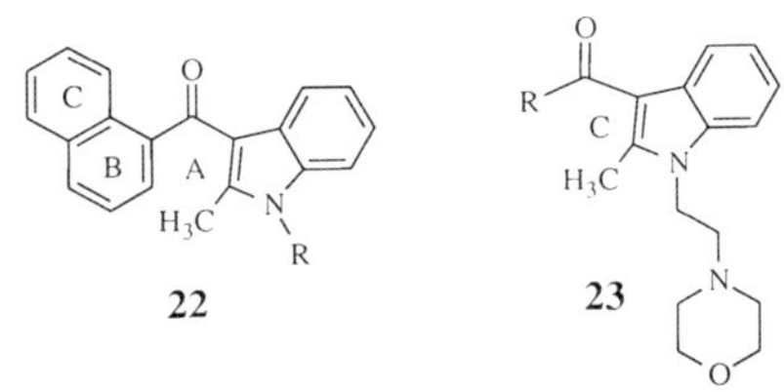

22 23

Analogue	Substituent	K_I(nM)	Mouse tetrad (μmol/kg)
22a	2-Phenylethyl	1250	>257
22b	2-Cyclohexylethyl	46	66
22c	*n*-Propyl	164	ND
22d	*n*-Butyl	22	ND
22e	*n*-Pentyl	10	2.6
22f	*n*-Hexyl	48	14
22g	*n*-Heptyl	>10,000	>189
22h	RS-2-Heptyl	33	36
22i	R-2-Heptyl	81	22
22j	S-2-Heptyl	72	ND
23a	Methyl	>10,000	>350
23b	1-Methylallyl	>10,000	>103

[a] Reported previously (Huffman *et al.*, 1994). Receptor affinity (K_I) was determined in the [^{3}H]CP 55,940 binding assay and expressed as nM. The results from the mouse tetrad assays were averaged and expressed as μmol/kg.
ND, not determined.

murine Ltk$^-$ cells transfected with the human cannabinoid receptor (Felder *et al.*, 1993). Additional compounds with changes in the fatty acid moiety have been evaluated by both groups (Felder *et al.*, 1993; Mechoulam *et al.*, 1994) and the results are presented in Table 9. The binding affinities obtained by these two groups are not the same, which is likely due to differences in methodologies and to differences between rat brain membranes and transfected cells.

Several tentative conclusions can be drawn from the limited results obtained so far: (1) at least three double bonds on the fatty acid chain are required for

Table 9 Receptor binding affinities of various anandamides[a]

Analogue	Nomenclature	Backbone	R	K_i[b]	K_i[c]	A.C.[d]
	ANANDAMIDE (20:4, *n*-6)	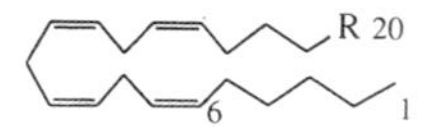	$CONHCH_2CH_2OH$	39	543	160
24	Anandamide (20:3, *n*-6)		$CONHCH_2CH_2OH$	54	598	390
25	Anandamide (22:4, *n*-6)	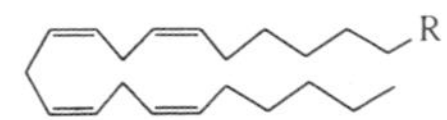	$CONHCH_2CH_2OH$	35	848	190
26	Anandamide (22:6, *n*-3)		$CONHCH_2CH_2OH$	383	12 200	6000
27	Anandamide (18:3, *n*-6)	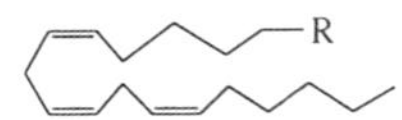	$CONHCH_2CH_2OH$	4600	>41 400	>24 000
28	Anandamide (18:2, *n*-6)	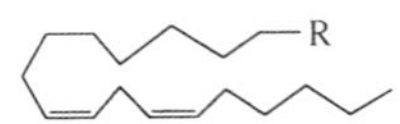	$CONHCH_2CH_2OH$	>1000		
29	Anandamide (18:3, *n*-3)	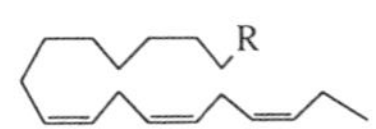	$CONHCH_2CH_2OH$	>25 000		
30	Anandamide (20:5, *n*-3)		$CONHCH_2CH_2OH$	200–400		
31	Anandamide (20:2, *n*-6)	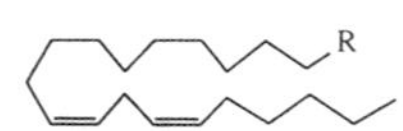	$CONHCH_2CH_2OH$	1500		
32	Anandamide (20:1, *n*-9)	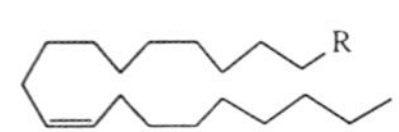	$CONHCH_2CH_2OH$	>1000		

Table 9 *(contd)*

Analogue	Nomenclature	Backbone	R	K_I[b]	K_I[c]	A.C.[d]
33	Anandamide (20:3, *n*-3)		$CONHCH_2CH_2OH$	>10 000		
34	Analogue (20:4, *n*-6)	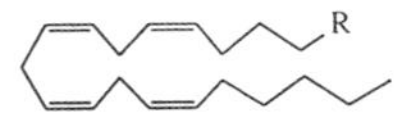	CONH**CH_2**CH_2CH_2OH		364	274
35	Analogue (20:4, *n*-6)		CONHCH(**CH_3**)CH_2OH		1300	507
36	Analogue (20:4, *n*-6)		$CONHCH_2CH(OH)$**CH_3**		4000	611
37	Analogue (20:4, *n*-6)	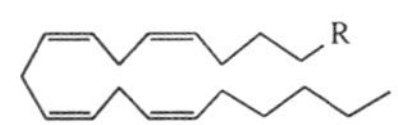	$CONH_2$		9600	10 100

[a]Applying the standard eicosanoid nomenclature to the above arachidonic acid derivatives means anandamide is written as anandamide (20:4, *n*-6), where 20 refers to the number of carbons in the backbone, 4 refers to the number of unsaturated bonds and *n*-6 signifies that the first double bond originates at C6. It is understood that the double bonds are each separated by one single bond.
[b]Mechoulam *et al.* (1994). K_I's were determined using [^{3}H]HU-243 binding in rat brain membranes and expressed as nM.
[c]Felder *et al.* (1993). K_I's were determined using [^{3}H]CP 55,940 binding to murine Ltk^- cells transfected with the human cannabinoid receptor, and the results are expressed as nM.
[d]Felder *et al.* (1993). Inhibition of forskolin-stimulated adenylate cyclase (A.C.) is presented as IC_{50} (nM).

pronounced activity; (2) activity is strongly reduced if the first double bond is on the third carbon atom from the non-acidic end of the fatty acid; (3) highest potencies are observed with the C-20 and C-22 polyunsaturated acids. It should be stressed that these are tentative proposals owing to the limited number of analogues which have been studied.

The National Institutes for Health (NIH) group (Felder *et al.*, 1993) have reported that substitution of a propanol for the ethanol does not eliminate activity and that the straight chain is more active than a branched chain. Replacement of the hydroxyl group with a fluorine atom increases the binding potency nearly 10-fold (R.K. Razdan, B.R. Martin and D.R. Compton, unpublished observations).

In studies of biological activity, anandamide has been examined in mice following peripheral administration and found to exhibit THC-like properties in producing hypomotility, hypothermia, antinociception and catalepsy (Fride and Mechoulam, 1993). When evaluated by i.v. administration in the mouse tetrad model, it is less potent than Δ^9-THC by four- to 20-fold, depending upon the pharmacological measure (Smith *et al.*, 1994). However, it should be pointed out that anandamides present the pharmacologist with new challenges in conducting SAR studies, particularly with comparison of *in vivo* and *in vitro* data. Anandamide has a shorter duration of action than Δ^9-THC, which is believed to be due to metabolic factors. Anandamide is rapidly degraded by an amidase which can be blocked by the enzyme inhibitor phenylmethylsulfonyl fluoride (PMSF) (Deutsch and Chin, 1993). PMSF also increases the binding affinity of anandamide almost 50-fold when the traditional rat brain membrane assay is used (Childers *et al.*, 1993). Efforts in our laboratory to produce THC-like effects in mice or rats after i.p. administration have proven to be difficult. However, we have just been able to obtain Δ^9-THC-like responding in rat drug discrimination with anandamide (our unpublished results).

2.4 Pharmacological activity in humans

The methods for evaluation of cannabinoids in humans have been well established (Perez-Reyes *et al.*, 1972; Hollister, 1974). Obviously, many of the most recently described analogues have not been evaluated for psychotomimetic effects because insufficient pharmacological and toxicological data are available. Moreover, the time and expense cannot be adequately justified in many cases. However, a strong case can be made for the anandamides. If anandamide is truly the endogenous cannabinoid ligand, then it is essential that this compound be evaluated in humans for its physiological effects.

2.5 Molecular modelling of the cannabinoid pharmacophore

Analysis of the SAR within a class of compounds often results in the delineation of a number of structural features governing pharmacological potency. Such has been the case in developing the very useful model of cannabinoid–receptor interaction depicted in Figure 1. However, lists of structural requirements usually treat structural modifications as isolated events (i.e. that modification of one group does not affect another) and the effect of modifying more than one feature of a molecule can often be difficult to predict. The development of theoretical pharmacophore models based on quantum mechanics and molecular modelling has extended the characterization of the SARs beyond a descriptive list of requirements. Typical molecular modelling procedures enable rapid calculation of three-dimensional conformations of molecules which are often in close agreement with structures determined by X-ray crystallography. In addition to providing a method for structural determination, the application of quantum mechanics to molecular modelling also allows a number of physical and chemical properties of molecules, such as molecular volumes and electrostatic potentials, to be readily determined. Since the interaction between a ligand and its site of action is frequently non-covalent in nature, a variety of the molecular properties involved in these non-covalent interactions can be accounted for by the use of molecular mechanics forcefields. Thus, molecular models, forcefields, and three-dimensional structure–activity analyses have enabled a more comprehensive understanding of the three-dimensional steric and electrostatic forces involved in the production of a particular pharmacological effect.

As mentioned earlier, studies (Johnson and Melvin, 1986) using a large number of synthetic 'non-classical' cannabinoid analogues, based on the structure of (−)-9-*nor*-9β-hydroxyhexahydrocannabinol (HHC) (Wilson *et al.*, 1976), clearly demonstrated that one would achieve extreme pharmacological potency in structures that were distinct from the classical cannabinoid ring system. Substitution of the pyran ring oxygen with a nitrogen (phenanthridines) or carbon (phenanthrenes) yielded active compounds, and complete elimination of the pyran ring, as in CP 47,497, did not eliminate activity (Weissman *et al.*, 1982; Johnson and Melvin, 1986). Using Alinger's Molecular Mechanics Program (MMI) calculations it was shown that a low-energy conformation of CP 47,497 closely approximated the conformation of HHC with the positions of the side-chain, phenol and alcohol groups nearly identical (Figure 2). These results suggest that the pyran ring in the 'classical' cannabinoid ring system anchors the phenol and cyclohexyl ring systems in an active conformation. Thus, it was concluded that both the substituent at C9, and its orientation in space, are important determinants of biological potency (Johnson and Melvin, 1986).

The geometry of the ring system and the orientation of the C9 substituent has continued to be implicated in structure–activity studies as an important determinant of potency. It has been suggested that differences in the potency of cannabinoid analogues could sometimes be explained by deviations in the planarity of the tricyclic ring systems (Kriwacki and Makriyannis, 1989). These investigators

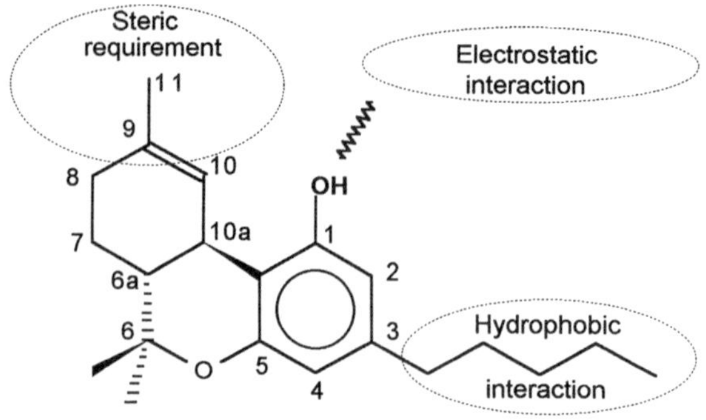

Figure 1 Δ^9-THC and the three-point requirement for cannabinoid receptor interaction.

reported that in Δ^9-THC the cyclohexene (C) ring assumes a flattened chair conformation, causing the methyl group at the C9 position to project above the plane of the aromatic ring. Conversely, the ring system of the less active structural isomer Δ^{9-11}-THC assumes a chair form, as found in cyclohexane, which projects the methylene into the plane of the aromatic ring, forming a coplanar conformation with all three rings. Similar findings were also reported with these and other cannabinoids when the torsion angle between C11–C9–C10 was correlated with activity (Reggio *et al.*, 1989).

In another comparative molecular modelling approach (Reggio *et al.*, 1993), the cumulative molecular volume of various active analogues was subtracted from the cumulative molecular volume of two inactive analogues. The remaining region in space occupied by the atoms of inactive analogues and not occupied by the atoms of active analogues was interpreted as the volume occupied by the receptor (or receptor essential volume, REV). Because of steric interference, the REV cannot be occupied by the ligand if favourable receptor interaction is to take place. The essential volume mapping of cannabinoid analogues indicated an area behind the plane of the aromatic ring and at the approximate location of the C9 and C11 carbons that was 'receptor essential'. From all of these observations, it was concluded that steric bulk, in or behind the plane of the aromatic ring at the C9 substituent region, could be associated with elimination, or with a substantial decrease, in activity. It was also clear from the molecular modelling and structure–activity analyses that the three-dimensional geometry (or steric parameters) of cannabinoid analogues is an extremely important property determining the potency of cannabinoids.

In addition to the aforementioned steric considerations, electrostatic parameters have also been considered to be important in influencing cannabinoid potency. The phenolic hydroxyl group is considered to be a particularly important cannabinoid functional group owing to the possibility of its hydrogen bonding with the receptor. The orientation of the phenolic hydroxyl group is energetically favoured at torsion angles about the C1-OH bond of approximately $-1°$ or approximately $150°$ (Reggio *et al.*, 1989; Semus and Martin, 1990). In addition to this conformational

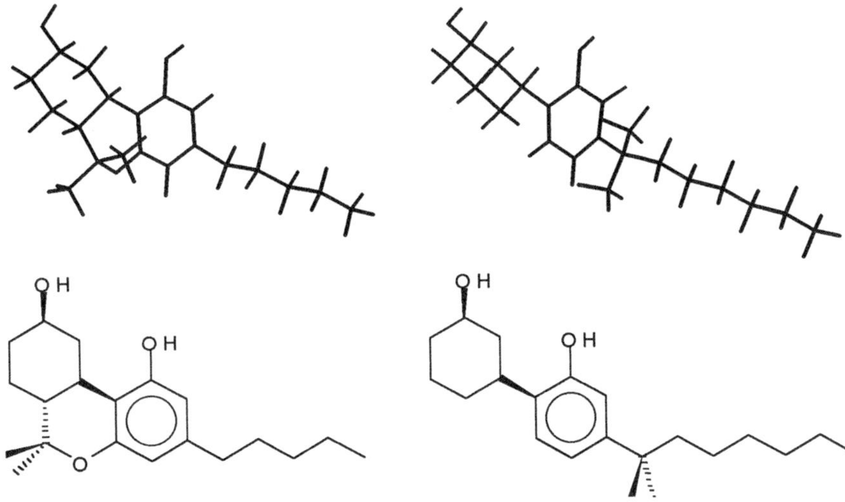

Figure 2 Low-energy conformations of HHC (left) and CP 47,497 (right).

freedom, the situation is complicated in that hydrogen bonding with the hydroxyl group can either be through the oxygen lone pairs or through interaction with the phenolic proton. The role of the hydroxyl group in hydrogen bonding and its orientation has, therefore, been the subject of molecular modelling studies (Reggio, 1987; Semus and Marin, 1990). In the Reggio studies, it was suggested that the active conformation of THC is one where the C1-OH torsion angle is 155° (Reggio, 1987). This conclusion was based on the observation that the relatively inactive 1-methoxy-Δ^9-THC analogue cannot obtain this conformation with favourable energy. However, this conclusion assumes that it is the phenolic oxygen, not the phenolic proton (which is absent in 1-methoxy-Δ^9-THC), that is involved in interaction with the receptor. Semus and Martin (1990) concluded that the phenolic proton was involved in receptor interaction, but no conclusion with regard to the active conformation was possible. Clearly, the phenolic hydroxyl group plays an important role in receptor interaction and further studies of the conformation and mode of interaction of this functional group with the receptor are required.

The previously described techniques of molecular modelling and SAR analysis often assume that changes in the steric, electrostatic or lipophilic properties of one functional group do not affect the properties of another. This approach can be problematic in that some molecules whose structures are markedly different from the prototypes are extremely difficult to assess in terms of their expected potency and the nature of their interaction with the receptor. For example, the AAIs, such as WIN-55212-2 (Table 6), differ from Δ^9-THC so dramatically that their alignment with the prototype cannabinoid is not clear (D'Ambra *et al.*, 1992). Not only has the superposition of these compounds with cannabinoids been debated, it has even been suggested that AAIs do not interact with the same binding site as cannabinoids. Indeed, molecular modelling and SAR studies performed on a series of 40 AAIs led the Sterling group (Estep *et al.*, 1991) to propose a three-point pharmacophore model for these compounds (based on the positions in space of two aromatic moieties and the nitrogen atom of the side chain) that is dramatically different from the cannabinoid pharmacophore. However, when radiolabelled ligand-binding analysis was performed the binding of radiolabelled AAI was selectively displaced by active cannabinoids (Haycock *et al.*, 1991; Kuster *et al.*, 1993). In addition, AAIs were shown to displace the binding of [^{3}H]CP 55,940 (Ward *et al.*, 1991) and [^{3}H](–)-11-OH-Δ^9-THC-DMH (Thomas *et al.*, 1992), indicating that the binding site for AAIs was close to or equivalent to the cannabinoid binding site. The increasingly wide variety of compounds that have been shown to produce cannabinoid effects and appear to interact at the same receptor have underscored the need for a more robust pharmacophore model that looks at molecules in terms of their entire molecular characteristics.

A recently developed molecular modelling and quantitative structure–activity relationship (QSAR) program, comparative molecular field analysis (or CoMFA, for review see Cramer *et al.*, 1988), utilizes computer-based forcefields to describe molecules in terms of the steric and electrostatic forces they would exert upon a probe atom placed at various points of a lattice surrounding the molecule (the theoretical receptor space). If the alignment of molecules is not easily discerned, a field-fitting program is available which aligns structurally diverse compounds so as to minimize the difference in their molecular fields. The steric and electrostatic molecular fields are then analysed for their ability to correlate with observed pharmacological potency through partial least-squares analysis (Dunn *et al.*, 1984; Cramer *et al.*, 1988). The results of the analysis can be displayed as contour plots around the molecules of interest, enabling the visualization of those regions where changes in the steric and electrostatic fields have the greatest influence on the predicted biological potency. The QSAR analysis can also be used to predict the potency of compounds, thereby facilitating rational drug design.

In a CoMFA study consisting of 33 cannabinoid compounds, the molecular fields of various 'classical' and 'non-classical' compounds were correlated with their potency in four behavioural assays designed to measure cannabinoid activity and their IC_{50} values in an *in vitro* binding assay (Thomas *et al.*, 1991). The QSAR analyses resulted in high correlation coefficients for fitting the potency of each compound in behavioural and binding-affinity assays, with final analysis r^2 values all exceeding 0.88. In addition, six compounds that were not contained in the training set were aligned to Δ^9-THC and their potency was predicted based on the model derived from the 33 compounds. The correlation coefficient between predicted and actual potencies of five true unknowns was determined to be 0.76. The unknowns that were predicted included the active and inactive enantiomers of the AAI compound WIN 55,212-2 shown in Table 6. As the training set did not include AAIs, the ability to predict the enantiospecificity within these two compounds indicates that the model is robust and capable of serving as a pharmacophore model for a wide variety of cannabinoid compounds.

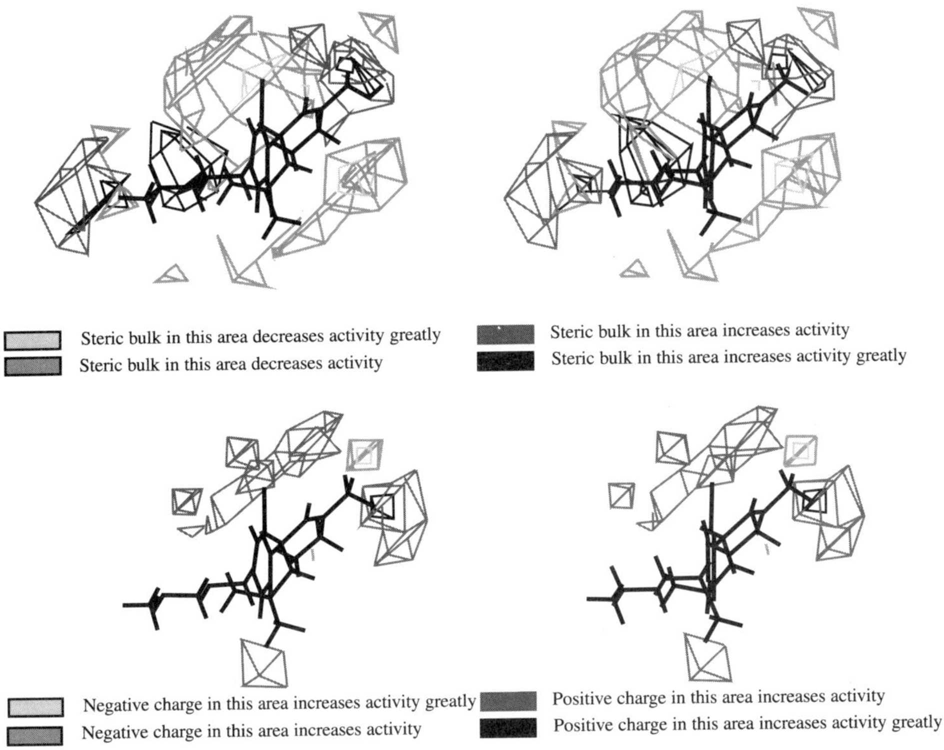

Figure 3 Contour plots of CoMFA data showing the steric and electrostatic pharmacophores as defined for cannabinoid-induced hypothermia in mice. Δ^9-THC is shown within the contours plots.

The contour plots (Figure 3) reveal a pharmacophore model that weights the contribution of the side chain, the phenolic hydroxyl group and the C9–C11 substituent heavily. In the steric map it is clear that the steric bulk of the side-chain is important for increasing the predicted potency. Depending upon its orientation, steric bulk at the C9 position leads to either decreased or increased predicted potency. Areas in the electrostatic map emphasize the role of the phenolic hydroxyl and the pyran oxygen. In addition to these functional groups there is a contoured area at the C11 position that indicates that a hydroxyl group at this position would lead to increased predicted potency. Thus, the pharmacophore model described by the CoMFA analysis agrees with previous hypothetical models attempting to account for differences in pharmacological potencies. The ability to rationalize both the *in vivo* behavioural potencies and the *in vitro* binding affinities of cannabinoids with steric and electrostatic force fields is in agreement with a specific, receptor-based mechanism of action in that these molecular properties are typically considered significant in drug–receptor interactions. Assuming that the pharmacological responses which are produced by drugs are the result of an interaction with a single receptor in a single binding conformation or orientation, the results of all of these SAR analyses can be interpreted with respect to the complementary molecular properties of the receptor.

The discovery of the putative endogenous cannabinoid (Devane *et al.*, 1992b), anandamide, has presented a unique challenge to molecular modelling and cannabinoid SAR. Although anandamide does possess a hydroxyl group and a lipophilic side chain, it is not clear at this time how the basic tenets of the cannabinoid pharmacophore are able to accommodate anandamide. As mentioned above, there is considerable interest in synthesizing and testing anandamide analogues in an attempt to understand the SAR within fatty acid ethanolamides, particularly with regard to how they relate to the previously determined cannabinoid SAR. Whether molecular modelling and quantitative SAR analysis will be able to accommodate all compounds possessing cannabinoid activity has not been determined. Clearly, any pharmacophore model will only be capable of accurately predicting molecules that are somewhat akin to the molecules in the training set, either in terms of size or functional groups present. Indeed, there may be non-cannabinoid compounds that would 'fit' into the pharmacophore model and wrongly be predicted to be cannabimimetic. Furthermore, the assumption of a common mode and conformation of interaction for anandamide and other cannabinoids may prove to be incorrect. Finally, bulk water and other solutes may be involved in the conformation or interaction of some ligands with the active site, a factor which is not typically included in a pharmacophore model. However, it seems likely that the combination of computer-assisted molecular modelling, three-dimensional quantitative structure–activity relationship analysis and protein structure determination will provide us with previously unattainable insights into the nature of cannabinoid–receptor interaction.

2.6 Evidence for and against receptor subtypes

Several strategies have been used to obtain evidence for subpopulations of receptors. In almost all cases, receptor subtypes are associated with ligands which produce multiple pharmacological/physiological effects. It is for this reason that receptor subtypes have been proposed for the cannabinoids. An impressive array of behavioural and other pharmacological effects are produced by the cannabinoids which makes it difficult to believe that a single receptor can be solely responsible. Typically, analogues are developed which exhibit selectivity and thereby implicate receptor subtypes. As we pointed out in the critique of cannabinoid models, there is an excellent correlation among all of the methods used to date. Unfortunately, this excellent correlation also means that no analogues have emerged which demonstrate a unique pharmacological profile. Occasionally an analogue is found which does not produce all of the pharmacological effects of Δ^9-THC or its potency ratios among all of the tests do not mimic those of Δ^9-THC. However, it has been difficult to translate these differences into a cohesive SAR for receptor subtypes. Indeed, it was disappointing that three structurally distinctive radiolabelled ligands (CP 55,940, 11-OH-Δ^9-THC-DMH, and WIN 55,212-2) provided identical binding profiles for the cannabinoids.

At present, it does not appear likely that a receptor subtype selective ligand will be developed first. It is more likely that molecular biology will provide the definitive evidence for multiple receptors, as it has for most other receptor systems in recent years. In 1990 the first reports on the molecular cloning of the cannabinoid receptor appeared (Matsuda *et al.*, 1990). They used an oligonucleotide probe based on the G-protein coupled receptor for substance K to isolate a clone from a rat brain library that had homology with other G-protein coupled receptors but was unique. When the sequence for the cannabinoid receptor was published another group quickly reported that they had isolated what appeared to be the human homologue of this receptor (Gérard *et al.*, 1990). The sequence of this human homologue was 90% identical at the nucleic acid level and 98% identical at the amino acid level to the rat clone. The extent of sequence conservation strongly suggested that the role of this protein would be identical to the rat cannabinoid receptor. They subsequently expressed the clone in COS cells and demonstrated specific binding with [^{3}H]CP 55,940 (Gérard *et al.*, 1991). The evidence indicated that these two groups had characterized an identical (CB_1) receptor.

Although multiple cannabinoid receptors have not been identified in brain, a peripheral receptor has been identified which is structurally different from the CB_1 receptor (Munro *et al.*, 1993). This cloned receptor (CB_2) is expressed in macrophages in the marginal zone of spleen and has 44% homology with the brain receptor (68% homology when only the transmembrane regions are considered). Based upon a limited pharmacological evaluation of this receptor, these investigators concluded that the recognition site of this peripheral receptor is different from that of the brain receptor. However, this assumption is based largely upon the observation that cannabinol and Δ^9-THC are nearly equipotent in binding to the peripheral receptor, whereas cannabinol has an affinity for the brain receptor which is much lower than that of Δ^9-THC. Other investigators have also demonstrated cannabinoid-receptor binding in mouse spleen cells as well as the presence of mRNA for the cannabinoid receptor (Kaminski *et al.*, 1992). However, the latter group has suggested that this receptor is associated with T-cells (Schatz *et al.*, 1993) and is similar to the receptor found in brain (Kaminski *et al.*, 1992). Additional characterization of these receptors in spleen will be needed in order to determine whether they are distinct entities and to ascertain their functional roles. However, the discovery of these cannabinoid receptors in spleen raises the possibility that other receptor subtypes with entirely unique functional roles may exist.

2.7 Strategies for developing antagonists

2.7.1 General considerations[1]

As yet a specific cannabinoid antagonist has not been found, although there have been some reports which have suggested that a variety of drugs (e.g. cannabidiol, cannabinol, phenitrone, imipramine and amphetamine) attenuate the effects of Δ^9-THC (Martin, 1986). However, subsequent studies have failed to find antagonism by these compounds and indicate that the attenuation of the THC-induced effect was simply the summation of opposite pharmacological responses rather than a direct effect (Spaulding *et al.*, 1972; Karniol *et al.*, 1974; Hollister and Gillespie, 1975; Lemberger *et al.*, 1976; Binder and Barlage, 1980; Fujiwara *et al.*, 1980; Browne and Weissman, 1981; Lew and Richardson, 1981). There has been one report (Burstein *et al.*, 1987) that the acid metabolite of Δ^9-THC attenuates the cataleptic effects produced by Δ^9-THC, and it was concluded that this metabolite is a cannabinoid antagonist. Unfortunately, we have been unable to replicate these findings in our laboratory. In another report, Δ^{9-11}-THC was found to significantly reduce the effects of Δ^9-THC on conditioned responding in monkeys (Beardsley *et al.*, 1987; Martin *et al.*, 1987), but it failed to alter the effects of Δ^9-THC in other behavioural models. Studies in our laboratory also demonstrated that $\Delta^{9,11}$-THC is not able to antagonize the discriminative cue of Δ^9-THC in monkeys.

Since there are no general guidelines for designing an antagonist in a particular class of compounds, the strategies which have been followed for cannabinoids are discussed below. The cannabimimetic properties of novel drugs is currently determined by using the mouse tetrad model as described above. This procedure can also be used to evaluate potential antagonistic properties of a compound by evaluating its ability to attenuate the pharmacological effects of Δ^9-THC (Little *et al.*, 1987; Martin *et al.*, 1987; Compton and Martin, 1990).

2.7.2 Ethers

One possible approach to the development of a specific cannabinoid antagonist is to modify the structure of Δ^9-THC sufficiently to prevent activation of the receptor site without altering recognition. Based on this strategy we prepared several ethers of Δ^8 and Δ^9 and Δ^{9-11}-THC. This seemed an attractive approach since it was known that both methyl ethers of Δ^8- and Δ^9-THC were inactive in the monkey at doses up to 10 mg/kg, while Δ^9-THC produced prominent effects at 0.10–0.25 mg/kg (Edery *et al.*, 1972). Furthermore, they had not been evaluated previously in rodents for agonist or antagonist properties. The various ethers that were synthesized and tested are shown in Table 10 (**38, 39, 40**). In addition to methyl ethers of Δ^8-, Δ^9- and Δ^{9-11}-THC, a few novel ethers were synthesized as preliminary probes to ascertain whether an ether linkage at the phenolic site could impart any antagonistic properties. The

[1] The recent discovery of a competitive cannabinoid antagonist SR141716A occurred after the completion of this chapter (Rinaldi-Carmona *et al.*, 1994). This antagonist, described in Chapter 1, will have a significant impact on cannabinoid research.

Table 10 Synthesized as potential cannabinoid antagonists

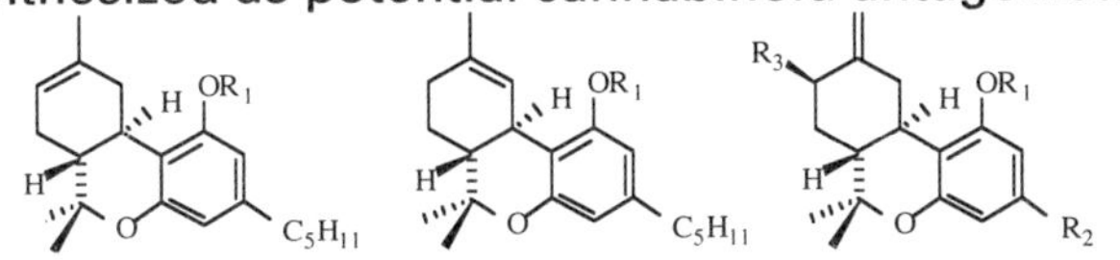

Δ^{8}-THC derivatives Δ^{9}-THC derivatives $\Delta^{9,11}$-THC derivatives

Compound	Structure	R_1	R_2	R_3	IC_{50}[a]
Δ^8-THC	Δ^8	H			179
38	Δ^8	CH_3			3200
41	Δ^8	CH_2 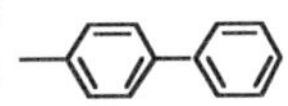			b
43	Δ^8	$(CH_2)_4NH_2$			b
44	Δ^8	$(CH_2)_3NH_2$			b
45	Δ^8	$(CH_2)_6NH_2$			2500
46	Δ^8	$(CH_2)_2$ 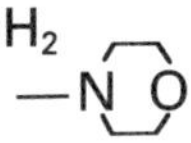			4300
Δ^9-THC	Δ^9	H			218
39	Δ^9	CH_3			10 000
$\Delta^{9,11}$-THC	$\Delta^{9,11}$	H	n-C_5H_{11}	H	334
49	$\Delta^{9,11}$	H	n-C_5H_{11}	OH	1200
40	$\Delta^{9,11}$	CH_3	n-C_5H_{11}	H	
42	$\Delta^{9,11}$	CH_2 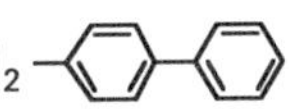	n-C_5H_{11}	H	
47	$\Delta^{9,11}$	H	n-C_3H_7	H	1400
48	$\Delta^{9,11}$	CH_3	n-C_3H_7	H	b

[a]IC_{50} values (nM, means±SEM) for displacement of 1 nM[^{3}H]CP 55,940.
[b]Even at concentrations of 10 μM, less than 50% displacement was obtained.

rationale for the synthesis of the biphenyl ethers (**41** and **42**) was based on the finding that this substituent (biphenyl ether) has been found to impart antagonistic properties in the prostaglandin class of drugs (Brittain *et al.*, 1984; Cross and Dickinson, 1987), and there is evidence to suggest that some cannabinoid effects may be mediated by the prostaglandins (Fairbairn and Pickens, 1980; Johnson *et al.*, 1982; Burstein, 1987). We also synthesized aminoalkyl ethers (**43–45**) because we thought that a nucleophilic site attached to the THC molecule (in place of a phenol) might possibly interact with the receptor. The length of the chain was varied to facilitate this potential interaction. A morpholinoalkyl ether (**46**) was synthesized because the presence of an electron rich site, such as the oxygen of the morpholine, could possibly result in a mixed agonist/antagonist, as morpholino alkyl esters of THCs are known to be very potent agonists (Razdan, 1986). None of these analogues was found to exhibit antagonist properties in the mouse behavioural assays.

The observation that Δ^{9-11}-THC significantly reduced the effect of Δ^9-THC in monkeys (Beardsley *et al.*, 1987; Martin *et al.*, 1987) served as a possible lead for developing an antagonist. However, Δ^{9-11}-THC exhibited depressant properties which precluded testing it as an antagonist at higher doses. Therefore, two structural modifications were made in the anticipation that these depressant properties might be eliminated while retaining any antagonist activity. First, the side-chain was shortened (from C_5 to C_3) to reduce the agonistic activity of $\Delta^{9,11}$-THC, resulting in compounds **47** and **48** (Table 10). Second, a hydroxyl group was introduced in the vicinity of the exocyclic group of Δ^{9-11}-THC, resulting in compound **49**. The rationale for this latter change is based upon analogy with the opiate field, where it is well known that the addition of a hydroxyl group (at C14) to morphine imparts antagonistic properties.

Unfortunately none of the above-mentioned compounds provided a lead for an antagonist. In another approach, novel THCs were prepared carrying structural groups such as sulfonamide, hydroxyl, bromine, etc. which have provided leads for antagonists in other fields. Thus, the sulfonamide of Δ^8-THC (**50**, Scheme 1) was synthesized based on the analogy with antagonists in the thromboxane A_2/prostaglandin field (Hamanaka *et al.*, 1989). Similarly, following the analogy of structural groups which have provided antagonists of β-adrenergic receptors, THC analogues **51a**, **51b**, and **51c** (Scheme 1) were synthesized. These analogues also failed to yield antagonistic activity.

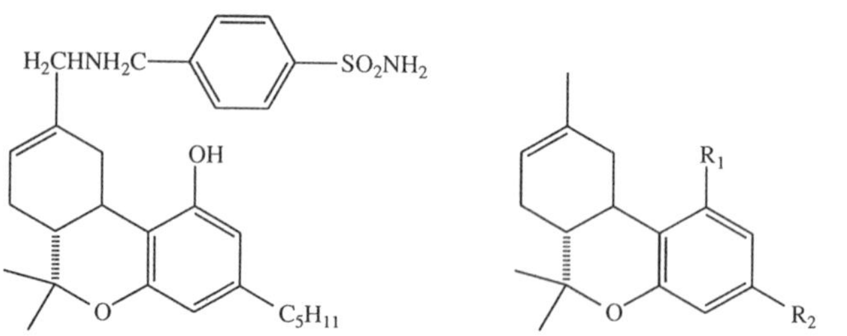

50 **51**

	R_1	R_2
a,	OH	$NHSO_2CH_3$
b,	$NHSO_2CH_3$	OH
c,	OH	Br

Scheme 1

2.7.3 Alkylating agents of weak or inactive cannabinoids

Nitrogen mustards have been used successfully in numerous instances to covalently alkylate receptors, thereby producing irreversible antagonism. Our initial strategy involved incorporation of the nitrogen mustard moiety into an inactive cannabinoid such as cannabidiol (CBD) and at a position not likely to be critical for receptor activation. However, the resulting analogue (**52**), depicted in Scheme 2, lacked antagonistic activity (Compton *et al.*, 1990). Out of concern that CBD might not be capable of acting at the receptor, a similar analogue was prepared using Δ^8-THC as the template (**53**). Failure of this analogue to antagonize Δ^9-THC led us to place the nitrogen moiety at

Scheme 2

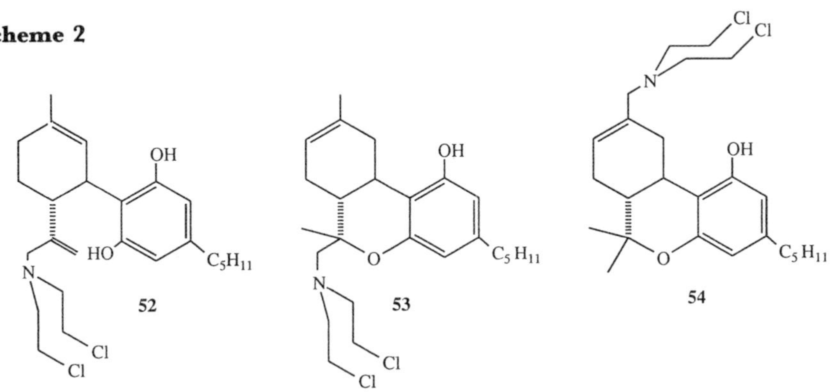

a position that we knew was critical for receptor interaction, i.e. substitution in the C_{11} position in Δ^8-THC. This nitrogen mustard analogue (**54**) also was unable to block the behavioural effects of Δ^9-THC in the mouse model (Compton *et al.*, 1990).

2.7.4 Alkylating agents of potent cannabinoids

In continuation of our search for a lead to an antagonist we recently synthesized two novel analogues of Δ^8-THC which show promise in the attempted development of cannabinoid antagonists. One analogue is a nitrogen mustard derivative, (**55**) possessing a 1′,1′-dimethylheptyl side-chain and the other is a cyano derivative (**56**) with a 1′,1′-dimethylpentyl side-chain (Scheme 3). The former resulted from our studies on various nitrogen mustard analogues of THCs which were found to be much less active than Δ^9-THC. To increase the binding affinity we incorporated a 1′,1′-dimethylheptyl side-chain and designed **55**. The cyano analogue **56** was designed following the lead for leukotriene B_4 receptor antagonists in which a nitrile group has imparted significant antagonist activity (Herron *et al.*, 1992).

Scheme 3

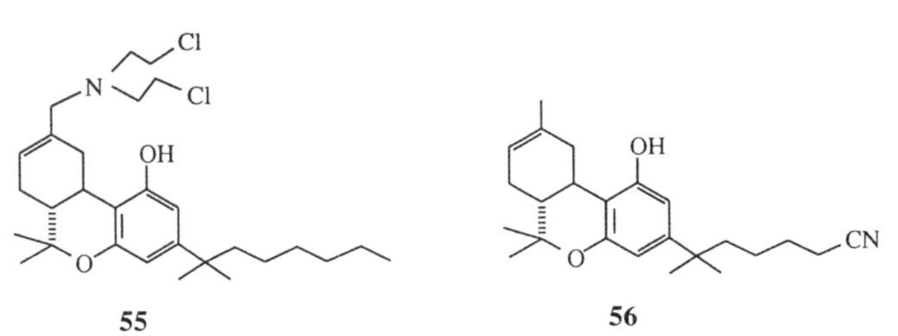

The nitrogen mustard analogue **55** possesses a K_I of 1.3±0.2 nM and ED_{50} values between 2.3 and 5.7 mg/kg in the four *in vivo* measures in the mouse tetrad model. At 100 mg/kg no fatalities were observed, and mice evaluated either 24 h or 48 h later still exhibited maximal agonistic effects in all measures. However, no agonistic effects remained at 72 h post-injection following a dose of 10 mg/kg (i.v.). After

72 h the antagonist properties of the mustard analogue were evaluated versus administration of Δ^9-THC (3 mg/kg, i.v.). Results on locomotor activity were inconclusive, but the effects of Δ^9-THC on rectal temperature and immobility were completely blocked, while the effect on the tail-flick measure was reduced from 73% MPE to 40% MPE. In the rat drug-discrimination model, the mustard analogue showed agonistic activity. Similarly, the cyano derivative **56** possesses a K_I of 0.67 nM and is extremely potent in the mouse tetrad model, with ED_{50} values between 0.007 and 0.33 mg/kg. At a dose of 100 mg/kg (i.v.) this analogue produced significant effects on locomotor activity (89% inhibition), rectal temperature (6.8°C decrease), tail-flick antinociception (71% MPE), and immobility (62%), while producing no fatalities. After 72 h these animals were administered 30 mg/kg Δ^9-THC (i.p.). The cyano analogue completely antagonized the effects of Δ^9-THC on temperature and immobility (the only measures evaluated). Subsequent experiments indicate that at a dose as low as 0.3 mg/kg it completely antagonized the effects of THC on temperature and immobility responses. However, the degree of antagonism of this drug on tail-flick (MPE) or locomotor responses was inconclusive.

These data indicate that these compounds possess acute agonist activity but have delayed antagonist properties, thus falling in the category of mixed agonist/antagonist. These compounds have a very interesting pharmacological profile and have provided a lead, for the first time, for a cannabinoid antagonist. In this context, it is noteworthy that among the recently discovered AAI cannabinoids, one analogue WIN 56,098 was reported to provide a lead for an antagonist. At a concentration of 10 mM, WIN 56,098 competitively antagonized the effects of AAIs, pravadoline and (+)-WIN 55,212-2 with dose ratios of 40 and 100, respectively, in the mouse vas deferens response (Pacheco *et al.*, 1991). However, the corresponding dose ratios were much less significant for other cannabinoids such as Δ^9-THC and levonantradol. Nevertheless, these data did suggest a lead for an antagonist. However, subsequent *in vivo* work in mice failed to confirm any antagonist activity for WIN 56,098 (Compton *et al.*, 1992a).

To summarize, to date, no specific antagonist for cannabinoids has been found. A promising lead has been provided with the compounds **55** and **56**. Unfortunately, the mixed agonist/antagonists effects of these compounds complicate their use. Hopefully, further structural modifications of THCs along these lines will lead to a specific antagonist which is devoid of agonist effects.

2.8 Promising leads

It is well documented that a variety of different structural entities possess cannabinoid activity. These include the 'classical THC' analogues which have the ABC ring system (Table 8), the 'non-classical THC' analogues (levonantradol and ring-opened AC and ACD compounds in Tables 4 and 5), aminoalkylindoles (Tables 6, 7, and 8) and anandamide analogues (Table 9). They are distinct and unique structures which provide entirely new templates for designing cannabinoid agonists/antagonists.

Over the years, several-hundred cannabinoids have been synthesized and several leads for potent agonists have been found. In the 'classical THC' group, (−)-11-OH-Δ^8-THC-DMH and (−)-11-OH-Δ^9-THC-DMH are several-hundred times more potent than Δ^9-THC which makes them two of the most potent cannabinoids known. Other potent compounds in this group are the 5′-bromo-Δ^8-THC-DMH and the hexahydro analogue of 11-OH-THC-DMH. In the 'non-classical THC' series, the most potent agonists are (−)-ACD (Table 5) and CP 55,940. It is noteworthy that all of these agonists show the full spectrum of pharmacological activity, similar to that of Δ^9-THC. Hence, to date, separation of activities to provide leads for selective agonists has been unsuccessful.

In the antagonist field, despite several claims that some cannabinoids are antagonists and attempts to design compounds as potential antagonists, the discovery of a specific THC antagonist has not been achieved. At best, the work from our laboratory has provided leads for mixed agonist/antagonists. Compounds **55** and **56** do reveal that the addition of reactive functional groups, such as nitrogen mustards and cyano groups, may produce antagonistic properties. The only other potential lead for an antagonist came from the Sterling research group who demonstrated that WIN 56,098, an aminoalkylindole, had antagonistic properties when evaluated *in vitro*. Subsequent *in vivo* work in mice, however, failed to confirm any antagonistic activity for this compound. To our knowledge there have been no reports of a potential lead either from the Pfizer group ('non-classical THCs') or from the anandamide analogues.

It is likely that a lead for an antagonist may come from the anandamide analogues since several antagonists are known in the prostanoid field. Hopefully, further structural modifications of THCs along the lines of compounds **55** and **56**, as well as with structural manipulations of aminoalkylindoles and anandamides, will lead us to THC antagonists. Conversely, an antagonist may emerge which has a structure distinct from all of the above templates.

2.9 Summary

It is clear that the lack of selective agonists and pure antagonists has arrested rapid progress in the cannabinoid field. It is therefore paramount that chemists and pharmacologists continue their search for these vital probes.

As stated earlier, it would be difficult to believe that a single receptor is solely responsible for the myriad effects of Δ^9-THC. If the examination of other centrally acting agents can be used as a guide, then it would be very surprising if multiple receptors or mechanisms are not found for the cannabinoids. The report on the discovery of peripheral cannabinoid receptors in the spleen points to the possibility that other receptor subtypes with entirely unique functional roles may exist. In this connection, it should be emphasized that the development of SAR in cannabinoids is complex. One reason for this is that analogues have been screened in various models over the years. As a result, many analogues have not been evaluated in every model, making comparisons difficult. Multiple assays are critical for verifying that a compound is truly a cannabinoid, since this class of compounds shares many pharmacological properties with other centrally acting agents. Furthermore, SAR must be established in different pharmacological models if multiple mechanisms of action are to be delineated. At present, there is an excellent correlation among all of the *in vivo* and *in vitro* assays, which is reassuring. However, this excellent correlation brings in to question whether multiple receptors or mechanisms exist for the cannabinoids.

One issue which has received relatively little attention is the criteria necessary for establishing a relationship between structure and activity. In the early literature, a typical SAR was established using Δ^9-THC, Δ^8-THC, cannabinol and cannabidiol, because they were the compounds that were readily available. However, the difficulty of establishing a correlation using only four points is obvious. These early investigators could usually deduce from the results whether the effect in question was likely to be associated with cannabinoids. However, it was almost impossible to distinguish between different cannabinoid actions. Although it is difficult to identify the minimum number of compounds needed to draw firm conclusions, a convincing argument for an SAR can be made if several analogues are chosen from each of the subsets which now include 'classical THCs', 'non-classical THCs', aminoalkylindoles and anandamides. Of course, many investigators face the practical difficulty of obtaining large numbers of analogues and some pharmacological procedures are too difficult to accommodate large numbers of compounds. As a generalization, an SAR conducted with 10–15 analogues is likely to result in definitive conclusions, whereas those carried out with 4–5 compounds should be interpreted cautiously. Despite these problems, there is general agreement among all of the assays used for SAR in cannabinoids and the conclusions about SAR described in various publications are quite informative.

In general, the most useful model of cannabinoid–receptor interaction is the three-point attachment depicted in Figure 1. Indeed, this model provided the basis for designing new analogues with potent activity. Refinement of this model, as described by the CoMFA model, has resulted in a pharmacophore for cannabinoids. However, there are many aspects which need further examination. For example, it is clear that the phenolic hydroxyl group plays an important role in receptor interactions, but further studies are required to delineate clearly the conformation and mode of interaction of this functional group. This model can accommodate the aminoalkylindoles, although further refinements are necessary.

The recent discovery of anandamide has, however, complicated this issue. Although anandamide possesses a hydroxyl group and a lipophilic side chain, it is unclear at present how it could be accommodated in the cannabinoid pharmacophore. Many more anandamide analogues will have to be synthesized and studied for their biological activity before this aspect will be clarified. This last point underscores the irony with the anandamides. Frequently, the initial ligands for a receptor are flexible, making it necessary to design conformationally restrained molecules to characterize a pharmacophore. In the cannabinoids, we began with a rigid structure (Δ^9-THC), then developed the more flexible bicyclic 'non-classical THCs' and finally discovered the very flexible anandamides.

The cannabinoid field is presently experiencing an exciting phase reminiscent of that of the opioids after the discovery of endorphins. Unlike opioids, however, no cannabinoid antagonist or selective agonist is known. A lead is provided by compounds **55** and **56** in that reactive functional groups may serve to convert agonists to irreversible antagonists. Rapid advances in this field can be envisioned if an antagonist is discovered. It would also prove to be a valuable tool in the investigation of dependence-producing properties of cannabinoids. The discovery of selective agonists should help in understanding the mechanism(s) of action of this class of compounds and in developing new cannabinoid drugs for therapeutic use. It seems very likely that these goals will be reached in the near future through the combined use of computer-assisted molecular modelling, three-dimensional quantitative SAR analysis and protein-structure determination.

Acknowledgements

Portions of this research were supported by NIDA grants DA-03672 and DA-05488 and the Commonwealth of Virginia Center on Drug Abuse Research.

Note added in proof

The recent discovery of a competitive cannabinoid antagonist SR141716A occurred after the completion of this chapter (Rinaldi-Carmona *et al.*, 1994). This antagonist, described in Chapter 1, will have a significant impact on cannabinoid research.

References

Balster, R.L. & Prescott, W.R. (1992) Δ^9-tetrahydrocannabinol discrimination in rats as a model for cannabis intoxication. *Neurosci. Biobehav. Rev.* **16**, 55–62.

Beardsley, P.M., Scimeca, J.A. & Martin, B.R. (1987) Studies on the agonistic activity of Δ^{9-11}-tetrahydrocannabinol in mice, dogs and rhesus monkeys and its interactions with Δ^9-tetrahydrocannabinol. *J. Pharmacol. Exp. Ther.* **241**, 521–526.

Bell, M.R., D'Ambra, T.E., Kumar, V., Eissenstat, M.A., Herrmann, J.L., Wetzel, J.R., Rosi, D., Philion, R.E., Daum, S.J., Hlasta, D.J., Kullnig, R.K., Ackerman, J.H., Haubrich,

D.R., Luttinger, D.A., Baizman, E.R., Miller, M.S. & Ward, S.J. (1991) Antinociceptive (aminoalkyl)indoles. *J. Med. Chem.* **34**, 1099–1110.
Bidaut-Russell, M., Devane, W.A. & Howlett, A.C. (1990) Cannabinoid receptors and modulation of cyclic AMP accumulation in the rat brain. *J. Neurochem.* **55**, 21–26.
Binder, M. & Barlage, U. (1980) Metabolic transformation of (3R,4R)-$\Delta^{1(7)}$-tetrahydrocannabinol by a rat liver microsomal preparation. *Helv. Chim. Acta* **63**, 255–267.
Binder, M., Witteler, F.-J., Schmidt, B., Franke, I. Bohnenberger, E. and Sandermann, H. (1984) Triglyceride/phospholipid partitioning and pharmacokinetics of some natural and semi-synthetic cannabinoids: further evidence for the involvement of specific receptors in the mediation of psychotropic effects of Δ^1-THC and Δ^6-THC. In *The Cannabinoids: Chemical, Pharmacologic and Therapeutic Aspects* (eds Agurell, S., Dewey, W.L. & Willette, R.E.), pp. 709–727. Academic Press, New York.
Brittain, R.T., Coleman, R.A., Collington, E.W., Hallett, P., Humphrey, P.P.A., Kennedy, I., Lumley, P., Sheldrick, R.L.G. & Wallis, C.J. (1984). AH 23848: a potent, selective thromboxane receptor blocking drug. *Br. J. Pharmacol.* **83**, 377P.
Browne, R.G. & Weissman, A. (1981). Discriminative stimulus properties of Δ^9-THC: mechanistic studies. *J. Clin. Pharmacol.* **21**, (Suppl.) 227s-234s.
Burstin, S., Hunter, S.A., Latham, V. & Renzulli, L. (1987) A major metabolite of Δ^1-tetrahydrocannabinol reduces its cataleptic effect in mice. *Experientia* **43**, 402–403
Burstein, S.H. (1987) Inhibitory and stimulatory effects of cannabinoids on eicosanoid synthesis. In *Structure–Activity Relationships of the Cannabinoids* (eds Rapaka, R.S. & Makriyannis, A.), NIDA Res. Monogr. 79, pp. 158–172. US Government Printing Office, Washington, DC.
Caulfield, M.P. & Brown, D.A. (1992). Cannabinoid receptor agonists inhibit Ca current in NG108–15 neuroblastoma cells via a pertussis toxin-sensitive mechanism. *Br. J. Pharmacol.* **106**, 231–232.
Charalambous, A., Lin, S., Marciniak, G., Banijamali, A., Friend, F.L., Compton, D.R., Martin, B.R. & Makriyannis, A. (1991) Pharmacological evaluation of halogenated Δ^8-THC analogs. *Pharmacol. Biochem. Behav.* **40**, 509–512.
Childers, S.R., Sexton, T. & Roy, M.B. (1993) Effects of anandamide on cannabinoid receptors in rat brain membranes. *Biochem. Pharmacol.* **47**, 711–715.
Compton, D.R. & Martin, B.R. (1990). Pharmacological evaluation of water soluble cannabinoids and related analogs. *Life Sci.* **46**, 1575–1585.
Compton, D.R., Little, P. J., Martin, B.R., Gilman, J.W., Saha, J.K., Jorapur, V.S., Sard, H.P. & Razdan, R.K. (1990). Synthesis and pharmacological evaluation of amino, azido, and nitrogen mustard analogues of 10-substituted cannabidiol and 11- or 12- substituted Δ^8-tetrahydrocannabinol. *J. Med. Chem.* **33**, 1437–1443.
Compton, D.R., Gold, L.H., Ward, S.J., Balster, R.L. & Martin, B.R. (1992a) Aminoalkylindole analogs: cannabimimetic activity of a class of compounds structurally distinct from Δ^9-tetrahydrocannabinol. *J. Pharmacol. Exp. Ther.* **263**, 1118–1126.
Compton, D.R., Johnson, M.R., Melvin, L.S. & Martin, B.R. (1992b) Pharmacological profile of a series of bicyclic cannabinoid analogs: classification as cannabimimetic agents. *J. Pharmacol. Exp. Ther.* **260**, 201–209
Compton, D.R., Rice, K.C., de Costa, B.R., Razdan, R.K., Melvin, L.S., Johnson, M.R. & Martin, B.R. (1993) Cannabinoid structure–activity relationships: correlation of receptor binding and in vivo activities. *J. Pharmacol. Exp. Ther.* **265**, 218–226.
Cramer, III. R.D., Bunce, J.D. & Patterson, D.E. (1988) Crossvalidation, bootstrapping, and partial least squares compared with multiple regression in conventional QSAR studies. *Quant. Struct.–Act. Relat.* **7**, 18–25.
Cross, P.W. & Dickinson, R.P. (1987) Thromboxane synthetase inhibitors and antagonists. In *Annual Reports in Medicinal Chemistry* 22, (ed. Bailey D.M.), pp. 95–106. Academic Press, Orlando, FL.

D'Ambra, T.E., Estep, K.G., Bell, M.R., Eissenstat, M.A., Josef, K.A., Ward, S.J., Haycock, D.A., Baizman, E.R., Casiano, F.M., Beglin, N.C., Chippari, S.M., Grego, J.D., Kullnig, R.K. & Daley, G.T. (1992) Conformationally restrained analogues of pravadoline – nanomolar potent, enantioselective, (aminoalkyl)indole agonists of the cannabinoid receptor. *J. Med. Chem.* **35**, 124–135

Deutsch, D.G. & Chin, S.A. (1993) Enzymatic synthesis and degradation of anandamide, a cannabinoid receptor agonist. *Biochem. Pharmacol.* **46**, 791–796

Devane, W.A., Dysarz, F.A., Johnson, M.R., Melvin, L.S. & Howlett, A.C. (1988) Determination and characterization of a cannabinoid receptor in rat brain. *Mol. Pharmacol.* **34**, 605–613

Devane, W.A., Breuer, A., Sheskin, T., Järbe, T.U.C., Eisen, M.S. & Mechoulam, R. (1992a) A novel probe for the cannabinoid receptor, *J. Med. Chem.* **35**, 2065–2069

Devane, W.A., Hanus, L., Breuer, A., Pertwee, R.G., Stevenson, L.A., Griffin, G., Gibson, D., Mandelbaum, A., Etinger, A. & Mechoulam, R. (1992b) Isolation and structure of a brain constituent that binds to the cannabinoid receptor. *Science* **258**, 1946–1949

Dunn, III. W.J., Wold, S., Edlund, U., Hellberg, S. & Gasteiger, J. (1984) Multivariate structure-activity relationships between data from a battery of biological tests and an ensemble of structure descriptors: the PLS method. *Quant. Struct.–Act. Relat.* **3**, 131–137.

Edery, H., Grunfeld, Y., Ben-Zvi, Z. & Mechoulam, R. (1971) Structural requirements for cannabinoid activity. *Ann. NY Acad. Sci.* **191**, 40–53.

Edery, H., Grunfeld, Y., Porath, G., Ben-Zvi, Z., Shani, A. & Mechoulam, R. (1972) Structure–activity relationships in the tetrahydrocannabinol series. *Arzneim-Forsch. (Drug Res.)* **22**, 1995–2003

Estep, K.G., D'Ambra, T.E., Olefirowicz, E.M., Bell, M.R., Eissenstat, M.A., Haycock, D.A. & Ward, S.J. (1991) Conformationally restrained aminoalkylindoles: potent, stereoselective ligands at the cannabinoid binding site. In *Problems of Drug Dependence 1990: Proceedings of the 52nd Annual Scientific Meeting* (ed. Harris, L.S.), NIDA Res. Monogr. 105, p. 300. US Government Printing Office, Washington, DC.

Fairbairn, J.W. & Pickens, J.T. (1980) The effect of conditions influencing endogenous prostaglandins on the activity of Δ^1-tetrahydrocannabinol in mice. *Br. J. Pharmacol.* **69**, 491–493

Felder, C.C., Briley, E.M., Axelrod, J., Simpson, J.T., Mackie, K. & Devane, W.A. (1993) Anandamide, an endogenous cannabimimetic eicosanoid, binds to the cloned human cannabinoid receptor and stimulates receptor-mediated signal transduction. *Proc. Natl Acad. Sci. USA* **90**, 7656–7660

Fride, E. & Mechoulam, R. (1993) Pharmacological activity of the cannabinoid receptor agonist, anandamide, a brain constituent. *Eur. J. Pharmacol.* **231**, 313–314

Fujiwara, M., Ibii, N., Kataoka, Y. & Ueki, S. (1980) Effects of psychotropic drugs on Δ^9-tetrahydrocannabinol-induced long-lasting muricide. *Psychopharmacol.* **68**, 7–13.

Gérard, C., Mollereau, C., Vassart, G. & Parmentier, M. (1990) Nucleotide sequence of a human cannabinoid receptor cDNA. *Nucl. Acids* Res. **18**, 7142.

Gérard, C.M., Mollereau, C., Vassart, G. & Parmentier, M. (1991) Molecular cloning of a human cannabinoid receptor which is also expressed in testis. *Biochem. J.* **279**, 129–134

Gold, L.H., Balster, R.L., Barrett, R.L., Britt, D.T. & Martin, B.R. (1992) A comparison of the discriminative stimulus properties of Δ^9-tetrahydrocannabinol and CP-55,940 in rats and rhesus monkeys. *J. Pharmacol. Exp. Ther.* **262**, 479–486.

Hamanaka, N., Seko, T., Miyazaki, T., Naka, M., Furuta, K. & Yamamoto, H. (1989) The synthesis of potent thromboxane A_2/prostaglandin endoperoxide receptor antagonists. *Tetrahedron Lett.* **30**, 2399–2402.

Hanuš, L., Gopher, A., Almog, S. & Mechoulam, R. (1993) Two new unsaturated fatty acid ethanolamides in brain that bind to the cannabinoid receptor, *J. Med. Chem.* **36**, 3032–3034.

Hardman, H.F., Domino, E.F. & Seevers, M.H. (1971) General pharmacological actions of some synthetic tetrahydrocannabinol derivatives. *Pharmacol. Rev.* **23**, 295–315.

Harris, L.S., Carchman, R.A. & Martin, B.R. (1978) Evidence for the existence of specific cannabinoid binding sites. *Life Sci.* **22**, 1131–1138.

Haycock, D.A., Kuster, J.E., Stevenson, J.I., Ward, S.J. & D'Ambra, T. (1991) Characterization of aminoalkylindole binding: selective displacement by cannabinoids. In *Problems of Drug Dependence 1990: Proceedings of the 52nd Annual Scientific Meeting* (ed. Harris, L.S.), NIDA Res. Monogr. 105, pp. 304–305. US Government Printing Office, Washington, DC.

Herron, D.K., Goodson, T., Bollinger, N.G., Swanson-Bean, D., Wright, I.G., Staten, G.S., Thompson, A.R., Froelich, L.L. & Jackson, W.T. (1992) Leukotriene B_4 receptor antagonists: the LY255283 series of hydroxyacetophenones. *J. Med. Chem.* **35**, 1818–1828.

Hollister, L.E. (1974) Structure–activity relationships in man of cannabis constituents, and homologs and metabolites of Δ^9-tetrahydrocannabinol. *Pharmacology* **11**, 3–11.

Hollister, L.E. & Gillespie, H. (1975) Interactions in man of Δ^9-tetrahydrocannabinol. II. Cannabinol and cannabidiol. *Clin. Pharmacol. Ther.* **18**, 80–83.

Howlett, A.C. (1984) Inhibition of neuroblastoma adenylate cyclase by cannabinoid and nantradol compounds. *Life Sci.* **35**, 1803–1810.

Howlett, A.C., Johnson, M.R., Melvin, L.S. & Milne, G.M. (1988) Nonclassical cannabinoid analgetics inhibit adenylate cyclase: development of a cannabinoid receptor model. *Mol. Pharmacol.* **33**, 297–302.

Howlett, A.C., Champion, T.M., Wilken, G.H. & Mechoulam, R. (1990). Stereochemical effects of 11-OH-Δ^8-tetrahydrocannabinol-dimethylheptyl to inhibit adenylate cyclase and bind to the cannabinoid receptor. *Neuropharmacol.* **29**, 161–165.

Howlett, A.C., Evans, D.M. & Houston, D.B. (1992) The cannabinoid receptor. In *Marijuana/Cannabinoids: Neurobiology and Neurophysiology* (ed. Murphy, L. & Bartke, A.), pp. 35–72. CRC Press, Boca Raton, FL.

Huffman, J.W., Dai, D., Martin, B.R. & Compton, D.R. (1994) Design, synthesis and pharmacology of cannabimimetic indoles. BioMed. Chem. Lett. in press.

Järbe, T.U.C. & Mathis, D.A. (1992) Dissociative and discriminative stimulus functions of cannabinoids/cannabimimetics. In *Marijuana/Cannabinoids: Neurobiology and Neurophysiology* (eds Murphy, L. and Bartke, A.), pp. 425–458. CRC Press, Boca Raton, FL.

Järbe, T.U.C. & McMillan, D.E. (1979) Discriminative stimulus properties of tetrahydrocannabinols and related drugs in rats and pigeons. *Neuropharmacol.* **18**, 1023–1024

Järbe, T.U.C. & McMillan, D.E. (1980) Δ^9-THC as a discriminative stimulus in rats and pigeons: generalization to THC metabolites and SP-111. *Psychopharmacol.* **71**, 281–289.

Järbe, T.U.C. & Ohlin, G.C. (1977) Stimulus effects of Δ^9-THC and its interaction with naltrexone and catecholamine blockers in rats. *Psychopharmacol.* **54**, 193–195.

Järbe, T.U.C., Hiltunen, A.J. & Mechoulam, R. (1989) Stereospecificity of the discriminative stimulus functions of the dimethylheptyl homologs of 11-hydroxy-Δ^8-tetrahydrocannabinol in rats and pigeons. *J. Pharmacol. Exp. Ther.* **250**, 1000–1005.

Johnson, M.R. & Melvin, L.S. (1986) The discovery of nonclassical cannabinoid analgetics. In *Cannabinoids as Therapeutic Agents* (ed. Mechoulam, R.), pp. 121–145. CRC Press, Inc., Boca Raton, FL.

Johnson, M.R., Melvin, L.S. & Milne, G.M. (1982) Prototype cannabinoid analgetics, prostaglandins and opiates – a search for points of mechanistic interaction. *Life Sci.* **31**, 1703–1706.

Kaminski, N.E., Abood, M.E., Kessler, F.K., Martin, B.R. and Schatz, A.R. (1992) Identification of a functionally relevant cannabinoid receptor on mouse spleen cells that is involved in cannabinoid-mediated immune modulation. *Mol. Pharmacol.* **42**, 736–742.

Karniol, I.G., Shirakawa, I., Kasinski, N. Pfeferman, A. & Carlini, E.A. (1974) Cannabidiol interferes with the effects of Δ^9-tetrahydrocannabinol in man. *Eur. J. Pharmacol.* **28**, 172–177.

Kriwacki, R.W. & Makriyannis, A. (1989) The conformational analysis of Δ^9- and $\Delta^{9,11}$-tetrahydrocannabinols in solution using high resolution nuclear magnetic resonance spectroscopy. *Mol. Pharmacol.* **35**, 495–503.

Kuster, J.E., Stevenson, J.I., Ward, S.J., D'Ambra, T.E. & Haycock, D.A. (1993) Aminoalkylindole binding in rat cerebellum: selective displacement by natural and synthetic cannabinoids. *J. Pharmacol. Exp. Ther.* **264**, 1352–1363

Lemberger, L., Dalton, B., Martz, R., Rodda, B. & Forney, R. (1976) Clinical studies on the interaction of psychopharmacologic agents with marihuana. *Ann. NY Acad. Sci.* **281**, 219–228.

Lew, E.O.H. & Richardson, J.S. (1981) Neurochemical and behavioral correlates of the interaction between amphetamine and Δ^9-tetrahydrocannabinol in the rat. *Drug Alc. Dependence* **8**, 93–101.

Little, P.J., Kaplan, N.C. & Martin, B.R. (1987) Pharmacological profile of Δ^9-THC carbamate. *Alc. Drug Res.* **7**, 517–523

Little, P.J., Compton, D.R., Johnson, M.R., Melvin, L.S. & Martin, B.R. (1988) Pharmacology and stereoselectivity of structurally novel cannabinoids in mice. *J. Pharmacol. Exp. Ther.* **247**, 1046–1051.

Little, P.J., Compton, D.R., Mechoulam, R. & Martin, B.R. (1989) Stereochemical effects of 11-OH-Δ^8-THC dimethylheptyl in mice and dogs. *Pharmacol. Biochem. Behav.* **32**, 661–666.

Little, P.J. & Martin, B.R. (1991) The effects of Δ^9-tetrahydrocannabinol and other cannabinoids on cAMP accumulation in synaptosomes. *Life Sci.* **48**, 1133–1141.

Mackie, K., Devane, W.A. & Hille, B. (1993) Anandamide, an endogenous cannabinoid, inhibits calcium currents as a partial agonist in N18 neuroblastoma cells. *Mol. Pharmacol.* **44**, 498–503.

Mackie, K. & Hille, B. (1992) Cannabinoids inhibit N-type calcium channels in neuroblastoma-glioma cells. *Proc. Natl Acad. Sci. USA* **89**, 3825–3829.

Martin, B.R. (1986) Cellular effects of cannabinoids. *Pharmacol. Rev.* **38**, 45–74.

Martin, B.R., Compton, D.R., Little, P.J., Martin, T.J. & Beardsley, P.M. (1987) Pharmacological evaluation of agonistic and antagonistic activity of cannabinoids. In *Structure–Activity Relationships of Cannabinoids* (eds Rapaka, R.S. & Makriyannis, A.), NIDA Res. Monogr. 79, pp. 108–122. US Government Printing Office, Washington, DC.

Martin, B.R., Compton, D.R., Thomas, B.F., Prescott, W.R., Little, P.J., Razdan, R.K., Johnson, M.R., Melvin, L.S., Mechoulam, R. & Ward, S.J. (1991) Behavioral, biochemical, and molecular modeling evaluations of cannabinoid analogs. *Pharmacol. Biochem. Behav.* **40**, 471–478.

Martin, B.R., Prescott, W.R. & Zhu, M. (1992) Quantitation of rodent catalepsy by a computer-imaging technique. *Pharmacol. Biochem. Behav.* **43**, 381–386.

Martin B.R., Compton, D.R., Semus, S.F., Lin, S.Y., Marciniak, G., Grzybowska, J., Charalambous, A. & Makriyannis, A. (1993) Pharmacological evaluation of iodo and nitro analogs of Δ^8-THC and Δ^9-THC. *Pharmacol. Biochem. Behav.* **46**, 295–301.

Matsuda, L.A., Lolait, S.J., Brownstein, M.J., Young, A.C. & Bonner, T.I. (1990) Structure of a cannabinoid receptor and functional expression of the cloned cDNA. *Nature* **346**, 561–564.

Mechoulam, R. & Edery, H. (1973) Structure-activity relationships in the cannabinoid series. In *Marijuana Chemistry, Pharmacology, Metabolism, and Clinical Effects* (ed. R. Mechoulam), pp. 101–136. Academic Press, New York.

Mechoulam, R., Feigenbaum, J.J., Lander, N., Segal, M., Järbe, T.U.C., Hiltunen, A.J. & Consroe, P. (1988) Enantiomeric cannabinoids: stereospecificity of psychotropic activity. *Experientia* **44**, 762–764.

Mechoulam, R., Lander, N., Srebnik, M., Breuer, A., Segal, M., Feigenbaum, J.J., Järbe, T.U.C. & Consroe, P. (1987) Stereochemical requirements for cannabimimetic activity. In *Structure–Activity Relationships of the Cannabinoids* (eds Rapaka, R.S. & Makriyannis, A.), NIDA Res. Monogr. 79, pp. 15–30, US Government Printing Office, Washington, DC.

Mechoulam, R., Hanuš, L. & Martin, B.R. (1994) The search for endogenous ligands of the cannabinoid receptor. *Biochem. Pharmacol.* in press.

Melvin, L.S. & Johnson, M.R. (1987) Structure-activity relationships of tricyclic and non-classical bicyclic cannabinoids. In *Structure–Activity Relationships of the Cannabinoids* (eds Rapaka, R.S. & Makriyannis, A.), NIDA Res. Monogr. 79, pp. 31–47. US Government Printing Office, Washington, D.C.

Melvin, L.S., Johnson, M.R., Harbert, C.A., Milne, G.M. & Weissman, A. (1984) A cannabinoid derived protypical analgesic. *J. Med. Chem.* **27**, 67–71.

Milne, G.M. & Johnson, M.R. (1981) Levonantradol: a role for central prostanoid mechanisms. *J. Clin. Pharmacol.* **21** (Suppl.), 367S–374S.

Munro, S., Thomas, K.L. & Abu-Shaar, M. (1993) Molecular characterization of a peripheral receptor for cannabinoids. *Nature* **365**, 61–65.

Pacheco, M., Childers, S.R., Arnold, R., Casiano, F. & Ward, S.J. (1991) Aminoalkylindoles: actions on specific G-protein-linked receptors. *J. Pharmacol. Exp. Ther.* **257**, 170–183.

Pacheco, M.A., Ward, S.J. & Childers, S.R. (1993) Identification of cannabinoid receptors in cultures of rat cerebellar granule cells. *Brain Res.* **603**, 102–110.

Paton, W.D.M., Pertwee, R.G. & Temple, D. (1972) The general pharmacology of cannabinoids. In *Cannabis and Its Derivatives* (ed. Paton, W.D.M. & Crown, J.), pp. 50–75. Oxford University Press, London.

Perez-Reyes, M., Timmons, M.C., Lipton, M.A. Davis, K.H. & Wall, M.E. (1972) Intravenous injection in man of Δ^9-tetrahydrocannabinol and 11-OH-Δ^9-tetrahydrocannabinol. *Science* **177**, 633–635.

Pertwee, R.G. & Greentree, S.G. (1988) Δ^9-Tetrahydrocannabinol-induced catalepsy in mice is enhanced by pretreatment with flurazepam or chlordiazepoxide. *Neuropharmacol.* **27**, 485–491.

Pertwee, R.G., Stevenson, L.A., Elrick, D.B., Mechoulam, R. & Corbett, A.D. (1992) Inhibitory effects of certain enantiomeric cannabinoids in the mouse vas deferens and the myenteric plexus preparation of guinea-pig small intestine. *Br. J. Pharmacol.* **105**, 980–984.

Prescott, W.R., Gold, L.H. & Martin, B.R. (1992) Evidence for separate neuronal mechanisms for the discriminative stimulus and catalepsy induced by Δ^9-THC in the rat. *Psychopharmacol.* **107**, 117–124.

Razdan, R.K. (1986) Structure–activity relationships in cannabinoids. *Pharmacol. Rev.* **38**, 75–149.

Reggio, P.H. (1987) Molecular determinants for cannabinoid activity: refinement of a molecular reactivity template. In *Structure–activity relationships of the Cannabinoids* (eds R.S. Rapaka & Makriyannis A.), NIDA Res. Monogr. 79, pp. 82–95. US Government Printing Office, Washington, D.C.

Reggio, P.H., Greer, K.V. & Cox, S.M. (1989) The importance of the orientation of the C9 substituent to cannabinoid activity. *J. Med. Chem.* **32**, 1630–1635.

Reggio, P.H., Panu, A.M. & Miles, S. (1993) Characterization of a region of steric interference at the cannabinoid receptor using the active analog approach. *J. Med. Chem.* **36**, 1761–1771.

Rinaldi-Carmona, M., Barth, F., Héaulme, M., Shire, D., Calandra, B., Congy, C., Martinez, S., Maruani, J., Néliat, G., Caput, D., Ferrera, P., Soubrié, P., Brelière, J.C. and Le Fur, G. (1994) SR141716A, a potent and selective antagonist of the brain cannabinoid receptor. FEBS Lett **350**, 240–244.

Rosell, S., Agurell, S. & Martin, B. (1976) Effects of cannabinoids on isolated smooth muscle preparations. In *Marihuana: Chemistry, Biochemistry, and Cellular Effects* (ed Nahas G.G.), pp. 397–406. Springer-Verlag, New York.

Roth, S.H. & Williams, P.J. (1979) The non-specific membrane binding properties of Δ^9-tetrahydrocannabinol and the effects of various solubilizers. *J. Pharm. Pharmacol.* **31**, 224–230.

Schatz, A.R., Koh, W.S. & Kaminski, N.E. (1993) Δ^9-Tetrahydrocannabinol selectively inhibits T-cell dependent humoral immune responses through direct inhibition of accessory T-cell function. *Immunopharmacol.* **26**, 129–137.

Semus, S.F. & Martin, B.R. (1990) A computergraphic investigation into the pharmacological role of the THC-cannabinoid phenolic moiety. *Life Sci.* **46**, 1781–1785.

Smith, P.B., Compton, D.R., Welch, S.P., Razdan, R.K., Mechoulam, R. & Martin, B.R. (1994) The pharmacological activity of anandamide, a putative endogenous cannabinoid, in mice. *J. Pharmacol. Exp. Ther.* **270**, 219–227.

Spaulding, T.C., Ford, R.D., Dewey, W.L., McMillan, D.E. & Harris, L.S. (1972) Some pharmacological effects of phenitrone and its interaction with Δ^9-THC. *Eur. J. Pharmacol.* **19**, 310–317.

Thomas, B.F., Compton, D.R., Martin, B.R. & Semus, S.F. (1991) Modeling the cannabinoid receptor: a three-dimensional quantitative structure-activity analysis. *Mol. Pharmacol.* **40**, 656–665.

Thomas, B. F., Wei, X. & Martin, B.R. (1992) Characterization and autoradiographic localization of the cannabinoid binding site in rat brain using [^{3}H]11-OH-Δ^9-THC-DMH. *J. Pharmacol. Exp. Ther.* **263**, 1383–1390.

Walton, R.P., Martin, L.F. & Keller, J.H. (1937) The relative activity of various purified products obtained from American grown hashish. *J. Pharmacol. Exp. Ther.* **62**, 239–251.

Ward, S.J., Baizman, E., Bell, M., Childers, S., D'Ambra, T., Eissenstat, M., Estep, K., Haycock, D., Howlett, A., Luttinger, D. & Miller, M. (1991) Aminoalkylindoles (AAIs): A new route to the cannabinoid receptor? In *Problems of Drug Dependence 1990: Proceedings of the 52nd Annual Scientific Meeting* (ed Harris L.S.), NIDA Res. Monogr. 105, pp. 425–426. US Government Printing Office, Washington, DC.

Weissman, A., Milne, G.M. & Melvin, L.S. (1982) Cannabimimetic activity from CP-47,497, a derivative of 3-phenylcyclohexanol. *J. Pharmacol. Exp. Ther.* **223**, 516–523.

Wilson, R.S., May, E.L., Martin, B.R. & Dewey, W.L. (1976) 9-Nor-9-hydroxyhexahydrocannabinols. Synthesis, some behavioral and analgesic properties, and comparison with tetrahydrocannabinols. *J. Med. Chem.* **19**, 1165–1167.

CHAPTER 3

THE ROLE OF CELL MEMBRANES IN CANNABINOID ACTIVITY

Alexandros Makriyannis

School of Pharmacy, Institute of Materials Science, University of Connecticut, CT 06268, USA

Table of Contents

Cannabinoid Receptors
ISBN 0-12-551460-3

3.1 Introduction

Until the early 1980s the predominant view on the molecular mechanism of cannabinoid action was that these molecules exert all of their pharmacological effects by interacting with cell membrane lipids. Such an interaction would induce a perturbation of the membrane lipid bilayer with a concomitant change in the function of several membrane associated proteins. This view was supported by several lines of evidence pointing to the ability of cannabinoids to interact with model and biological membranes (Makriyannis and Rapaka, 1990; Martin, 1986). Thus, it had been pointed out that the natural biologically active tetrahydrocannabinols, which are tricyclic terpenes, are highly lipophilic molecules that preferentially partition in the cell membrane when allowed to interact with biological preparations. This all-important property was found to be shared by all of those synthetic analogues possessing cannabimimetic activity, including a group of analogues developed at Pfizer as potential analgesic agents (Johnson *et al.*, 1981; Melvin *et al.*, 1984). These molecules lacked the tetrahydropyran structure and were referred to as non-classical cannabinoids (NCCs) to distinguish them from the classical cannabinoid tricyclic analogues (CCs) in which the benzopyran ring system is intact (see Chapters 1 and 2). The highly lipophilic properties of cannabinoids are also shared by a recently discovered group of cannabimimetic molecules developed at Sterling Winthrop (Bell *et al.*, 1991; D'Ambra *et al.*, 1992), which are generally designated aminoalkylindoles (AAIs) and by arachidonylethanolamide, commonly known as anandamide, the very recently discovered native cannabimimetic ligand (Devane *et al.*, 1992; see also Chapters 1 and 2).

The physicochemical property shared by all analogues possessing cannabimimetic activity is most appropriately described as amphipathic to account for the presence of one or more polar groups attached to the highly hydrophobic molecular framework of these molecules. For the CCs and NCCs the polar groups include a required phenolic hydroxyl and possibly one to three additional alcohol hydroxyl groups. Conversely, the AAIs have polar groups which are not hydroxyls, while the anandamide structure includes only one aliphatic hydroxyl. As we shall discuss later, it is this amphipathic property of cannabimimetic molecules which is primarily responsible for their ability to affect the thermotropic properties and the fluidity of the lipids in natural and model membranes.

Although a plethora of experimental data has convincingly demonstrated that cannabinoids and related analogues can effectively perturb model and biological membranes, the evidence that such perturbation is responsible for the biological properties of these molecules is less convincing. The most compelling arguments favouring the lipid over the receptor hypothesis for cannabinoid activity have been the lack of any evidence (a) for a receptor or other functional protein with which cannabinoids interact or (b) for a native ligand resembling the cannabinoids. However, today both of the above uncertainties relative to the molecular mechanism of cannabinoid activity have been answered. The first direct evidence for a cannabinoid receptor came from experiments in which [^{3}H]CP 55,940, a tritiated NCC, was shown to bind stereospecifically to a brain membrane preparation (Devane *et al.*, 1988). Also, the relative abilities of a representative group of cannabimimetic analogues to inhibit [^{3}H]CP 55,940 binding were found to parallel their potencies for producing behavioural effects in animals (Little *et al.*, 1988). Finally, conclusive evidence was obtained when the clone for a cannabinoid receptor was recognized and expressed in several cell systems (Matsuda *et al.*, 1990; Gérard *et al.*, 1991). These transfected cells demonstrated all the key features of a cannabinoid-responsive G-protein coupled receptor (Matsuda *et al.*, 1990). Over the past 5 years this cannabinoid receptor was identified in a diverse number of cell systems, including cells from the central nervous system (Herkenham *et al.*, 1990), the lung (Burstein *et al.*, 1982, 1983) and the immune system (Bouaboula *et al.*, 1993), and very recently the first cannabinoid receptor subtype was discovered (Munro *et al.*, 1993). Also very recently, a putative native cannabinoid ligand, anandamide, was isolated and shown to possess all of the biochemical and pharmacological properties of cannabinoids (Devane *et al.*, 1992).

3.2 Membrane-based cannabinoid activity

The discovery of cannabinoid receptors in different organs and cells has helped to explain many of the pharmacological effects of cannabinoids. It is, therefore, tempting to attribute all of the pharmacological properties of cannabinoids to their interactions with one or more cannabinoid receptors and, thus, settle the long-standing argument of lipid- versus receptor-mediated cannabinoid activity in favour of the receptor hypothesis. However, such a conclusion is somewhat premature as the molecular basis for much of the pharmacology of the cannabinoids remains unexplained. Indeed, the wide variety of pharmacological effects induced by cannabinoids and the great diversity of membrane-associated proteins whose functions are affected by cannabimimetic agents argues for a more complex mechanism of action. There is also evidence that cannabinoids exert many of their biochemical effects at the membrane level. Which cannabinoid effects are thus receptor-mediated and which of their pharmacological properties can be attributed to membrane interactions? What role do cannabinoid–membrane interactions play in determining the pharmacological properties of this structurally diverse group of drug molecules? The evidence is still incomplete and sometimes contradictory. In this chapter we shall attempt to examine the available experimental evidence and place it into perspective.

Let us pose the questions that can be directly related to the interactions of cannabinoids with membranes as will be discussed in this chapter. First, how do cannabinoids diffuse across biological membranes; how do they cross the blood–brain barrier and what molecular features facilitate this passive transport? Second, do cannabinoids have the ability to perturb membranes, thus affecting the function of membrane-associated proteins? What are the molecular features of this perturbation; and how can membrane perturbation be related to their pharmacological effects? Third, what are the molecular features of the cannabinoid–membrane interactions and in what manner do these interactions determine the drug's ability to interact productively with the cannabinoid receptor? Before discussing these questions let us first start with a brief introduction outlining the role of the cell membrane in drug action.

3.2.1 The role of membrane lipids in drug action

Although it is generally recognized that drug lipophilicity plays an important role in determining pharmacological activity, the molecular features of the drug–membrane interaction have received less attention. In which instances do drug–lipid interactions affect drug activity? First, the membrane bilayer is a selective barrier that many drug molecules need to penetrate in order to reach their sites of action. For example, drugs acting on the central nervous system (CNS) must be able to cross the blood–brain barrier, while orally administered drugs need to cross the gastric mucosa or the intestinal walls to enter the blood circulation. To penetrate these barriers, the drug molecules must have certain structural features which may vary depending on the type of membrane the drug needs to cross. This type of drug selectivity can serve as an appropriate basis for drug design or drug targeting. Second, the membrane lipid bilayer serves as a matrix into which numerous functional proteins are embedded and a lipophilic drug may produce some of its effects by perturbing the lipid bilayer. This event, which is manifested as a change in membrane fluidity, may modulate the function of one or more membrane-associated proteins and the extent of this modulation appears to be dependent on the composition of the lipids surrounding the protein (Jain and White, 1977; Sandermann, 1978; Loh and Law, 1980).

A third instance of membrane involvement in drug activity is the critical role of the lipid bilayer in facilitating the interaction of the drug molecule with its site of action on the target protein. Thus, it is expected that a lipophilic drug will partition preferentially in the lipid component of the membrane where it maintains a favoured conformation, possibly different from the one in solution (Behling and Jelinski, 1990). In the membrane, the drug also assumes a preferred orientation and location (depth) and undergoes fast lateral diffusion within the bilayer leaflet. It has been argued that, through this two-dimensional diffusion within the lipid phase, the drug may gain access to an intrabilayer receptor site (Rhodes *et al.*, 1985; Schwyzer, 1986). According to this view such an approach of the drug via the membrane to its site of action on the receptor will be highly favoured over one in which the drug reaches the receptor by diffusion through the bulk extracellular fluid which would act as a solvent. Thus, this two-step pathway for a successful drug–receptor interaction involves initial partitioning of the drug from the extracellular fluid to an energetically favourable location, orientation and conformation within the bilayer, followed by a rapid lateral diffusion to the receptor site. Correspondingly, the model imposes two sets of structural criteria which the drug must fulfil in order to undergo an optimal interaction with its site of action. First, the drug must possess suitable structural features which would allow it to locate in the appropriate depth and in the proper orientation within the membrane, thus enabling it to interact productively with the receptor site. Second, the drug must possess the stereoelectronic properties required for optimal binding with the active site.

3.2.2 Diffusion of cannabimimetic drugs across the blood–brain barrier

Biologically active cannabinoids are known to be able to enter from the peripheral circulation into the brain where they produce their CNS effects. Early experiments, in which radiolabelled (–)-Δ^9-tetrahydrocannabinol (Δ^9-THC) was used, established that after administration the drug could be detected in the brains of animals as the unmodified cannabinoid or its metabolites (Roth and Williams, 1979).

3.2.2.1 Positron emission tomography (PET)

A more recent approach for studying the distribution of a cannabinoid in live animals involved the use of PET (Charalambous *et al.*, 1991). This method requires the drug molecule to be labelled with a short-lived positron-emitting isotope, which in this instance was an ^{18}F label, introduced at the end side-chain carbon of the parent (–)-Δ^8-tetrahydrocannabinol (Δ^8-THC) to give (–)-5′-[^{18}F]Δ^8-THC (Marciniak *et al.*, 1991). The radiolabelled cannabinoid was administered to live baboons, and the presence and distribution of the drug in the brain was followed using computerized tomography. The experiment showed that the drug reaches its maximum concentration during the first 5 min and persists up to 2 h after injection (Figure 1), with high drug concentrations being found in the striatum, thalamus and cerebellum. The same radio-analogue was also studied in mice where it was shown that approximately 1% of the drug was distributed in the brain (Table 1).

3.2.2.2 Blood–brain barrier model

A very recent experiment aimed at uncovering the structural features that determine the ability of cannabimimetic drugs to diffuse into the brain, used an *in vitro* blood–brain barrier model system. This preparation, which involves a monolayer of bovine brain endothelial cells, was found to be a good model for the blood–brain barrier. When used to test a group of analogues and rank them according to their abilities to transport across the monolayer, this *in vitro* test provides results that

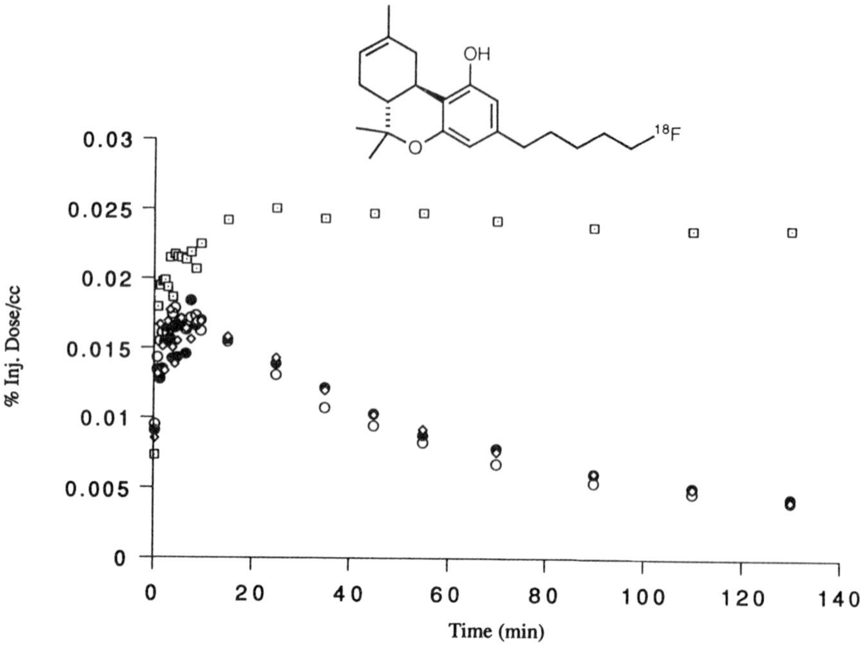

Figure 1 Time courses for (−)-5′-^{18}F-Δ^8-THC distribution in baboon brain: ○, cerebellum; ●, left striatum; ⊡, sella turcica; ◇ thalamus (Charalambous *et al.*, 1991).

generally mirror those obtained from *in vivo* transport experiments. Ayotte *et al.* (1993) tested a group of cannabimimetic analogues in such an endothelial cell monolayer system which was appropriately adapted for highly lipophilic molecules. The experiment (Figure 2) provided some interesting results. First, it demonstrated that both the biologically active (−)-Δ^8-THC and its inactive (+)-enantiomer transported equally across the model blood–brain barrier with no measurable difference in their rates of transport. This observation suggested a passive diffusion mechanism for the transport of cannabinoids into the brain – a process which, unlike protein-mediated transport, is not expected to exhibit significant enantioselectivity. Second, two biologically inactive (−)-Δ^8-THC analogues, in which the phenolic group is methylated, exhibited much slower rates of transport. Of the two analogues, the one with a bromo-substitution in the 5′-position of the side-chain was slower to diffuse, perhaps reflecting the hindrance imposed by the bulky bromine atom. Finally, the same study confirmed recent data that the putative native ligand anandamide can diffuse into the brain. Indeed, *R*-methanandamide (AM-356), a recently developed, metabolically more stable anandamide analogue (Abadji *et al.*, 1994), was shown to diffuse across the monolayer at a rate only somewhat slower than that of Δ^8-THC (Ayotte *et al.*, 1993).

Table 1 Tissue distribution of (–)-5′-^{18}F-Δ^8-THC in mice

	Mean±SEM (*n*=4) Time After Injection					
	5 min		30 min		60 min	
Organ	%/Organ	%/Gram	%/Organ	%/Gram	%/Organ	%/Gram
Brain	1.07±0.14	2.23±0.34	0.52± 0.05	1.07±0.09	0.24± 0.01	0.52± 0.02
Blood	—	2.65±0.21	—	2.27±0.69	—	1.38± 0.12
Heart	0.88±0.17	6.45±1.10	0.19± 0.03	1.53±0.15	0.10± 0.01	0.88± 0.06
Lungs	0.81±0.12	4.82±0.74	0.39± 0.17	1.72±0.23	0.19± 0.05	1.15± 0.15
Liver	25.31±2.73	20.07±2.06	13.49± 7.90	9.08±5.20	9.72± 0.60	7.44± 0.33
Spleen	0.41±0.07	2.85±0.43	0.16± 0.04	1.07±0.09	0.10± 0.01	0.67± 0.09
Kidneys	3.17±0.46	7.33±0.85	0.75± 0.16	1.86±0.26	0.41± 0.02	1.04± 0.08
Small Intestine	6.84±2.74	3.42±1.04	30.89±13.6	16.08±5.02	42.66±20.2	22.92±10.1
Femur	—	1.38±0.31	—	1.50±0.19	—	1.65± 0.26
Muscle	—	2.57±0.30	—	1.06±0.23	—	0.68± 0.19

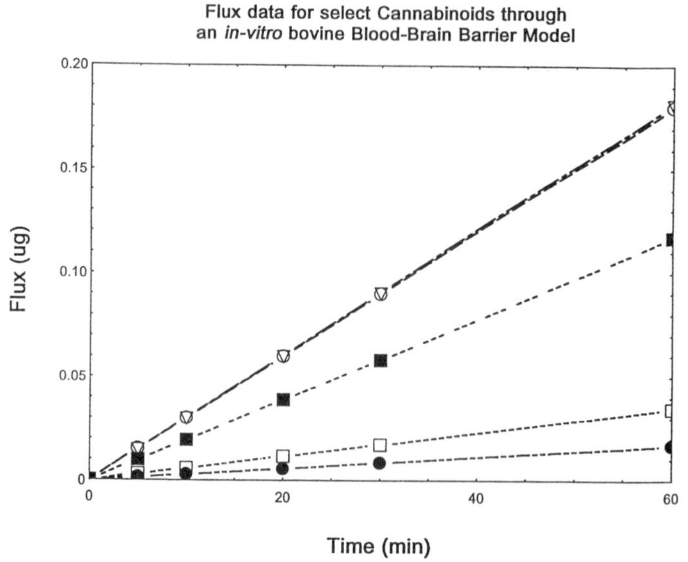

Figure 2 Flux measurements across the bovine brain microvessel endothelial cell monolayer from the apical to the basolateral side of (−) Δ^8-THC(∇), (+)Δ^8-THC(○), (-)1-O-methyl-5′Br-Δ^8-THC(●), (-)1-O-methyl-Δ^8-THC(□) and AM-356 (■) (Abadji *et al.*, 1994) which is a chiral metabolically stable anandamide analogue (*R*-methanandamide) whose potency is four- to 10-fold that of anandamide. Data shown are the mean± SD of three separate experiments (Ayotte *et al.*, 1993).

3.2.2.3 Molecular features of cannabinoid diffusion across cell membranes

The above results provide good evidence that cannabinoids and anandamides reach their sites of action in the brain through a passive diffusion process which does not appear to be governed by requirements for enantiomeric stereoselectivity. Consequently, the observed large differences in potencies between cannabinoid enantiomers when measured from *in vivo* experiments cannot be accounted for by preferential transport into the brain but must be explained on the basis of stereo-selectivity requirements at the active site.

Although there is considerable information on the kinetics of passive transport of drug molecules into the brain, the molecular mechanism of this diffusion process is not yet fully understood. To obtain some understanding of how cannabinoids are able to traverse the amphipathic membrane bilayer, a number of biophysical studies aimed at exploring the interactions of these molecules with model and biological membranes were undertaken. As a result of such work, it has become possible to characterize the conformation and location of a number of cannabinoids in membranes and to obtain detailed information on the manner with which these molecules orient within the membrane bilayer. This latter aspect of cannabinoid–

membrane interactions appears to play an important role in determining cannabimimetic activity. Based on the above studies, a mechanism was proposed for the transbilayer diffusion of highly lipophilic cannabinoids with no hydrogen donor polar groups, as for example, the *O*-methyl-tetrahydrocannabinols (Makriyannis and Yang, 1989). These molecules appear to enter the outer leaflet of the membrane bilayer and localize at the level of the inner segment of the lipid alkyl chains from where they can diffuse into the cell. Such a molecular model for passive diffusion across the membrane seems to be valid for all lipophilic, non-amphipathic small or middle-sized organic molecules such as hormones, neuro-transmitters, nutrients and the biologically less active or inactive cannabinoids, in which the phenolic hydroxyl hydrogen is substituted. Conversely, cannabinoids with free hydroxyl groups enter the membrane and, like other amphipaths, preferentially reside at the interface between the hydrophilic head groups and hydrophobic tails of the membrane phospholipids (i.e. at the level of the glycerol moieties of the phospholipid molecules). Subsequently, these cannabinoid molecules diffuse across the membrane by a different mechanism, the molecular features of which have not yet been fully elucidated.

3.2.3 Cannabimimetic effects through membrane perturbation

Although the existencc of cannabinoid receptors can account for some of the known physiological effects of cannabimimetic drugs, many other effects cannot yet be linked to these receptors. Cannabinoids are thus known to produce biochemical changes in a variety of membrane preparations at concentrations which are arguably 'physiological' (Leuschner *et al.*, 1984; Hillard *et al.*, 1986, 1990) and to affect numerous membrane-related functions. Their effects on membrane-bound enzymes include Na^{+}-K^{+}-, Mg^{2+}- and Ca^{2+}-ATPases (Gilbert *et al.*, 1977; Hershkowitz *et al.*, 1977; Collins and Haavik, 1979), NADH dehydrogenase (Bartova and Birmingham, 1976), lysophosphatidylcholine acyl transferase (Greenberg *et al.*, 1977), phospholipase A_2 (Burstein and Hunter, 1981), monoamine oxidase (Schurr and Livne, 1976) as well as those effects on adenylate cyclase which are non-receptor mediated (Howlett *et al.*, 1989). Cannabinoids also affect the function of neurotransmitter uptake systems (Hershkowitz *et al.*, 1977), blood platelets (Nathan *et al.*, 1982), human lung fibroblasts (Burstein *et al.*, 1983), neutrophils (Naccache *et al.*, 1982) and certain non-cannabinoid receptors (Hillard and Bloom, 1982; Bloom, 1984; Bloom and Hillard 1985). A common feature with all of these effects is that the drug molecules do not seem to interact directly with the sites of action of specific proteins and do not compete with any ligands known to interact with these sites.

3.2.3.1 Cannabinoid effects on membrane lipids

It is generally accepted that the lipid micro-environment surrounding the membrane-associated enzymes (Sandermann, 1978), receptors (Jain and White, 1977) or other functional proteins, also referred to as boundary lipids or lipid annulus, can influence protein function. The question of whether certain groups of drug molecules can induce some or all of their effects by perturbing the lipid component of the membrane has received a great deal of attention over a period of at least three decades. In its most current form this hypothesis postulates that lipophilic drugs can affect the conformational and dynamic properties of boundary lipids surrounding membrane proteins and modulate the function of one or more of these proteins. As boundary lipids vary from protein to protein, their composition may be the source of any observed selectivity in the pharmacological properties of these different groups of lipophilic drugs.

3.2.3.2 Experimental uncertainties

Although considerable effort has been directed at designing appropriate experiments, there is still no method that can unequivocally link the ability of a drug molecule to perturb membrane lipids with its pharmacological activity. Because of this, the extent to which membrane perturbation plays a role in determining cannabimimetic activity is still unclear. The reasons for the encountered difficulties are twofold. The first of these is in relating effective drug concentrations at the sites of action to the concentrations required to induce measurable conformational and dynamic changes in natural or model membranes. Such direct comparisons are subject to the experimental limitations of the biophysical experiments used to measure membrane perturbations that require relatively high drug concentrations. The uncertainty in correlating *in vitro* drug concentrations with *in vivo* pharmacological levels is further compounded by difficulties in estimating drug concentrations at the site of action. For example, in most biochemical and biophysical experiments in which membrane preparations are used, drug concentrations are reported relative to total sample volume, although in many instances virtually all of the drug is expected to reside in the membrane where it is present in considerably higher concentrations. The second uncertainty which must be considered when the membrane-perturbing properties of a drug molecule are linked to its ability to induce a pharmacological response is a lack of detailed understanding on what constitutes 'significant perturbation'. Although it is generally accepted that membrane 'fluidity' is a major parameter governing membrane protein activity, the direct correlation between membrane 'fluidity' and function is not well understood. A major reason for this is that the concept of 'fluidity' presupposes changes in bilayer dynamics but does little to reveal their nature.

3.2.3.3 Experimental evidence

A number of investigators have advanced the view that cannabinoids may exert some of their effects through the membrane lipid component. Support for a 'membrane perturbation' mechanism to explain some aspects of cannabinoid activity has come from experiments aimed at correlating induced changes in membrane fluidity or bilayer chain dynamics with pharmacological activity. However, because of the above-mentioned uncertainties in interpreting the experimental data, some of the results have been inconclusive and occasionally contradictory.

There is considerable experimental evidence showing that cannabinoids affect the thermotropic (Bach *et al.*, 1976; Bruggeman and Melchior, 1983) and dynamic properties of model (Lawrence and Gill, 1975; Makriyannis *et al.*, 1990b) and biological membranes (Hillard *et al.*, 1990) at concentrations claimed to be 'physiological' (Hillard *et al.*, 1985, 1990; see also Chapter 6). Thus, these drug molecules have been shown to decrease both the phase transition temperature and the enthalpy of pure phospholipids and calf brain lipids (Bruggeman and Melchior, 1983; Makriyannis *et al.*, 1990b; Paton, 1975). They have also been shown to affect the 'microviscosity' of phospholipids in experiments with phospholipid unilamellar vesicles using electron spin resonance (Lawrence and Gill, 1975) and with multilamellar liposomes using ^2H solid-state nuclear magnetic resonance (NMR) (Makriyannis *et al.*, 1990b). In such experiments, (–)-Δ^9-THC and other biologically active cannabinoids were shown to perturb membrane bilayers at the level of the fatty acid chains and at the membrane interface, while the biologically inactive (–)-*O*-methyl-Δ^8-THC produced only marginal effects (Makriyannis *et al.*, 1988; Yang *et al.*, 1992).

Previously, Lawrence and Gill (1975), using electron spin resonance phospholipid probes, reported that the biologically active (–)-Δ^9-THC and (–)-11-hydroxy-Δ^9-THC both increased phospholipid chain mobility in egg lecithin/cholesterol model membranes, whereas the biologically less active cannabinol and the inactive cannabidiol decreased chain mobility. Later, in a similar study using fluorescence measurements, Hillard and coworkers (1986) demonstrated that the hydrophobic core of brain synaptic membranes is sensitive to the disordering effect of (–)-Δ^9-THC and its 11-hydroxy analogue at low micromolar concentrations, while the same concentrations of cannabidiol and cannabinol produced no demonstrable change in the mobility of the membrane hydrophobic chains. However, these differences could not be duplicated in model lipid bilayers made from a variety of different phospholipids.

While using a different approach to explore the role of membrane perturbation in cannabinoid activity, Howlett *et al.* (1989) studied the effects of (–)-Δ^9-THC and cannabidiol on the Arrhenius plots of hormone-stimulated adenylate cyclase activity in neuroblastoma cell membranes and compared them to those of benzyl alcohol, a known membrane-perturbing agent (see also Chapter 6). The studies led to the conclusion that, while the effects of cannabidiol on this enzyme could be explained by the observed increased lipid fluidity in the membrane system, the effects of (–)-Δ^9-THC were more congruent with a cannabinoid receptor-based mechanism. However, a similar experiment by Hillard *et al.* (1990), directed at investigating the effect of (–)-Δ^9-THC on adenylate cyclase stimulated by either isoprenaline or glucagon in cardiac ventricular membranes, led to a different conclusion. In contrast to the previous study, Hillard *et al.* (1990) found their results to be consistent with the hypothesis that the effects of this biologically active cannabinoid are mediated by a change in membrane lipid order in the region of the receptor and/or the G_s protein. In an analogous study, cannabinoids were shown to eliminate the discontinuity in the Arrhenius plot for microsomal *O*-demethylase activity (Bach *et al.*, 1977), an effect attributed to a change in the phase-transition properties of the lipids associated with this enzyme (Collins and Haavik, 1979).

3.2.3.4 The molecular features of membrane perturbation

Although there is still considerable uncertainty regarding the extent to which perturbation of cellular membranes by cannabinoids plays a role in determining their pharmacological properties, more information is available regarding the molecular features of membrane perturbation as well as the structural features required in a drug molecule for it to be a membrane-perturbing agent. Much of the information related to the above two questions has been gathered from biophysical experiments designed to study the conformational and dynamic properties of membranes as well as the interactions of drug molecules or native ligands with membranes.

3.2.3.4.1 Methods for studying drug–membrane interactions: Because of their complexity and inherent instability, membranes do not lend themselves to a detailed analysis of their structure and dynamics by a single biophysical method. It is, therefore, preferable to use jointly several experimental techniques for such studies. Makriyannis and coworkers (1991b) have developed an approach which combines the use of differential scanning calorimetry (DSC), small-angle X-ray or neutron diffraction and solid-state NMR. Each method has distinct advantages and limitations and, when used independently, can give only partial information. However, when combined, the methods are complementary and lead to a more detailed molecular representation of drug–membrane interactions. DSC is a relatively inexpensive technique which allows us to study quantitatively the thermotropic properties of membranes and the effects of drugs on these properties. Conversely, the diffraction methods give information on the thickness and the packing of the membrane bilayer. Solid-state NMR (Figure 3) gives the most detailed molecular information on membrane conformation and dynamics and on the effects of drugs on these properties. Furthermore, diffraction and NMR data when used jointly, can identify the location, orientation and conformation of the drug molecule in the membrane.

3.2.3.4.2 Role of drug structure in membrane perturbation: Traditionally, the potency of drugs that are believed to produce their effects by interacting with the lipid component of the membrane has been correlated with their lipid solubility, as reflected by their measured or calculated partition coefficients. In such correlations, no consideration is given to variations in their conformational properties or to their relative and absolute configurations. However, there is good evidence showing that in many groups of drugs, small variations in structure with no significant effect on their partitioning properties produce major changes in pharmacological potency. This has been well demonstrated for anaesthetic steroids, there being evidence for a correlation between the anaesthetic potency of such drugs and their ability to

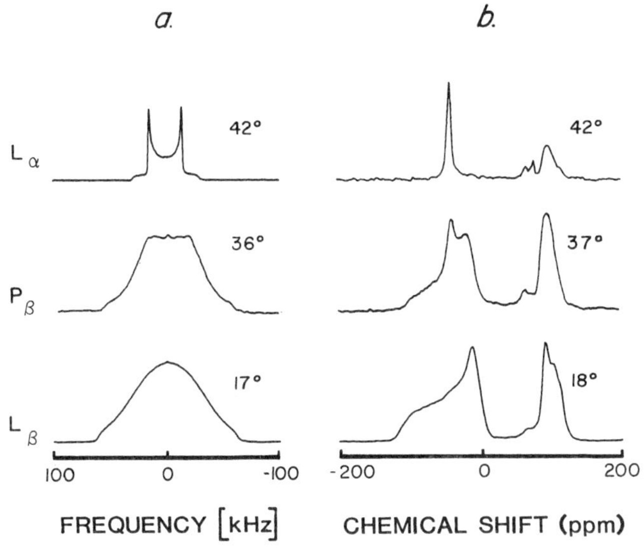

Figure 3 ^{2}H- and ^{13}C solid-state NMR spectra representing the three phases for dipalmitoylphosphatidyl choline (DPPC) aqueous dispersions within the temperature range at which the studies were conducted. (a) Solid-state ^{2}H NMR spectra from the 2[7′, 7′-C^{2}H$_2$] segment of DPPC obtained using a $90°_x$-τ-$90°_y$ quadrupolar echo sequence at 45.3 MHz. (b) Solid-state ^{13}C spectra from the 2[1′-^{13}C] carbonyl group of DPPC obtained using a cross-polarization experiment with a 180° refocusing pulse (Makriyannis *et al.*, 1991b).

perturb model membranes (Lawrence and Gill, 1975; Makriyannis and Fesik, 1980; 1983; Fesik and Makriyannis, 1985; Makriyannis *et al.*, 1986; 1991a). Analogous studies have also been reported showing some correlation between the psychotropic potencies of cannabinoids and their effects on model membrane preparations as measured by ESR (Lawrence and Gill, 1975), fluorescence polarization (Niemi, 1979; Kriwacki, 1986) and DSC (Burstein *et al.*, 1986). In an attempt to correlate the stereoelectronic properties of anaesthetic steroids and cannabinoids with their different abilities to perturb model and biological membranes, molecular explanations have been provided showing how subtle changes in the geometry of anaesthetic steroids (Fesik and Makriyannis, 1985; Makriyannis *et al.*, 1991a) (Figure 4) and cannabinoids (Makriyannis *et al.*, 1987; Van der Schyf *et al.*, 1988) (Figure 5) can significantly affect their interactions with membrane bilayers.

3.2.3.4.3 Role of lipid structure and composition: The observed influence that the stereochemistry of a drug can have on its ability to perturb membrane lipids can be understood if the defined geometric parameters of the phospholipid bilayer are considered. These include the natural asymmetry of the phospholipid and cholesterol molecules as well as the conformational preferences and dynamic properties of the lipid hydrocarbon chains and/or polar head groups. These requirements

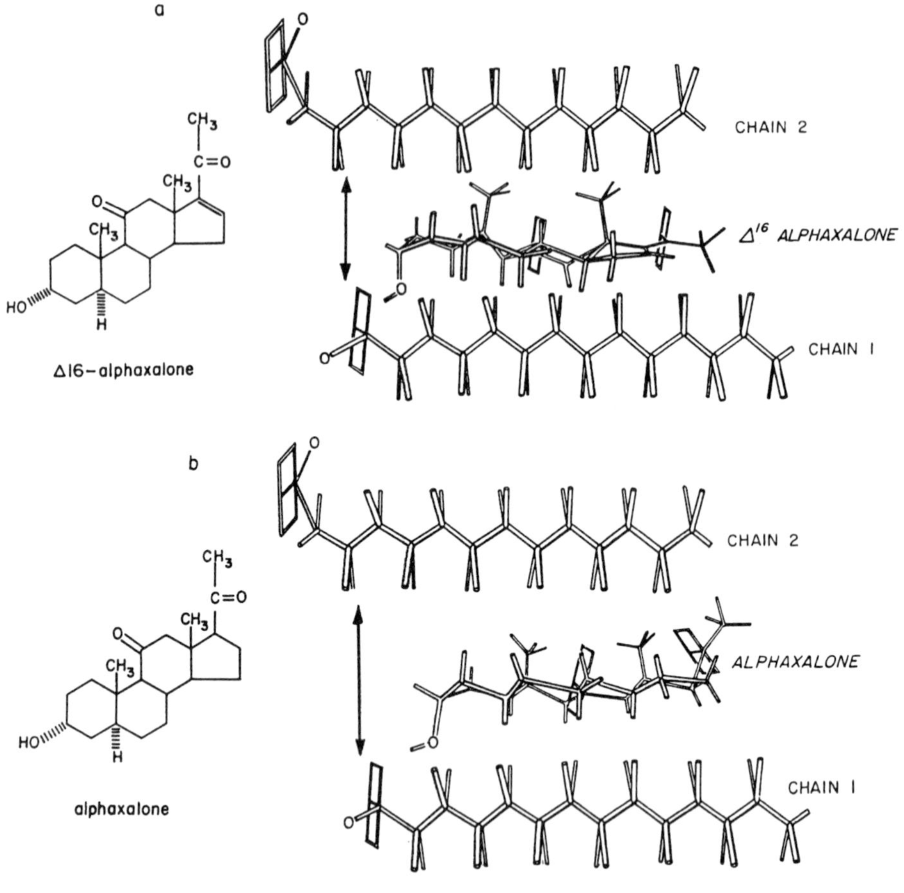

Figure 4 Molecular models for steroid:phospholipid chain interactions. Each of the two steroids is placed between chain 2 (top) and chain 1 (bottom) of two adjacent dimyristoylphosphatidyl choline molecules. The steroids are represented as being all *trans*. The model shows that Δ^{16}-alphaxalone (a) accommodates easily between the two chains while alphaxalone (b) requires more space (Fesik and Makriyannis, 1985).

will determine whether the drug can insert into, and create changes in the conformational and dynamic properties of the bilayer. The effects of a cannabinoid on a membrane will then, presumably, vary according to the nature of the lipid(s) composing the membrane. When viewed from such a perspective, DSC and solid-state NMR results showing that the two optical isomers of Δ^9-THC affect differently the thermotropic and dynamic properties of model membrane preparations, can be understood (Makriyannis *et al.*, 1987 and Figure 6). The varied response to the cannabinoids of different membrane systems is congruent with the observations that different lipids modulate functions of different proteins and that the phase

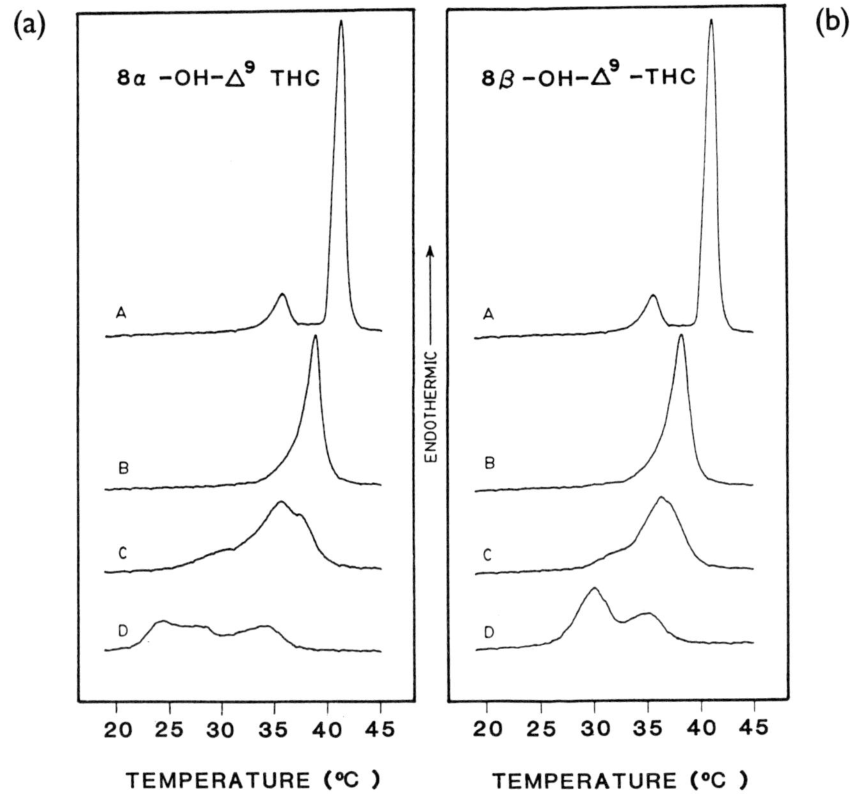

Figure 5 Thermograms of dipalmitoylphosphatidyl choline (DPPC) with 8α-OH-Δ^9-THC (a): (A) DPPC alone, (B) X_{THC2}=0.05, (C) X_{THC2}=0.1, and (D) X_{THC2}=0.2; and 8β-OH-Δ^9-THC (b): (A) DPPC alone, (B) X_{THC1}=0.05, (C) X_{THC1}=0.1, and (D) X_{THC1}=0.2 (Burstein *et al.*, 1986). X_{THC1} is the molar ratio of 8β-OH-Δ^9-THC to DPPC and X_{THC2}, the molar ratio of 8α-OH-Δ^9-THC to DPPC.

characteristics and dynamic properties of a biomembrane arise from the coexistence of several phases (Loh and Law, 1980).

3.2.3.4.4 The molecular mechanism of cannabinoid-induced membrane perturbation: Through a series of detailed studies combining ^{2}H and ^{13}C solid-state NMR with DSC experiments, Makriyannis and coworkers were able to elucidate the molecular details involved in the perturbation of a membrane bilayer by (−)-Δ^9-THC and other cannabinoids. The experiments, which involved model and biological membranes, clearly showed that the ability of these drug molecules to perturb membranes can be linked to their amphipathic properties. In CCs and NCCs

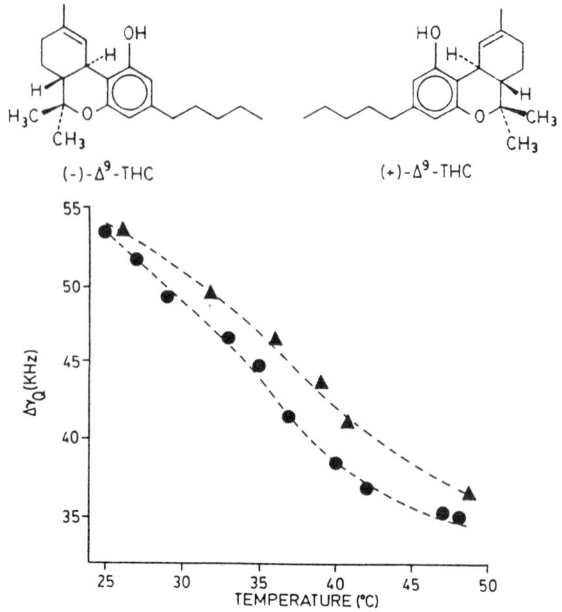

Figure 6 Temperature dependence of the ^{2}H quadrupolar splittings in the ^{2}H-NMR spectra of the 2(7′,7′-C^{2}H$_2$) segment of dipalmitoylphosphatidyl choline (DPPC). (●) DPPC + 0.15 M cholesterol + 0.15 M (−)-Δ^9-THC; (▲) DPPC + 0.15 M cholesterol + 0.15 M (+)-Δ^9-THC (Makriyannis *et al.*, 1987).

(Section 3.1), the phenolic hydroxyl is the minimum requirement for a polar component which provides the necessary polar/hydrophobic molecular mix characteristic of amphipathic molecules. Cannabinoids in which the phenolic OH hydrogen is substituted with an alkyl group lose their amphipathic properties and perturb the bilayer only marginally (Makriyannis *et al.*, 1988; Yang *et al.*, 1992) (Figure 7). The emerging picture from these studies with cannabinoids (Makriyannis and Yang, 1989; Makriyannis *et al.*, 1990b) is that membrane perturbation involves an increase in the ratio of *gauche:trans* conformers in the phospholipid bilayer chains and a decreased cooperativity between the chains. The critical interaction occurs at the membrane interface and is associated with a conformational change in the glycerol backbone of the membrane phospholipids. In all likelihood, perturbation of the lipid chains results from an imperfect packing of the drug in the bilayer (Makriyannis *et al.*, 1990a). Parallel experiments have also shown that similar molecular features are involved in the perturbation of membranes by anaesthetic steroids (Makriyannis *et al.*, 1990a, 1991a). It is likely that the principal features of the above-described model for membrane perturbation have general applicability and may describe the conformational and dynamic changes induced on the membrane by all amphipathic molecules. In a separate study, the same investigators have shown that the geometry of ring C (see Figure 1 in Chapter 1) is an important factor in determining

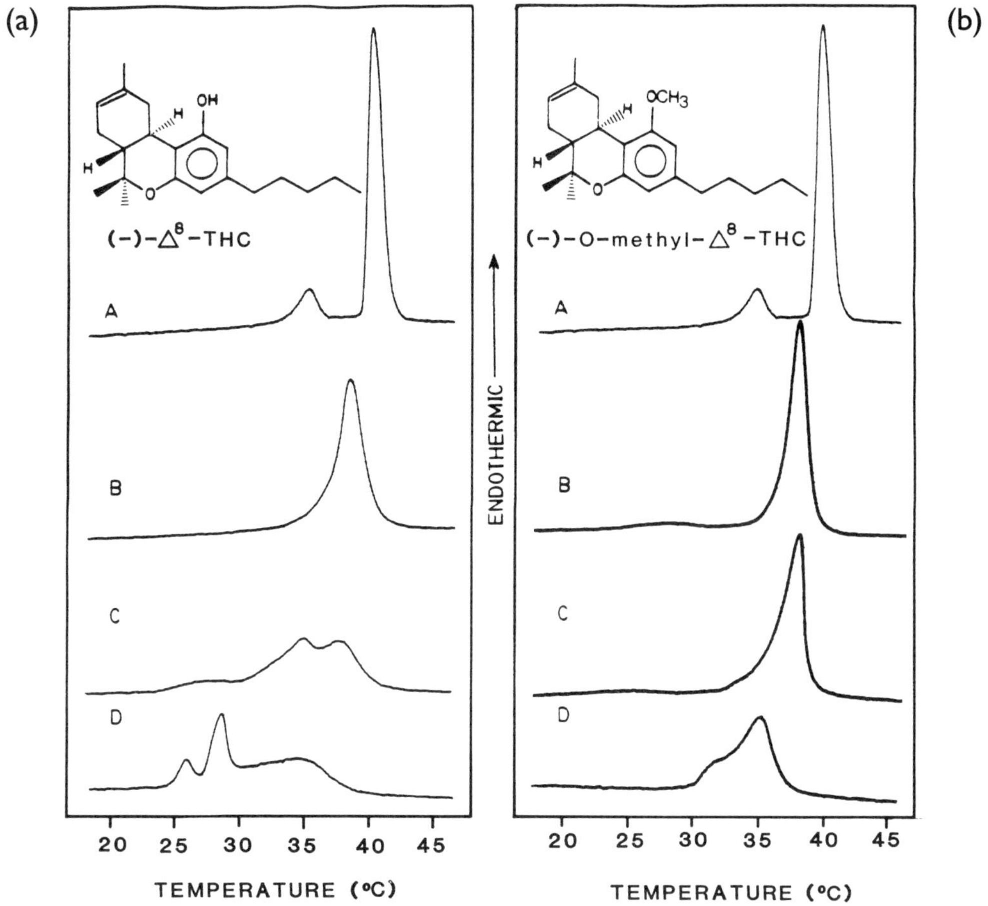

Figure 7 Thermograms of dipalmitoylphosphatidyl choline (DPPC) with (-)-Δ^8 THC (a) and (-)-O-methyl-Δ^8-THC (b). (A) DPPC alone; (B) x=0.05; (C) x=0.10; (D) x=0.20; where x is the molar fraction of cannabinoid in the preparation, e.g., THC:DPPC=0.1:0.9 M for x=0.1 (Makriyannis *et al.*, 1991b).

the ability of cannabinoids to perturb membranes (Makriyannis *et al.*, 1987). Cannabinoids in which the C ring is coplanar with the other two (e.g. $\Delta^{9,11}$-THC) perturb membranes less effectively than those in which the C ring deviates from planarity (e.g. Δ^9-THC).

3.2.3.5 Specific amphipathic cannabinoid interactions with non-catalytic sites

Interactions of lipophilic drug molecules, such as cannabinoids, with membrane lipids are generally discussed in terms of changes in the bulk properties of the lipid component and described as changes in 'fluidity'. However, there is increasing evidence from a variety of sources for specific drug–membrane interactions. For example, evidence has been presented for a stoichiometric Δ^9-THC–lecithin complex (Bruggeman and Melchior, 1983). The same authors have also shown that in a mixed phospholipid membrane preparation, Δ^9-THC exhibits a preference for lower melting point lipid.

The possibility exists that similar amphipathic interactions of cannabinoids may also occur with non-catalytic sites in certain membrane-associated proteins, leading to changes in their function(s).

3.2.4 Membrane-dependent interactions of cannabinoids with receptor sites

Several researchers have argued that drug conformation, orientation and location in the membrane play an important role in determining the drug's ability to interact productively with its site of action on the receptor. According to this argument, lateral diffusion of a drug molecule in the membrane is a very fast process and is kinetically the most effective route to the drug's target protein, which is also undergoing a fast lateral movement within the bilayer. In such a process, the probability for a productive drug–receptor interaction via the membrane pathway will be enhanced if the preferred location for both the drug molecule and the receptor active site are at the same depth in the membrane, with their motions being confined to the same two-dimensional space within the bilayer. The probability for such an interaction is also enhanced if the drug molecule assumes within the membrane an orientation and conformation which allow its key pharmacophoric groups to be appropriately aligned with regard to the respective binding components of the active site. Conversely, if during its lateral motion, the drug molecule assumes a preferred two-dimensional path which is at a different depth in the membrane bilayer than that of the active site, the probability for a productive drug–receptor collision is reduced (Figure 8). Similarly, the likelihood for a productive interaction is decreased if the drug exists in an unfavourable orientation and/or conformation in the membrane.

Biophysical studies by Makriyannis and coworkers have provided detailed information on the topography, the stereochemistry and the dynamic features of the cannabinoid–membrane interactions. The emerging picture is that the biologically active natural cannabinoid, (–)-Δ^9-THC, uses its phenolic hydroxyl group to anchor itself at the interface of the amphipathic bilayer through hydrogen bonding (Makriyannis *et al.*, 1989; Mavromoustakos *et al.*, 1990; Martel *et al.*, 1993). To achieve this effect, THC assumes an 'awkward' orientation in the bilayer with the

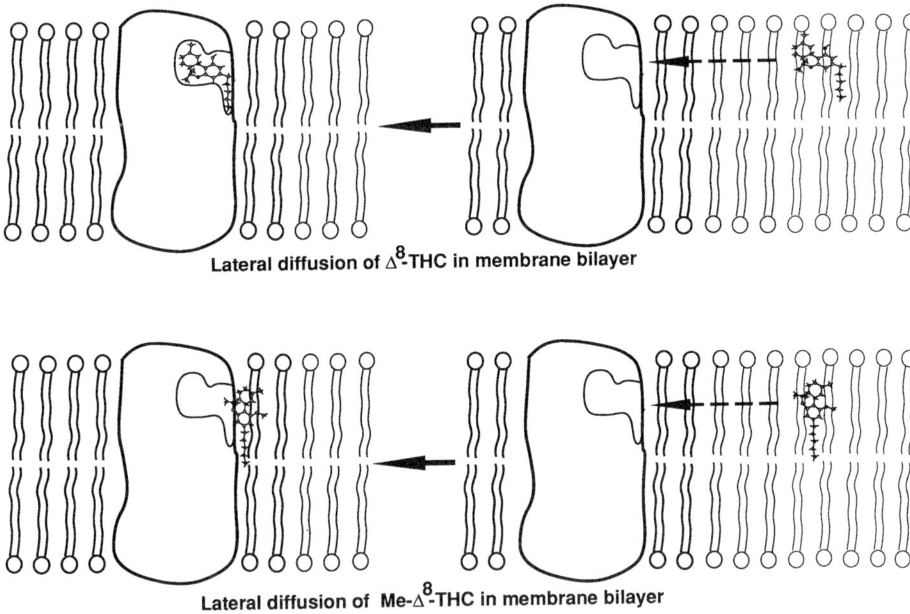

Figure 8 A model for cannabinoids reaching their active site by laterally diffusing in the membrane bilayer. Δ^8-THC and Me-Δ^8-THC assume different orientations and locations within the lipid bilayer. Δ^8-THC assumes an orientation in which the long axis of its tricyclic ring system is perpendicular to the bilayer chains and the phenolic hydroxyl is near the membrane interface. Conversely, Me-Δ^8-THC aligns with its long axis parallel to the chains and its average location is deeper and closer to the centre of the bilayer. According to this model, the cannabinoid active site on the membrane-associated receptor protein has a location and orientation that allows the biologically active Δ^8-THC to undergo a productive collision with it after diffusing laterally in the lipid bilayer. The inactive Me-Δ^8-THC orients and positions itself in an unfavourable orientation with regard to the site of action, thus decreasing the likelihood for a favourable interaction.

long axis of its tricyclic system perpendicular to the bilayer chains (Makriyannis *et al.*, 1991b) (Figures 9 and 10), while its aliphatic five-carbon side-chain orients parallel with the chains of the membrane phospholipids (Mavromoustakos *et al.*, 1990) (Figure 10). Cannabinoids with two hydroxyl groups, for example the two (−)-11-hydroxyhexahydrocannabinol diastereoisomers, orient in a manner that allows both hydroxyls to be near the interface facing the polar surface of the membrane (Makriyannis and Yang, 1989) (Figure 11). However, if the phenolic hydrogen of THC is replaced by a methyl group, the cannabinoid having been deprived of its anchoring hydroxyl group, now orients with its natural long axis parallel to the chains and deep towards the centre of the bilayer (Yang *et al.*, 1991) (Figure 9).

(a)
(b)
(c)

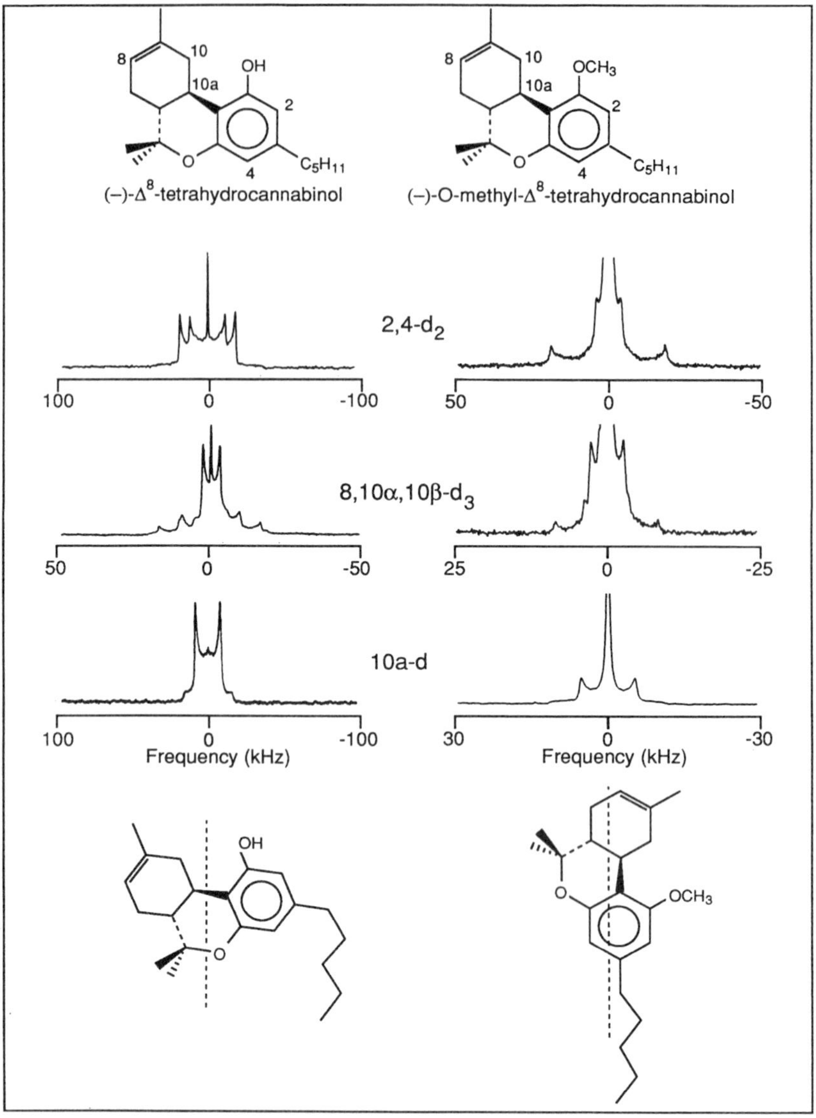

Figure 9 (a) Structures of Δ^8-THC and Me-Δ^8-THC with the deuterated positions numbered. (b) Solid state ^{2}H-NMR spectra from deuterated cannabinoids in fully hydrated dipalmitoylphosphatidyl choline (DPPC) bilayers at 42°C. Δ^8-THC and Me-Δ^8-THC were specifically deuterated in 2, 4, 10α, 10β and 10a positions. (c) The orientations of the cannabinoid analogues in DPPC bilayers, where the lipid chain direction is represented by the dashed lines.

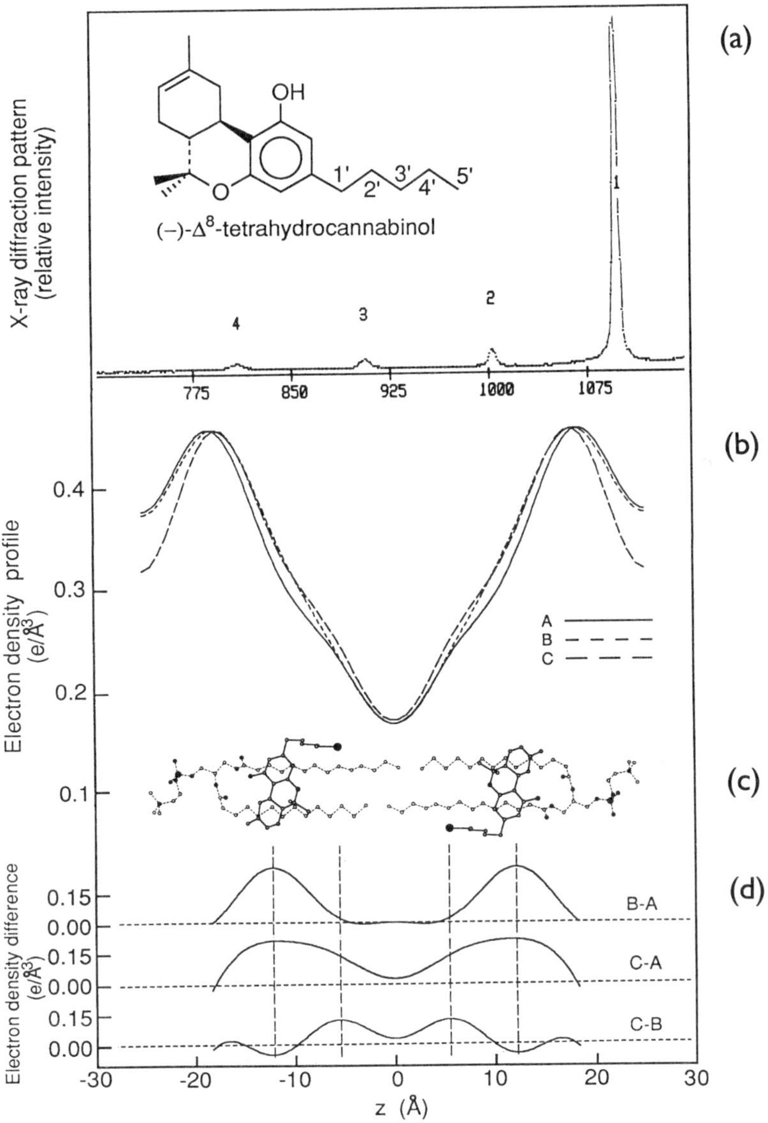

Figure 10 (a) Typical small angle X-ray diffraction pattern of dimyristoylphosphatidyl choline (DMPC) at 35°C equilibrated at a relative humidity of 98%, with relative diffraction pattern intensity versus detector channel number. (b) Superimposed electron density profiles of (A) DMPC alone, (B) DMPC + Δ^8-THC (x=0.15) and (C) DMPC + 5′-I-Δ^8-THC (x=0.15). (c) Molecular graphic representation of a DMPC bilayer with 5′-I-Δ^8-THC molecules in their proper orientation and location. (d) Electron density profile differences showing the location of the THC tricyclic ring system (outer pair of vertical dashed lines) and that of the iodine atom (inner pair of vertical dashed lines) (Makriyannis *et al.*, 1991b).

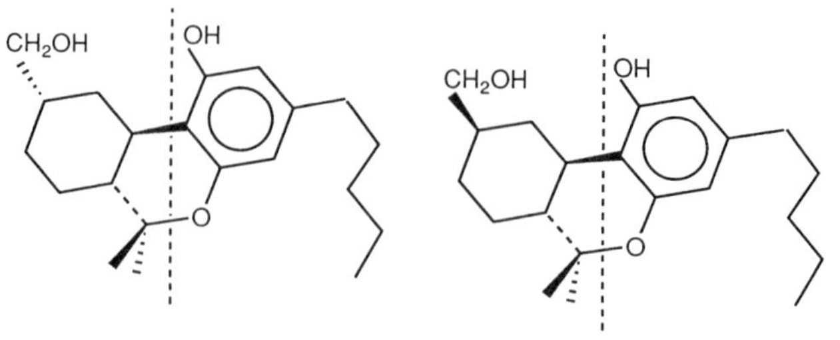

Figure 11 Orientations of 11-OH-9α- (on the left) and 11-OH-9β-hexahydrocannabinols in hydrated dipalmitoylphosphatidyl choline bilayers. The dashed lines represent the direction of the lipid acyl chains in the bilayer.

The evidence obtained to date supports the hypothesis that the ability of cannabinoids to orient with all their hydroxyl groups facing the polar surface in the amphipathic membrane system, seriously affects their pharmacological properties. It appears that the relative stereochemistry of the polar hydroxyl groups in the cannabinoid determines its orientation and location in the bilayer. If we assume that the cannabinoid diffuses across the membrane bilayer to reach its receptor site, then its orientation and location in the bilayer will, in turn, determine the relative alignment of the cannabinoid molecular fragments with regard to the receptor site and, consequently, the drug's ability to interact productively with the receptor. This would mean that there are two sequential criteria for cannabinoid activity: (a) 'proper' orientation of the drug in the membrane bilayer and (b) 'appropriate' fit between the drug and the receptor.

To evaluate how cannabinoid–membrane interactions vary depending on the type of membrane used, studies were carried out with a variety of membrane preparations of progressively increasing complexity. A comparison of the data from the different membrane preparations, showed no significant change in the manner in which the cannabinoid analogues tested oriented in the membrane (Mavromoustakos *et al.*, 1990; Yang *et al.*, 1991). However, the X-ray data revealed that there is a 2–3Å variation in the location of the drug in the bilayer. Indeed, we found that Δ^8-THC and its halogenated analogue preferentially reside 2–3Å deeper in the membrane preparations obtained from brain synaptosomal plasma membranes (Mavromoustakos *et al.*, 1991) compared with model membrane preparations (Mavromoustakos *et al.*, 1990). Molecular modelling used to compare the data from model and biological membranes, revealed that, in the synaptosomal preparation, the cannabinoid phenolic OH is still capable of engaging in a hydrogen-bonding interaction either with one of the membrane phospholipid carbonyls or with a water molecule located slightly below the interface between the hydrophilic headgroups and hydrophobic tails of the membrane phospholipids (Mavromoustakos *et al.*, 1991) (Figure 12). The reasons for the seemingly small, but significant differences in cannabinoid topography in the different membrane preparations are not known. It is also unclear what implications, if any, these intriguing observations may have on the molecular mechanism(s) of cannabinoid activity. It is tempting to speculate that the observed variation in drug topography, depending on the membrane sample used, may help in explaining observed differences in drug specificity with different membrane preparations (Martin, 1986). A similar hypothesis for another group of lipophilic drug molecules has been suggested (Mason *et al.*, 1994).

3.3 Conclusion

The discovery of a cannabinoid receptor in mammalian brains and subsequently in other organs has provided a sound basis for explaining many of the pharmacological properties of cannabinoids. Also, the horizons for a receptor-based hypothesis for cannabimimetic activity were recently broadened, first with the isolation of the putative cannabinoid ligand, anandamide, and soon after with the identification of the first cannabinoid receptor subtype. How do these new developments then affect the earlier ideas of an exclusively membrane-based mechanism? Clearly, the old ideas need to be revised. Do cellular membranes play an important role in cannabinoid activity? Clearly yes, although the extent to which cannabinoid–membrane interactions determine the very diverse pharmacological properties of these drug molecules has not been determined.

By virtue of their amphipathic/lipophilic properties, cannabinoids preferentially partition in the cellular membrane. As a consequence, we expect them to affect the properties of several membrane-associated proteins. To assess the role of these membrane effects in cannabinoid activity they need to be dissected from the receptor-based effects. According to the most current thinking, cannabimimetic agents reach their receptor sites preferentially by laterally diffusing within the amphipathic membrane bilayer rather than through the aqueous environment of the extracellular space. While in the membrane, the drug molecule will assume a preferred orientation and conformation and will engage in fast motions within a thermodynamically favourable two-dimensional space at a certain depth in the bilayer. All of these aspects of the drug–membrane interaction will be determined by the drug's stereo-electronic features and will affect its pharmacological properties. Furthermore, differences in membrane composition may affect cannabinoid-membrane interactions and could be the basis for some observed tissue or cell-type selectivity in cannabimimetic activity. Finally, most evidence seems to support a passive diffusion mechanism for cannabinoid transport across the membranes constituting the blood–brain barrier. Consequently, the ability of the cannabimimetic agents either to enter the CNS or to remain in the periphery is also subject to structural requirements which could serve as a basis for the design of useful centrally or peripherally active therapeutic agents.

It thus appears that although the cellular membrane may not be the principal target for cannabinoid activity, it nevertheless plays a key role in the mechanism

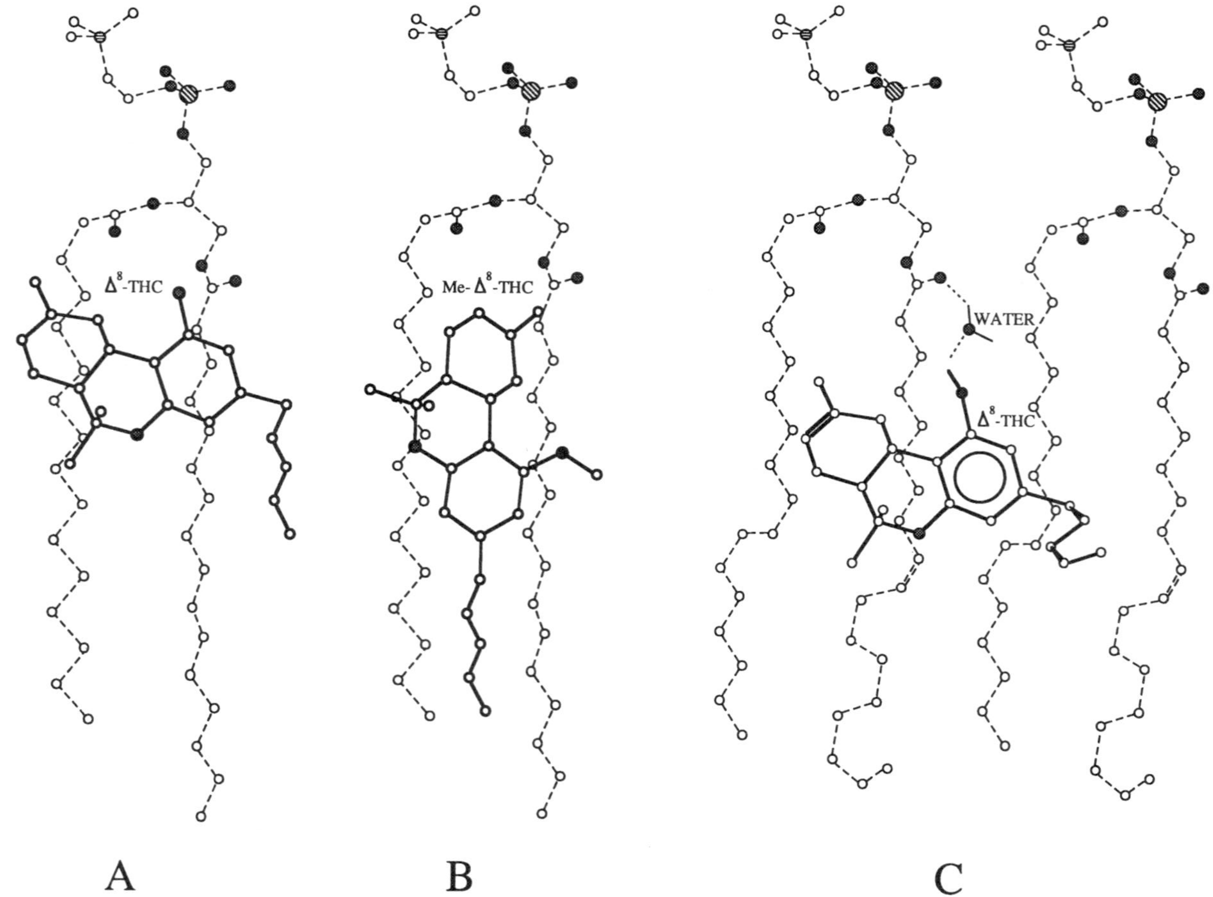

Figure 12 Graphical representation of the topography of (A) Δ^8-THC in a dipalmitoylphosphatidyl choline (DPPC) bilayer, (B) Me-Δ^8-THC in a DPPC bilayer and (C) Δ^8-THC in a biological membrane (Yang *et al.*, 1991).

of action of cannabimimetic agents. For this reason, detailed knowledge of cannabinoid–membrane interactions may contribute significantly in the design and development of novel therapeutic agents.

References

Abadji, V., Lin, S., Taha, G., Griffin, G., Stevenson, L.A., Pertwee, R.G. & Makriyannis, A. (1994) *R*-methanandamide: a chiral novel anandamide possessing higher potency and metabolic stability. *J. Med. Chem.* **37**, 1889–1893.

Ayotte, L.A., Makriyannis, A. & Fournier, D.J. (1993) Blood-brain barrier uptake and transport of Δ^8-tetrahydrocannabinol through a primary bovine brain microvessel endothelial cell culture monolayer. Am. Assoc. Pharm. Sci. 8th Annual Meeting. *Pharm. Res.* **10**, S182.

Bach, D., Raz, A. & Goldman, R. (1976) The effect of hashish compounds on phospholipid phase transition. *Biochim. Biophys. Acta* **436**, 889–894.

Bach, D., Bursuker, I. & Goldman, R. (1977) Differential scanning calorimetry and enzymic activity of rat liver microsomes in the presence and absence of Δ^1-tetrahydrocannabinol. *Biochim. Biophys. Acta* **469**, 171–179.

Bartova, A. & Birmingham, M.K. (1976) Effect of Δ^9-tetrahydrocannabinol on mitochondrial NADH-oxidase activity. *J. Biol. Chem.* **251**, 5002–5006.

Behling, R.W. & Jelinski, L.W. (1990) Importance of the membrane in ligand–receptor interactions. *Biochem. Pharmacol.* **40**, 49–54.

Bell, M.R., D'Ambra, T.E., Kumar, V., Eissenstat, M.A., Herrmann, J.L., Wetzel, J.R., Rosi, D., Philion, R.E., Daum, S.J., Hlasta, D.J., Kullnig, R.K., Ackerman, J.H., Haubrich, D.R., Luttinger, D.A., Baizman, E.R., Miller, M.S. & Ward, S.J. (1991) Antinociceptive (aminoalkyl)indoles. *J. Med. Chem.* **34**, 1099–1110.

Bloom, A.S. (1984) Effects of cannabinoids on neurotransmitter receptors in the brain. In *The Cannabinoids: Chemical, Pharmacologic and Therapeutic Aspects* (eds. Agurell, S., Dewey, W.L. & Willette, R.E.), pp. 575–589. Academic Press, Orlando.

Bloom, A.S. & Hillard, C.J. (1985) Cannabinoids, neurotransmitter receptors and brain membranes. In *Marihuana '84* (ed. Harvey, D.J.), pp. 217–231. IRL, Oxford Press.

Bouaboula, M., Rinaldi, M., Carayon, P., Carillon, C., Delpech, B., Shire, D., Le Fur, G. & Casellas, P. (1993) Cannabinoid-receptor expression in human leukocytes. *Eur. J. Biochem.* **214**, 173–180.

Bruggeman, E.P. & Melchior, D.L. (1983) Alterations in the organization of phosphatidylcholine/cholesterol bilayers by tetrahydrocannabinol. *J. Biol. Chem.* **258**, 8298–8303.

Burstein, S. & Hunter, S.A. (1981) Prostaglandins and Cannabis – VIII. Elevation of phospholipase A_2 activity by cannabinoids in whole cells and subcellular preparations. *J. Clin. Pharmacol.* **21** (Suppl.), 240S–248S.

Burstein, S., Hunter, S.A., Sedor, C. & Shulman, S. (1982) Prostaglandins and Cannabis – IX. Stimulation of prostaglandin E_2 synthesis in human lung fibroblasts by Δ^1-tetrahydrocannabinol. *Biochem. Pharmacol.* **31**, 2361–2365.

Burstein, S., Hunter, S.A. & Ozman, K. (1983) Prostaglandins and Cannabis XII. The effect of cannabinoid structure on the synthesis of prostaglandins by human lung fibroblasts. *Mol. Pharmacol.* **23**, 121–126.

Burstein, S., Hunter, S.A., Latham, V., Mechoulam, R., Melchior, D.L., Renzulli, L. & Tefft, R.E. (1986) Prostaglandins and Cannabis XV. Comparison of enantiomeric cannabinoids in stimulating prostaglandin synthesis in fibroblasts. *Life Sci.* **39**, 1813–1823.

Charalambous, A., Marciniak, G., Shiue, C.-Y., Dewey, S.L., Schlyer, D.J., Wolf, A.P. & Makriyannis, A. (1991) PET studies in the primate brain and biodistribution in mice using (–)-5′-^{18}F-Δ^8-THC. *Pharmacol. Biochem. Behav.* **40**, 503–507.

Collins, F.G. & Haavik, C.O. (1979) Effects of cannabinoids on cardiac microsomal CaATPase activity and calcium uptake. *Biochem. Pharmacol.* **28**, 2303–2306.

D'Ambra, T.E., Estep, K.G., Bell, M.R., Eissenstat, M.A., Josef, K.A., Ward, S.J., Haycock, D.A., Baizman, E.R., Casiano, F.M., Beglin, N.C., Chippari, S.M., Grego, J.D., Kullnig, R.K. & Daley, G.T. (1992) Conformationally restrained analogues of pravadoline: nanomolar potent, enantioselective, (aminoalkyl)indole agonists of the cannabinoid receptor. *J. Med. Chem.* **35**, 124–135.

Devane, W.A., Dysarz, F.A., Johnson, M.R., Melvin, L.S. & Howlett, A.C. (1988) Determination and characterization of a cannabinoid receptor in rat brain. *Mol. Pharmacol.* **34**, 605–613.

Devane, W.A., Hanuš, L., Breuer, A., Pertwee, R.G., Stevenson, L.A., Griffin, G., Gibson, D., Mandelbaum, A., Etinger, A. & Mechoulam, R. (1992) Isolation and structure of a brain constituent that binds to the cannabinoid receptor. *Science* **258**, 1946–1949.

Fesik, S.W. & Makriyannis, A. (1985) Geometric requirements for membrane perturbation and anesthetic activity. Conformational analysis of alphaxalone and Δ^{16}-alphaxalone and ^{2}H-NMR studies on their interactions with model membranes. *Mol. Pharmacol.* **27**, 624–629.

Gérard, C.M., Mollereau, C., Vassart, G. & Parmentier, M. (1991) Molecular cloning of a human cannabinoid receptor which is also expressed in testis. Biochem. J. **279**, 129–134.

Gilbert, J.C., Pertwee, R.G. & Wyllie, M.G. (1977) Effects of Δ^9-tetrahydrocannabinol and cannabidiol on a Mg^{2+}-ATPase of synaptic vesicles prepared from rat cerebral cortex. *Br. J. Pharmacol.* **59**, 599–601.

Greenberg, J.H., Saunders, M.E. & Mellors, A. (1977) Inhibition of a lymphocyte membrane enzyme by Δ^9-tetrahydrocannabinol *in vitro*. *Science* **197**, 475–477.

Herkenham, M., Lynn, A.B., Little, M.D., Johnson, M.R., Melvin, L.S., de Costa, B.R. & Rice, K.C. (1990) Cannabinoid receptor localization in brain. *Proc. Natl Acad. Sci. USA* **87**, 1932–1936.

Hershkowitz, M., Goldman, R. & Raz, A. (1977) Effects of cannabinoids on neurotransmitter uptake, ATPase activity and morphology of mouse brain synaptosomes. *Biochem. Pharmacol.* **26**, 1327–1331.

Hillard, C.J. & Bloom, A.S. (1982) Δ^9-tetrahydrocannabinol-induced changes in β-adrenergic receptor binding in mouse cerebral cortex. *Brain Res.* **235**, 370–377.

Hillard, C.J., Harris, R.A. & Bloom, A.S. (1985) Effects of the cannabinoids on physical properties of brain membranes and phospholipid vesicles – fluorescence studies. *J. Pharmacol. Exp. Ther.* **232**, 579–588.

Hillard, C.J., Bloom, A.S. & Houslay, M.D. (1986) Effects of Δ^9-tetrahydrocannabinol on glucagon receptor coupling to adenylate cyclase in rat liver plasma membranes. *Biochem. Pharmacol.* **35**, 2797–2803.

Hillard, C.J., Pounds, J.J., Boyer, D.R. & Bloom, A.S. (1990) Studies of the role of membrane lipid order in the effects of Δ^9-tetrahydrocannabinol on adenylate cyclase activation in heart. *J. Pharmacol. Exp. Ther.* **252**, 1075–1082.

Howlett, A.C., Scott, D.K. & Wilken, G.H. (1989) Regulation of adenylate cyclase by cannabinoid drugs. Insights based on thermodynamic studies. *Biochem. Pharmacol.* **38**, 3297–3304.

Jain, M.K. & White, H.B. (1977) Long-range order in biomembranes. *Adv. Lipid Res.* **15**, 1–60.

Johnson, M.R., Melvin, L.S., Althuis, T.H., Bindra, J.S., Harbert, C.A., Milne, G.M. & Weissman, A. (1981) Selective and potent analgetics derived from cannabinoids. *J. Clin. Pharmacol.* **21**, 271S–282S.

Kriwacki, R. (1986) *Studies on the Conformation of two Cannabinoids and their Interactions with Model Membranes Using High-resolution NMR Techniques.* M.S. Thesis, University of Connecticut, CT.

Lawrence, D.K. & Gill, E.W. (1975) The effects of Δ^1-tetrahydrocannabinol and other

cannabinoids on spin-labeled liposomes and their relationship to mechanisms of general anesthesia. *Mol. Pharmacol.* **11**, 595–602.
Leuschner, J.T.A., Wing, D.R., Harvey, D.J., Brent, G.A., Dempsey, C.E., Watts, A. & Paton, W.D.M. (1984) The partitioning of Δ^1-tetrahydrocannabinol into erythrocyte membranes *in vivo* and its effect on membrane fluidity. *Experientia* **40**, 866–868.
Little, P.J., Compton, D.R., Johnson, M.R., Melvin, L.S. & Martin, B.R. (1988) Pharmacology and stereoselectivity of structurally novel cannabinoids in mice. *J. Pharmacol. Exp. Ther.* **247**, 1046–1051.
Loh, H.H. & Law, P.Y. (1980) The role of membrane lipids in receptor mechanisms. *Annu. Rev. Pharmacol. Toxicol.* **20**, 201–234.
Makriyannis, A. & Fesik, S.W. (1980) Effects of anesthetics on sulfate transport in the red cell. *J. Neurosci. Res.* **5**, 25–33.
Makriyannis, A. & Fesik, S. (1983) Mechanism of steroid anesthetic action: interactions of alphaxalone and delta 16-alphaxalone with bilayer vesicles. *J. Med. Chem.* **26**, 463–465.
Makriyannis, A. & Rapaka, R.S. (1990) The molecular basis of cannabinoid activity. *Life Sci.* **47**, 2173–2184.
Makriyannis, A. & Yang, D.P. (1989) Solid-state NMR spectroscopy in the study of drug: membrane interactions: potential applications with antiarrhythmic agents. In *Molecular and Cellular Mechanisms of Antiarrhythmic Agents* (ed. Hondeghem, L.), pp. 287–299. Futura Publishing Co, Mount Kisco.
Makriyannis, A., Siminovitch, D.J., Das Gupta, S.K. & Griffin, R.G. (1986) Studies on the interaction of anesthetic steroids with phosphatidylcholine using ^{2}H and ^{13}C solid state NMR. *Biochim. Biophys. Acta* **859**, 49–55.
Makriyannis, A., Banijamali, A., Van der Schyf, C. & Jarrell, H. (1987) Interactions of cannabinoids with membranes. The role of cannabinoid stereochemistry and absolute configuration and the orientation of Δ^9-THC in the membrane bilayer. In *Structure–Activity Relationships of Cannabinoids, National Institute on Drug Abuse Research Monograph No. 79* (eds. Rapaka, R.S. & Makriyannis, A.), pp. 123–133. US Department of Health and Human Services, Rockville.
Makriyannis, A., Mavromoustakos, T., Jarrel, H. & Banijamali, A. (1988) Role of the phenolic hydroxyl group in psychoactive cannabinoids. Studies on the interaction of (−)-Δ^8-tetrahydrocannabinol and its O-methyl ether with model membranes. In *Marijuana: An International Research Report*, Monograph Series No. 7 (eds Chesher, G., Consroe, P. & Musty, R.), pp. 321–326. Australian Government Publishing Service, Canberra.
Makriyannis, A., Banijamali, A., Jarrell, H.C. & Yang, D.-P. (1989) The orientation of (−)-Δ^9-tetrahydrocannabinol in DPPC bilayers as determined from solid-state ^{2}H-NMR. *Biochim. Biophys. Acta* **986**, 141–145.
Makriyannis, A., Yang, D.-P. & Mavromoustakos, T. (1990a) The molecular features of membrane perturbation by anaesthetic steroids: a study using differential scanning calorimetry, small angle X-ray diffraction and solid state ^{2}H NMR. In *Steroids and Neuronal Activity, Ciba Foundation Symposium No. 153* (ed. Simmonds, M.), pp. 172–189. Wiley, Chichester.
Makriyannis, A., Yang, D.-P., Griffin, R.G. & Das Gupta, S.K. (1990b) The perturbation of model membranes by (−)-Δ^9-tetrahydrocannabinol. Studies using solid-state ^{2}H-NMR and ^{13}C-NMR. *Biochim. Biophys. Acta* **1028**, 31–42.
Makriyannis, A., DiMeglio, C.M. & Fesik, S.W. (1991a) Anesthetic steroid mobility in model membrane preparations as examined by high-resolution ^{1}H and ^{2}H NMR spectroscopy. *J. Med. Chem.* **34**, 1700–1703.
Makriyannis, A., Yang, D.-P. & Mavromoustakos, T. (1991b) Combined use of solid-state nuclear magnetic resonance spectroscopy, small-angle X-ray diffraction and differential scanning calorimetry in studies of cannabinoid:membrane interactions. In *Emerging Technologies in Drug Abuse Research, National Institute on Drug Abuse Research Monograph No. 112*

(eds Rapaka, R.S., Makriyannis, A. & Kuhar, M.), pp. 106–128. US Department of Health and Human Services, Rockville.

Marciniak, G., Charalambous, A., Shiue, C.-Y., Dewey, S.L., Schyler, D.J., Makriyannis, A. & Wolf, A.P. (1991) ^{18}F-Labeled tetrahydrocannabinol: synthesis and PET studies in a baboon. *J. Lab. Comp. Radiophys.* **30**, 413–415.

Martel, P., Makriyannis, A., Mavromoustakos, T., Kelly, K. & Jeffrey, K.R. (1993) Topography of tetrahydrocannabinol in model membranes using neutron diffraction. *Biochim. Biophys. Acta* **1151**, 51–58.

Martin, B.R. (1986) Cellular effects of cannabinoids. *Pharmacol. Revs.* **38**, 45–74.

Mason, R.P., Moring, J. & Herbette, L.G. (1994) An examination of the chemical and physical interactions of lipid soluble drugs with membranes in the heart and brain. In *New Methods in Drug Research*, Vol. 4 (ed. Makriyannis, A.), in press. J.R. Prous Scientific Publishing, Barcelona.

Matsuda, L.A., Lolait, S.J., Brownstein, M.J., Young, A.C. & Bonner, T.I. (1990) Structure of a cannabinoid receptor and functional expression of the cloned cDNA. *Nature* **346**, 561–564.

Mavromoustakos, T., Yang, D.-P., Charalambous, A., Herbette, L.G. & Makriyannis, A. (1990) Study of the topography of cannabinoids in model membranes using X-ray diffraction. *Biochim. Biophys. Acta* **1024**, 336–344.

Mavromoustakos, T., Yang, D.-P., Broderick, W., Fournier, D. & Makriyannis, A. (1991) Small angle X-ray diffraction studies on the topography of cannabinoids in synaptic plasma membranes. *Pharmacol. Biochem. Behav.* **40**. 547–552.

Melvin, L.S., Johnson, M.R., Harbert, C.A., Milne, G.M. & Weissman, A. (1984) A cannabinoid derived prototypical analgesic. *J. Med. Chem.* **27**, 67–71.

Munro, S., Thomas, K.L. & Abu-Shaar, M. (1993) Molecular characterization of a peripheral receptor for cannabinoids. *Nature* **365**, 61–65.

Naccache, P.H., Volpi, M., Becker, E.L., Makriyannis, A. & Sha'afi, R.I. (1982) Cannabinoid induced degranulation of rabbit neutrophils. *Biochem. Biophys. Res. Communs.* **106**, 1286–1290.

Nathan, I., Yarom, A., Dvilansky, A., Lander, N. & Livne, A. (1982) Cannabinoids and serotonin uptake by blood platelets: evidence for multiple sites of action. *Biochem. Pharmacol.* **31**, 439–441.

Niemi, W.D. (1979) Effect of Δ^9-tetrahydrocannabinol on synaptic transmission in the electric eel electroplaque. *Res. Communs. Chem. Pathol. Pharmacol.* **25**, 537–546.

Paton, W.D.M. (1975) Pharmacology of marijuana. *Annu. Rev. Pharmacol.* **15**, 191–220.

Rhodes, D.G., Sarmiento, J.G. & Herbette, L.G. (1985) Kinetics of binding of membrane-active drugs to receptor sites. Diffusion-limited rates for a membrane bilayer approach of 1,4- dihydropyridine calcium channel antagonists to their active site. *Mol. Pharmacol.* **27**, 612–623.

Roth, S.H. & Williams, P.J. (1979) The non-specific membrane binding properties of Δ^9-tetrahydrocannabinol and the effects of various solubilizers. *J. Pharm. Pharmacol.* **31**, 224–230.

Sandermann, H. (1978) Regulation of membrane enzymes by lipids. *Biochim. Biophys. Acta* **515**, 209–237.

Schurr, A. & Livne, A. (1976) Differential inhibition of mitochondrial monoamine oxidase from brain by hashish components. *Biochem. Pharmacol.* **25**, 1201–1203.

Schwyzer, R. (1986) Molecular mechanism of opioid receptor selection. *Biochemistry* **25**, 6335–6342.

Van der Schyf, C.J., Mavromoustakos, T. & Makriyannis, A. (1988) The conformation of (–) 8 a and (–) 8 b-hydroxy-Δ^9-tetrahydrocannabinols and their interactions with model membranes. *Life Sci.* **42**, 2231–2239.

Yang, D.-P., Banijamali, A., Charalambous, A., Marciniak, G. & Makriyannis, A. (1991) Solid state ^{2}H-NMR as a method for determining the orientation of cannabinoid analogs in membranes. *Pharmacol. Biochem. Behav.* **40**, 553–557.

Yang, D.-P., Mavromoustakos, T., Beshah, K. & Makriyannis, A. (1992) Amphipathic interactions of cannabinoids with membranes. A comparison between Δ^8THC and its *O*-methyl analog using differential scanning calorimetry, X-ray diffraction and solid state ^{2}H-NMR. *Biochim. Biophys. Acta* **1103**, 25–36.

CHAPTER 4

MOLECULAR BIOLOGY OF THE CANNABINOID RECEPTOR

Lisa A. Matsuda[1] and Tom I. Bonner[2]

[1]*Department of Psychiatry and Behavioral Sciences, Medical University of South Carolina, Charleston, SC 29425-0742, USA*
[2]*Laboratory of Cell Biology, NIMH, Bethesda, MD 20892-4090, USA*

Table of Contents

4.1 Introduction

A receptor-mediated mechanism of action for cannabinoids was only recognized within the last decade, and within that time the cannabinoid receptor has been characterized pharmacologically, localized in brain and its complementary DNA (cDNA) cloned. The cloning of this receptor was the serendipitous result of a non-targeted strategy to obtain cDNAs encoding novel receptors. As described in Chapter 6, the receptor for cannabinoids interacts with guanine nucleotide-binding proteins (G proteins) as part of its second messenger–transduction system. Receptors characterized by this type of coupling (G protein-coupled receptors) have been extensively studied and, structurally, are composed of a single polypeptide subunit with seven hydrophobic regions which appear to be membrane-

Cannabinoid Receptors
ISBN 0-12-551460-3

spanning domains (Probst *et al.*, 1992). Within these domains or immediately adjacent to them are approximately 30 mostly non-hydrophobic residues that are conserved in most G protein-coupled receptors. There are, however, receptors such as the metabotropic glutamate receptors (Houamed *et al.*, 1991; Masu *et al.*, 1991; Abe *et al.*, 1992; Tanabe *et al.*, 1992; Nakajima *et al.*, 1993) and hormone receptors (calcitonin, vasoactive intestinal polypeptide, glucagon, secretin and others; Ishihara *et al.*, 1991, 1992; Jüppner *et al.*, 1991; Lin *et al.*, 1991; Thorens, 1992; Jelinek *et al.*, 1993) that, while having a high degree of sequence similarity to their own small group of receptors, have essentially none of the residues that are typically present in most other G protein-coupled receptors that have been sequenced to date.

The identical amino acid sequence motifs found within different receptor proteins reflect a substantial amount of conserved nucleic acid sequence in their respective genes, and these conserved sequence motifs are the basis for various strategies that have successfully cloned receptors for various types of neurotransmitters or hormones. One of these cloned G protein-coupled receptors, SKR6, was ultimately found to be a marijuana or cannabinoid receptor. Expression of the SKR6 cDNA essentially reproduced the second messenger responses and binding properties previously described for the native cannabinoid receptor, and the corresponding gene for this receptor is particularly active in brain. Three years after the description of the first receptor clone a second cDNA (CX5) was discovered, which encodes a clearly different receptor which also interacts with cannabinoids and cannabimimetics (Section 4.3.1). While the predicted protein product of the CX5 cDNA also belongs to the G protein-coupled receptor superfamily, a clear distinction between the two receptors appears to be their tissue distribution; the gene encoding the CX5 receptor does not appear to be active in brain.

Although the progress in characterizing the cannabinoid receptor(s) has been rapid, a great deal regarding structure, function and significance remains unexplored. Following the sections of this chapter that describe various aspects of the cloned cannabinoid receptors (with emphasis on SKR6 clone), we discuss some of the implications of the information gained to date from these proteins and the questions that remain to be addressed. Issues dealing with the behaviour of the receptor at the cell surface (drug design and intracellular coupling) are dealt with in Section 4.4.1. Issues regarding the protein itself and its regulation are addressed in Section 4.4.2 while issues that apply mainly to the human condition, such as potential diseases involving the cannabinoid receptor and its overall functional significance are discussed in Section 4.4.3.

4.2 The cloned cannabinoid receptor

4.2.1 Cloning and identification of the rat cannabinoid receptor, SKR6

The non-targeted cloning strategy that isolated SKR6 involved the hybridization of an oligonucleotide probe to sequences in a rat cerebral cortex cDNA library. The 56-base probe was based on the sequence that encodes the second hydrophobic domain of the bovine substance K receptor, the first neuropeptide receptor cloned (Masu *et al.*, 1987). This domain contains an amino acid sequence that is highly conserved among many different types of G protein-coupled receptors (adrenergic, muscarinic, serotonergic, dopaminergic). By using appropriate salt concentrations and temperatures (low-stringency conditions) during the hybridization procedure, the probe was able to interact with sequences that encode numerous members of this receptor superfamily and was not limited to interactions with members of the tachykinin receptor family.

Potential receptor clones were identified within the library by hybridization of the radiolabelled oligonucleotide probe to Southern blots of subpools of cDNAs. Potentially different cDNAs were distinguished by the size of their hybridizing restriction fragments, were enriched by further subdividing pools of interest, and were isolated by direct hybridization to bacterial colony filter lifts. From several isolated cDNAs two (SKR6 and SKR14) contained sequences indicative of a receptor clone. The SKR6 cDNA was 5.7 kilobases (kb) in length and its sequence suggested that its protein product would have the seven hydrophobic domains and conserved amino acid residues which are characteristic of G protein-coupled receptor proteins (Matsuda *et al.*, 1990). The SKR14 cDNA contained a sequence that was identical to that of SKR6, however, this clone was truncated at both its 5′ and 3′ ends. Clones that are truncated in the 5′ region are usually the result of premature termination of cDNA synthesis that occurs early during the preparation of the library. In oligo(dT) primed cDNAs, 3′ truncation often occurs because of priming at A-rich stretches within the mRNA instead of at the poly(A) tail. However, in the case of SKR14, the poly(A) tail occurs at a point where no A-rich sequence is found in the otherwise identical SKR6 cDNA, and follows a canonical polyadenylation signal sequence. Thus the 3′ truncation of SKR14 appears to be due to an alternative polyadenylation site occurring *c.* 1.3 kb from the translation stop codon. It is probable, therefore, that at least two alternative forms of the RNA encoding this receptor exist and that the sizes of these transcripts differ as a result of variability in the lengths of their 3′ untranslated sequence.

The predicted amino acid sequence of the SKR6 clone clearly differed from all other previously cloned G protein-coupled receptors. However, SKR6 did not resemble any one receptor or receptor subfamily in particular, and its identity could not be predicted from the cDNA sequence. Consequently, the cDNA was introduced into various hosts for expression and 30–40 ligands (peptides, drugs, hormones, various transmitters) were tested directly for either their ability to bind specifically to the receptor or to elicit a functional response. Introduction of the SKR6 clone into a variety of expression systems was required since information regarding the coupling of this receptor could also not be deduced from its cDNA sequence. In mammalian cell lines (monkey COS-7, Chinese hamster ovary, baby hamster kidney cells) or *Xenopus* oocytes expressing the SKR6 clone, no ligands demonstrated specific binding or elicited a functional response that could be ascribed to the cloned receptor.

Clues leading to the identification of the SKR6 cDNA were obtained from the tissue- and cell-specific expression of the SKR6 gene. *In situ* hybridization histochemistry had shown that SKR6 RNA was widely distributed in the central nervous system of the rat, however, this information was not immediately useful. The observation, which initially suggested that SKR6 encoded a cannabinoid receptor, was obtained from Northern blot analysis of RNAs isolated from N18TG-2 neuroblastomas and NG108-15 neuroblastoma–glioma cells (Matsuda *et al.*, 1990). These data showed that the length of the hybridizing RNA molecule (≈6 kb) was consistent with the length of the SKR6 cDNA and, more importantly, that the SKR6 gene was active in two cell lines known to contain cannabinoid receptors (Howlett and Fleming, 1984; Devane *et al.*, 1986). Moreover, a glioma cell line (C6) which is negative for cannabinoid receptors (Howlett *et al.*, 1986) also appeared to be negative for the SKR6 RNA. The hypothesis that SKR6 might encode a cannabinoid receptor was strengthened by comparisons made between the localization pattern of the SKR6 RNA and that of cannabinoid receptors in rat brain (Herkenham *et al.*, 1990). Both the RNA and receptor proteins (localized by autoradiographic ligand binding) were present within numerous brain regions (Figure 1); however, some mismatch between binding sites and RNA was obvious. Because of the efferent projection patterns of the axons of particular populations of neurons some mismatch between these results should be expected (see 5.3). Generally, however, the similar distributions of cannabinoid receptors and the SKR6 RNA were consistent with the possibility that SKR6 encoded a cannabinoid receptor. Consequently, cannabinoid compounds were tested for the functional activation of the SKR6 clone transfected into an immortal mammalian cell line.

Expression of the SKR6 clone in Chinese hamster ovary cells (CHO cells) that are normally insensitive to cannabinoids rendered the cells responsive to cannabinoids. In neural cell lines and brain tissues analysed *in vitro*, psychoactive cannabinoids inhibit adenylate cyclase activity resulting in decreased accumulation of cyclic AMP (Howlett *et al.*, 1990a). A cannabinoid-induced reduction in cyclic AMP concentrations in SKR6-expressing CHO cells, therefore, would be consistent with the SKR6 cDNA encoding a functional cannabinoid receptor. In the presence of forskolin (to stimulate basal levels of cyclic AMP production), the naturally occurring Δ^9-tetrahydrocannabinol (Δ^9-THC) clearly decreased cyclic AMP production in these cells (Matsuda *et al.*, 1990). This cannabinoid-induced effect was dose-dependent, stereoselective and was not observed in CHO cells that did not express the SKR6 cDNA or in CHO cells that expressed other cloned G protein-coupled receptors (muscarinic or α-adrenergic receptors). In addition, CP 55,940, a potent,

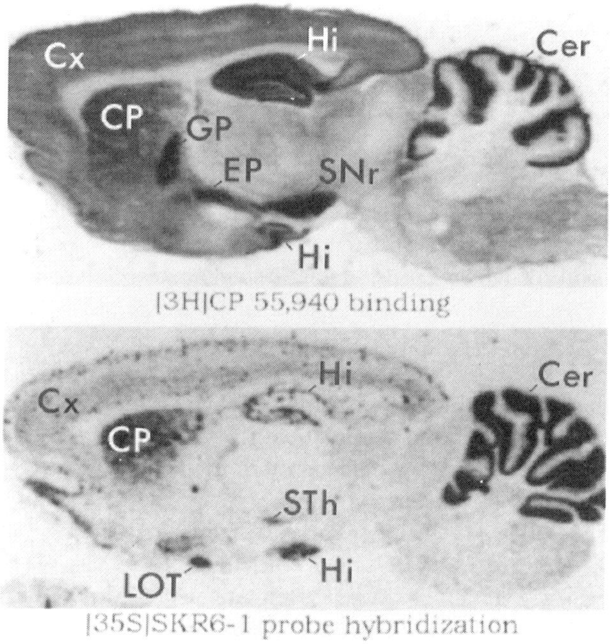

Figure 1 Sagittal sections through adult rat brain showing localization of cannabinoid receptor (CB_1) binding sites and RNA. Upper panel: Autoradiographic image of receptors bound to [^{3}H]CP 55,940. Lower panel: Localization of RNA hybridized to [^{35}S]SKR6-1 oligonucleotide probe. Both the specifically bound ligand and hybridized probe appear as dark images on a light background. Abbreviations: Cer, cerebellum; CP caudate putamen; Cx, cerebral cortex; EP, entopeduncular nucleus; GP, globus pallidus; Hi, hippocampus; LOT, nucleus of the lateral olfactory tract; SNr, substantia nigra reticulata; STh, subthalamic nucleus.

non-classical cannabinoid analogue that has been instrumental in the pharmacological and anatomical characterization of the native cannabinoid receptor (Devane *et al.*, 1988; Howlett *et al.*, 1988; Herkenham *et al.*, 1991a), also decreased the forskolin-stimulated synthesis of cyclic AMP in the SKR6-expressing CHO cells. These results demonstrated that the protein product of the SKR6 cDNA was a G-protein-coupled cannabinoid receptor.

4.2.2 Features of the cloned receptor

Translation of the SKR6 sequence predicts that the primary structure of the rat cannabinoid receptor contains 473 amino acids. Like other members of this receptor superfamily potential *N*-linked glycosylation sites can be identified in the amino acid sequence (residues 78 and 84) and it is probable that the actual mass of this protein exceeds the 52,844 daltons that is predicted by the cDNA sequence. Unlike most other G protein-coupled receptors, however, the cannabinoid receptor lacks a proline residue in the fifth hydrophobic domain. Prolines in the fourth, fifth, sixth and seventh domains are thought to introduce a bend in the protein that may influence the ligand-binding sites of these receptors (Probst *et al.*, 1992).

One other notable feature of the cannabinoid receptor is that it lacks an extracellular disulfide bond which appears to bridge two extracellular loops in most receptors in this superfamily (Savarese and Fraser, 1992). In the cannabinoid receptor, two cysteines reside between the fourth and fifth hydrophobic domains (residues 258 and 265, both of which reside within a single extracellular loop). Most other receptors, however, contain an additional cysteine between the second and third hydrophobic domains. This extra cysteine allows a disulfide link between extracellular loops, which likely serves to position the extracellular ends of the third and fifth hydrophobic domains in close proximity to each other. Should a disulfide bond link residues 258 and 265 in the cannabinoid receptor, it is possible that the fourth and fifth hydrophobic domains would be more closely associated with each other than in other G protein-coupled receptors. Indeed, it appears that high-affinity binding of CP 55,940 requires an intact disulfide linkage either within the cannabinoid receptor or in a protein that influences the binding activity of the receptor (i.e. G protein) (Lu, *et al.*, 1993). This type of alignment of hydrophobic domains may also be the case with the melanocortin receptors (Mountjoy, *et al.*, 1992), *mas* oncogene (Young, *et al.*, 1986) and three receptor-like clones that have similarly positioned cysteine residues in their extracellular loops (edg-1, R334 and RTA; Hla and Maciag, 1990; Eidne, *et al.*, 1991; Ross, *et al.*, 1990; Bonner and Brownstein, unpublished). It is possible that these receptors may be members of a subclass of G protein-coupled receptors that diverged from the prototypical neurotransmitter receptors (Savarese and Fraser 1992).

Shortly after the initial report of SKR6 as the cloned rat cannabinoid receptor, the sequence of the analogous human cDNA was reported (Gérard *et al.*, 1990). Although the human receptor contains only 472 amino acids, the identity between the rat and human receptors is very high (>97%) (Gérard *et al.*, 1991). When expressed in mammalian cells, both the rat and human cloned receptors demonstrate ligand selectivity and G protein coupling that is similar to that of native cannabinoid receptors. Radiolabelled-ligand binding studies reveal that the naturally occurring cannabinoid receptors most readily interact with psychoactive cannabinoids or cannabimimetics (Devane *et al.*, 1988; Herkenham *et al.*, 1990; Compton *et al.*, 1992). This ligand selectivity is also true for the cloned receptor. For example, in [^{3}H]CP 55,940 membrane filtration binding competition studies, relative binding affinities are HU-210>CP 55,244>CP 55,940>RM-8[1]>Δ^9-THC> nabilone>CP 50,556>2-iodo-Δ^8-THC>WIN 55,212-2 for the cloned human

[1]RM8 refers to the 1S enantiomer of 1,3-methanodibenzofuran-9-ol,7-(1,1-dimethylheptyl)-1,2,3,4,4a,9b-hexahydro-2,2-dimethyl-4-methylene, which is also designated HU-250 (Järbe *et al.*, 1993). This compound is the opposite enantiomer of RM7 (also designated HU-249), which is incorrectly defined in Felder *et al.* (1992).

cannabinoid receptor (Felder *et al.*, 1992). Binding studies conducted in membranes prepared from either CHO, COS-7 or L cells that express the cDNAs or the gene encoding this receptor all displayed, saturable and specific binding with [^{3}H]CP 55,940 and the binding dissociation constants (K_ds) of the cloned receptors were also similar to those observed with brain tissues (Gérard *et al.*, 1991; Felder *et al.*, 1992).

In tissues and neural cell lines, cyclic AMP production is also most efficiently reduced by psychoactive compounds compared with cannabinoids such as cannabinol and cannabidiol – compounds that appear to have little to no activity at the native receptor in a variety of systems (Howlett, 1987; Howlett *et al.*, 1990b; Bidaut-Russell *et al.*, 1990). As shown in Figure 2a, psychoactive compounds are also the most potent in affecting cyclic AMP concentrations in cells expressing the cloned receptors. Moreover, since these effects on cyclic AMP production are completely prevented in cells pretreated with pertussis toxin (Figure 2b) the cloned receptors also appear to couple with G proteins that are similar to those that interact with naturally occurring cannabinoid receptors.

4.2.3 Heterologous expression

The introduction of cloned receptors in a form that allows their expression in heterologous host cells, i.e., ones that would not otherwise express the receptors, is an essential technique of molecular biology. Usually, the cloned cDNA or gene is expressed in cells that are sufficiently similar to those in which it is naturally expressed, thereby, allowing the clone's function to be studied. For G protein-coupled receptors this means that, ideally, one tries to express the clones in cells that have the complete cellular transduction machinery through which the receptor acts, e.g., G proteins and effector molecules (enzymes or ion channels). The most powerful use of this technique arises from comparing the host cells expressing the cloned gene to the host cells into which the gene has not been introduced; the properties of the former that differ from those of the latter can then be ascribed to the introduced gene. In the case of a cloned receptor gene, the cell line with and without the receptor can be viewed as the biological equivalent of a pharmacological antagonist, which is particularly important for the cannabinoid receptor because antagonists have only recently been characterized (Pacheco *et al.*, 1991; Rinaldi-Carmona *et al.*, 1994). A general description for the use of clonal cell lines in the analysis of receptors is provided by Buckley *et al.* (1990).

Two heterologous expression systems have been used for the identification and characterization of the cannabinoid receptor. One system that was used during the initial screening of potential ligands involved the engineering of the clone into a vector that allowed RNA to be transcribed *in vitro* using a phage RNA polymerase. The homogeneous RNA product was then injected into *Xenopus* oocytes where it was translated into receptor protein. These injected eggs were then exposed to various ligands and monitored for an electrophysiological response that would signal that the receptor had been activated. The other heterologous expression system which has been used more extensively is that of mammalian cells that have been

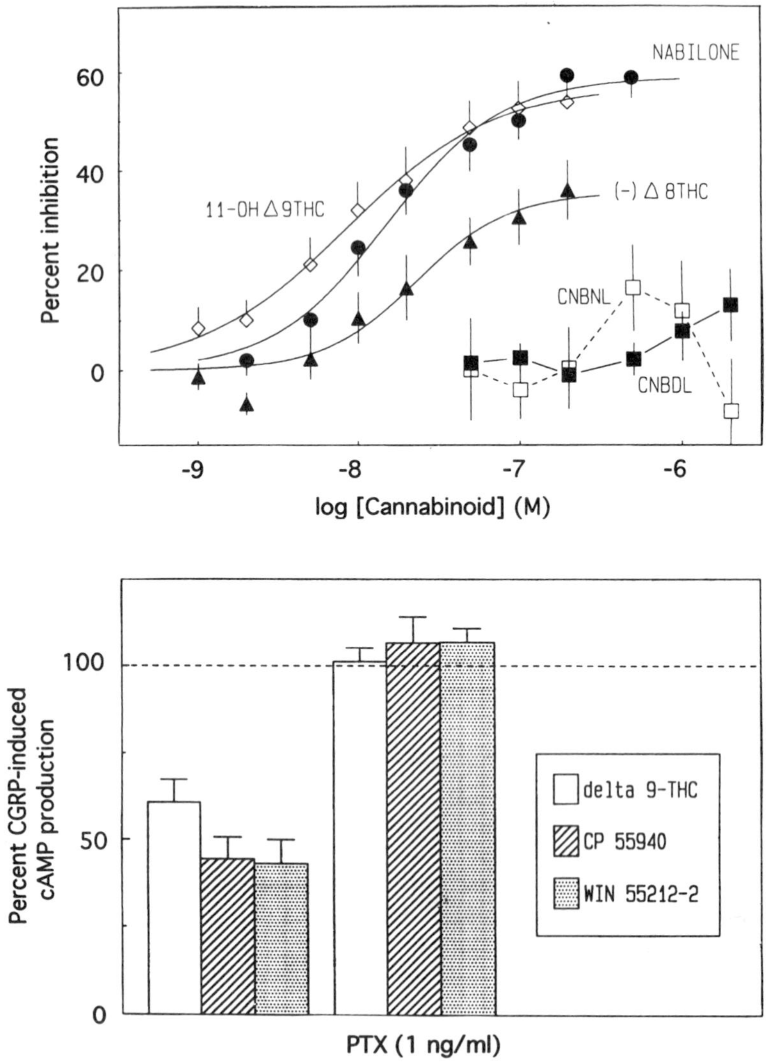

Figure 2 Cannabinoid-induced inhibition of cyclic AMP accumulation in Chinese hamster ovary (CHO) cells transfected with the cloned cannabinoid receptor (CB_1). Upper panel: dose response curves of various cannabinoids and nabilone. Data points represent the average percent inhibition ± SEM of forskolin (500 nM)-stimulated cyclic AMP production. Abbreviations: CNBNL, cannabinol; CNBDL, cannabidiol. Lower panel: effect of pertussis toxin (PTX) on cannabinoid-induced inhibition of cyclic AMP accumulation. Bars represent the averaged results ± SEM from 3–4 experiments (performed in triplicate); the amount of cyclic AMP produced is shown as a percentage of the calcitonin gene-related peptide (CGRP)-induced response (100%, dashed horizontal line). PTX-treated cells were cultured with the toxin for 18–20 h prior to exposure to CGRP ± cannabinoids. Concentrations of the various reagents were: CGRP, 250 nM; Δ^9-THC, 100 nM; CP 55,940, 5 nM; WIN 55,212–2, 100 nM.

phenotypically altered by a cloned gene engineered into an expression vector. Such a vector provides a promoter, typically of viral origin, that enables the host to transcribe RNA from the clone and, subsequently, translate that RNA into receptor proteins (Kaufman, 1990). Once engineered into the expression vector, the clone must then be introduced into the heterologous host such that (1) a significant fraction of the host cells successfully receive the vector (usually in the form of plasmid DNA), and (2) the clone is expressed at a detectable level. While eukaryotic cells can acquire foreign genes by a variety of methods (i.e., cell fusion, infection, transfection and direct injection), this chapter will focus on the transfection of cDNAs or genes encoding cannabinoid receptors, since this method has been used more extensively to study cloned receptors.

The introduction of exogenous DNA into eukaryotic cells (transfection) can be accomplished in numerous ways. In general, however, the manner in which new genetic material is ultimately processed by the heterologous host determines whether a transfection is transient or stable. Transient transfection is performed to obtain short-term (several days) expression of the cloned gene. Although the foreign gene is not stably incorporated into the host genome, typically a large fraction of the cells exposed to the plasmid DNA will express the clone, which results in readily detectable amounts of its protein product. The host cell population, however, will not be homogenous in terms of the levels of the protein product expressed. Some cell lines, such as COS-7, express foreign genes transiently and at a high level if the vector contains an SV40 origin of replication. The cannabinoid receptor was transfected into COS-7 cells for the initial determination of the ligand-binding properties of the cloned human cannabinoid receptor (cDNA) (Gérard *et al.*, 1991).

Stable transfections have the advantage of providing a homogeneous and continuous cell line that eliminates the need to perform transient transfections whenever more cells are needed. Transfections of this type are significantly more labour intensive and time-consuming (months), because incorporation of the clone into the host genome takes time and one must 'select' those cells that have successfully retained the cloned gene by eliminating the untransfected cells with an antibiotic. The stably transfected cells survive the antibiotic treatment because of the simultaneous expression of the clone and the appropriate antibiotic resistance gene. These cells are then made homogenous by establishing a clonal cell line that originates from a single tranfected cell. The CHO cells used to initially characterize the functional responses of the cloned cannabinoid receptors were stably transfected (Matsuda *et al.*, 1990; Gérard *et al.*, 1991) (see Table 1) as were the L cells that were used to examine the binding properties of the receptor encoded by the human cannabinoid receptor gene (Felder *et al.*, 1992).

The questions that can be answered by the heterologous expression of a receptor are limited to those of a relatively simple biochemical nature (i.e., examining the direct effects of a ligand on the receptor or the interaction of the receptor with its second messenger transduction system). In experiments involving stably transfected cell lines expressing cloned cannabinoid receptors, two unknowns that are unique to the cannabinoids have been addressed; the first unknown is the

Table 1 Effective concentrations (IC_{50}, nM) of various agonists for reducing cyclic AMP production in Chinese hamster ovary (CHO) cells expressing the cloned cannabinoid receptor (CB_1)

Receptor clone	(–)Δ^9-THC	CP 55,940	WIN 55,212–2	HU-210	Anandamide
Rat cDNA	9–14	0.93	24	0.02–0.9	201
Human cDNA	13	0.99	–	–	–
Human gene	–	–	–	–	160

Receptor clones were all stably transfected into CHO cells and basal levels of cyclic AMP production were increased by forskolin stimulation. Data obtained from Matsuda *et al.* (1990), Gérard *et al.* (1991), Felder *et al.* (1992, 1993) and Vogel *et al.* (1993).

mechanism of action of these compounds and the second unknown is the endogenous substance(s) which interacts with the cannabinoid receptor(s).

The long absence of an antagonist has hindered the determination of the mechanism of action (receptor- or non-receptor mediated) responsible for numerous cannabinoid-induced effects. In a recent study, the effects of two potent cannabimimetics (CP 55,940 and HU-210) were found to influence signal transduction systems by both receptor- and non-receptor mediated mechanisms (Felder *et al.*, 1992). In stably transfected CHO cells, a cannabinoid-induced reduction in cyclic AMP production was, as expected, dependent on the presence of the cloned cannabinoid receptor. Both compounds, however, also increased extracellular concentrations of [^{3}H]arachidonic acid and intracellular concentrations of Ca^{2+}. These responses were non-receptor mediated since they were found in both transfected and in non-transfected cells. Similarly, both effects were evident at only the highest concentrations of agonists tested and displayed no stereoselectivity. These potent cannabimimetics, however, did not activate another second messenger's effector enzyme, phospholipase C. These data suggest that not all second messenger systems are affected by the non-receptor-mediated actions of these compounds, and that the Ca^{2+} and arachidonic acid second messenger systems are specifically altered.

The discovery of the cannabinoid receptor initiated efforts to isolate and identify its functional correlate or endogenous ligand. Numerous investigators participating in this search have relied upon radiolabelled-ligand binding assays to test various preparations of tissue extracts. It appears, however, that tissue isolates that are capable of displacing a ligand in a binding assay or that evoke the functional response of interest often produce false-positive results. Studies involving transfected and non-transfected cell lines as a source of cannabinoid receptors can determine whether a compound that appears to bind to the receptor is also capable of activating the receptor. The important task of evaluating candidates for the physiological 'cannabinoid,' therefore, falls heavily upon the heterologously expressed cannabinoid receptor. Although transfected cell lines were not used during the initial evaluation of the candidate endogenous ligand, anandamide, this compound has subsequently been tested and found to activate the cloned cannabinoid receptor (Felder *et al.*, 1993; Vogel *et al.*, 1993). Interestingly, in the study conducted by Felder *et al.* (1993) anandamide significantly altered non-receptor-mediated cellular functions in a manner similar to that previously observed for other cannabimimetic compounds.

Expression of a cloned receptor in a heterologous host can also provide a system that is better defined than a protein's native environment. Therefore, another benefit of heterologous expression of a cloned gene is that its protein product can be studied in a population of cells that is far more, if not completely, homogeneous compared with the cellular makeup of many tissues. In addition, the cellular environment can also be manipulated, to a limited extent, such that subcellular components can be specifically absent (similar receptor subtypes) or present (a specific ion channel). Future research will undoubtedly utilize the benefits of heterologously expressed receptors to examine specific issues regarding cannabinoid receptor function.

Although heterologous expression is an effective way to control for the presence and absence of a cloned protein, the ability of the cloned gene and its product to function normally when expressed in this manner is by no means guaranteed. The requirement that the heterologous host does not express the gene to be introduced means that the host, in many instances, is different enough from the cells that normally express the cloned gene that the function of the clone may not exactly mimic that of the native protein. In the case of the G protein-coupled receptors, if a component of the second messenger transduction system varies significantly from that found in the receptor's normal environment, the functional properties of the cloned receptor will likely vary. For example, in baby hamster kidney cells that were transfected with the SKR6 cDNA and found to express the corresponding RNA, no cannabinoid-induced effects on cyclic AMP were ever detected (L.A. Matsuda, unpublished). Moreover, in stably transfected L cells that clearly express receptors, as detected by radioligand binding, no cannabinoid-induced effects on cyclic AMP production are observed. The reasons for a lack of functional responses in these cell lines are unknown, but serve to illustrate an important limitation to the application of transfected cells in the study of functional responses of cloned receptors. Indeed, the response initiated by the formation of a receptor–ligand complex can be interrupted at numerous points along the signal transduction cascade, and the composition of the relevant biomolecules in the expression host can readily affect the apparent functionality of the cloned receptor. The subtypes of the G proteins, the effector molecules and the processing of the clone itself need to be carefully considered before drawing conclusions regarding the functional responses of an expressed receptor, particularly if the results appear negative.

One other application of heterologous expression of the cloned cannabinoid receptor is to obtain the protein in relatively large amounts, and in an environment that allows the protein to be readily purified. Such studies would enable researchers to produce adequate quantities of protein to carry out immunizations to obtain receptor-specific antibodies or for biophysical and biochemical analysis of the protein. The large-scale synthesis of proteins for these purposes can be accomplished by expressing the clone in insect cells using baculovirus (Luckow and Summers, 1988) or a prokaryotic expression system. Although these expression systems are advantageous for procedures that require a specific protein in large quantity, the expressed protein may not be identical to the native receptor since mammalian post-translational modifications of the expressed protein are not reproduced in bacteria and may vary slightly in insect cells.

4.3 The receptor gene and its expression

4.3.1 Gene structure and chromosomal location

The gene encoding the cannabinoid receptor has been cloned from the human, rat and mouse genomes (Bonner, unpublished). The mouse gene encodes a 473 amino acid protein which is very similar to that encoded by the rat gene (>99% amino acid identity and *c.* 95% homology between their nucleic acid sequences). Sequences obtained from genomic DNA isolated from human (cosmid) or rat (phage) libraries indicate that the coding sequence and about 3.8 kilobases of 3′ untranslated sequence is contained within a single exon. This exon also contains 63 bases of the 5′ untranslated sequence (leader), with the remainder of the leader sequence residing in two additional 5′ exons. One of these exons accounts for the remainder of the rat cerebral cortex cDNA leader (Matsuda *et al.*, 1990) while the other corresponds to the remainder of the human brainstem cDNA leader (Gérard *et al.*, 1990). It appears that most cannabinoid receptor RNAs contain one of these two exons, but not both. This situation could arise in two ways. First, there could be an unidentified third exon at the 5′ end of the gene that is joined to one of the two exons (by alternative splicing) and then to the coding exon. Second, the two 5′ exons could result from transcriptional start sites that are under the control of different promoters (Schibler and Sierra, 1987). In this situation alternative splicing would not occur and neither 5′ exon would contain a splice acceptor site. This second case would provide some of the regulatory flexibility that is often provided by having multiple genes for very similar proteins.

Among the growing list of G protein-coupled receptors the majority of these receptors are typically grouped into smaller families based on the identity of their respective ligands. The primary structures of these ligand-dependent receptor subtypes are notably similar to each other and the number of subtypes ranges from two to eight (Watson and Girdlestone, 1994) or more. Attempts at identifying a second gene encoding a cannabinoid receptor by low-stringency hybridization of Southern blots have revealed only clones that do not resemble receptor genes (Lautens and Bonner, unpublished). This apparent lack of genes that closely resemble the initially described receptor, however, does not exclude the possibility that a more distantly related cannabinoid receptor gene exists. Recently, a cDNA (CX5) encoding a putative peripheral cannabinoid receptor (human) has been obtained from a promyelocytic leukaemic cell line (HL60) by the use of degenerate primers and polymerase chain reaction (PCR) (Munro *et al.*, 1993). This clone is notably smaller than the initially described human cannabinoid receptor (360 vs. 472 amino acids) and, while it clearly binds cannabinoids and cannabimimetic compounds, this peripheral binding site has not yet been shown to functionally respond to any of these ligands. To facilitate further discussion of the receptors encoded by CX5 and SKR6 in this communication, the protein products of these two cDNAs will be referred to as CB_2 and CB_1, respectively[1]. The rank potencies of various ligands in competition-binding assays differs between CB_1 and CB_2. However, whether or not these two receptors share a common endogenous ligand cannot yet be determined.

The human CB_1 gene has been genetically mapped to chromosome 6 based on a restriction fragment length polymorphism (RFLP) found within the intron preceding the coding exon (Caenazzo *et al.*, 1991). Physical mapping by *in situ* hybridization to metaphase chromosomes has confirmed this assignment to chromosome 6, and has further localized this gene to 6q14–q15 (Hoehe *et al.*, 1991).

4.3.2 Transcription of the receptor gene

The RNA for CB_1 has been identified in six tissues and various cell lines. In addition to the brain, receptor RNA has been found in lung and testes using Northern blot analysis. Receptor RNA has also been detected in spleen, thymus and various blood cell types as a PCR amplification product (Table 2). In immortal cell lines, this receptor's RNA has also been found in neural cell lines, lung fibroblasts and immune cells (Table 3). Northern blots hybridized with various probes (cDNA probes, oligonucleotide probes) specific for the receptor sequence reveal *c.* 6 kb RNA for this receptor. In some instances, other RNAs of *c.* 2.2 and 3.4 kb lengths can also be found, but the abundance of these smaller RNAs is usually much lower when compared with that of the larger transcript. The 2.2 kb RNA found in testes may correspond to the transcripts that are polyadenylated at the alternative polyadenylation site described for the SKR14 clone (Section 4.2.1).

CB_1 RNA has been detected by PCR amplification and Northern blot analysis in cerebellar tissues obtained from different mammals including rat, guinea-pig, dog, pig, common marmoset and human (Table 2). In addition, CB_1 RNA has been localized in specific populations of neurons by *in situ* hybridization histochemistry performed on brain tissue sections (Mailleux and Vanderhaeghen, 1992a; Matsuda *et al.*, 1993a). Cannabinoid receptor RNAs are found in neurons that comprise nearly all systems (sensory, associative, autonomic, motor, limbic, cognitive). The RNA for CB_1 is most abundant in anterior olfactory nuclei, the lateral nucleus of the olfactory tract, the dorsal lateral caudate nucleus and the granular layer of the cerebellar cortex. Interestingly, a number of forebrain areas (anterior olfactory nuclei, the lateral nucleus of the olfactory tract, the horizontal limb of the diagonal band, the basal lateral amygdala, hippocampus and cerebral cortex) contain small numbers of neurons that individually display very high levels of receptor RNA (Figure 3), although the majority of cells in these regions usually express the RNA to a more moderate level.

[1]The official human gene locus names for CB_1 and CB_2 receptors are *CNR1* and *CNR2*, respectively.

Table 2 Cannabinoid receptor (CB_1) RNA distribution in tissue

Tissue	Species	Method	mRNA size (kb)	Reference
Brain	Human	Northern blot	6.0	Matsuda *et al.* (1993b)
		Polymerase chain reaction		Bouaboula *et al.* (1993)
		In situ hybridization histochemistry		Mailleux *et al.* (1992)
Spleen		Polymerase chain reaction		Bouaboula *et al.* (1993)
Thymus		Polymerase chain reaction		Bouaboula *et al.* (1993)
Monocytes		Polymerase chain reaction		Bouaboula *et al.* (1993)
T4 cells		Polymerase chain reaction		Bouaboula *et al.* (1993)
T8 cells		Polymerase chain reaction		Bouaboula *et al.* (1993)
B cells		Polymerase chain reaction		Bouaboula *et al.* (1993)
NK cells		Polymerase chain reaction		Bouaboula *et al.* (1993)
Brain	Common marmoset	Northern blot	6.0	Matsuda *et al.* (1993b)
Brain	Pig	Northern blot	6.0	Matsuda *et al.* (1993b)
Brain	Dog	Northern blot	6.0	Matsuda *et al.* (1993b)
Testes		Northern blot	6.0, 2.2	Gérard *et al.* (1991)
Brain	Guinea-pig	Northern blot	6.0	Gérard *et al.* (1991)
Brain	Rat	Northern blot	5.8–6.0	Matsuda *et al.* (1993b)
				Gérard *et al.* (1991)
		Polymerase chain reaction		Kaminski *et al.* (1992)
		In situ hybridization histochemistry		Mailleux and Vanderhaeghen (1992a)
				Matsuda *et al.* (1993a)
Lung		Northern blot	5.8	Cabral *et al.* (1993)
Spleen	Mouse	Polymerase chain reaction		Kaminski *et al.* (1992)

Table 3 Cannabinoid receptor (CB_1) RNA expression in cell lines

Cell line	Species	Cell type	Method	mRNA size (kb)	Reference
HEL	Human	Lung fibroblast	Northern blot	5.8, 3.4	Pettit and Cabral (unpublished)
HFL		Lung fibroblast	Northern blot	5.8, 3.4	Pettit and Cabral (unpublished)
MRC-5		Lung fibroblast	Northern blot	5.8, 3.4	Pettit and Cabral (unpublished)
Ichikawa		T leukaemia	Polymerase chain reaction		Bouaboula *et al.* (1993)
Jurkat[a]		T leukaemia	Polymerase chain reaction		Daaka and Klein (unpublished)
JM		T leukaemia	Polymerase chain reaction		Bouaboula *et al.* (1993)
Raji[a]		B lymphoblast	Polymerase chain reaction		Daaka and Klein (unpublished)
Daudi		B lymphoblast	Polymerase chain reaction		Bouaboula *et al.* (1993)
IM9		B lymphoblast	Polymerase chain reaction		Bouaboula *et al.* (1993)
Bronchier		B lymphocyte	Polymerase chain reaction		Bouaboula *et al.* (1993)
U937		Monocyte	Polymerase chain reaction		Bouaboula *et al.* (1993)
THP1[a]		Monocyte	Polymerase chain reaction		Bouaboula *et al.* (1993) Daaka and Klein (unpublished)
B103 cells	Rat	Neuroblastoma	Northern blot	5.8	Cabral *et al.* (1993)
NG108–15	Mouse/rat	Neuroblastomaglioma	Northern blot	6.0	Matsuda *et al.* (1990)

Table 3 *(Contd)*

Cell line	Species	Cell type	Method	mRNA size (kb)	Reference
N18TG–2	Mouse	Neuroblastoma	Northern blot	6.0	Matsuda *et al.* (1990)
CTLL–2		T cell	Polymerase chain reaction		Daaka and Klein (unpublished)
NKB61A2		Natural killer-like	Polymerase chain reaction		Daaka and Klein (unpublished)
RAW276.4[a]		Macrophage	Polymerase chain reaction		Daaka and Klein (unpublished)

[a] Polymerase chain reaction amplification products increase after 2–18 h treatment with various inducing agents.

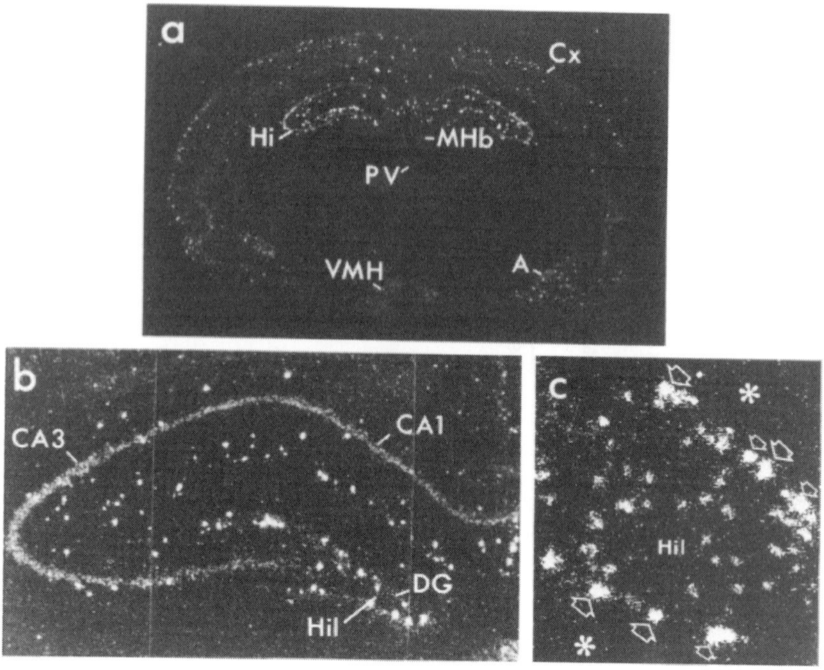

Figure 3 Cannabinoid receptor (CB_1) RNA in adult rat hippocampus. (a) Low magnification negative image of a coronal section hybridized with [^{35}S] tailed SKR6-1 probe and exposed to film for 16 days; labelling is white upon a dark background. Abbreviations: A, amygdala; Cx, cerebral cortex; Hi, hippocampus; MHb, medial habenula; PV, paraventricular thalamic nucleus; VMH, ventral medial nucleus of the hypothalamus. (b) Dark-field photomicrograph taken from a section similar to that shown in (a). Uniform labelling by the SKR6-1 probe is evident in the pyramidal cell layer of Ammon's horn (CA1, CA3). Very intensely labelled cells are found throughout the CA fields and subjacent to the granule cell layer in the dentate gyrus (DG); Hil, hilar region of the dentate gyrus. (c) Higher magnification photomicrograph of the hilar region of the dentate gyrus. Intensely labelled cells (open arrows) are closely associated with the granule cell layer (asterisk) which shows little to no specific labelling by the SKR6-1 probe. Other moderately labelled cells are found in the hilar region.

It is clear that certain neurons express substantially more CB_1 RNA than others. It is not clear, however, whether these neurons produce a substantially greater number of receptors. The localization of cannabinoid receptors by radio-labelled ligand autoradiography (film) does not permit the visualization of receptors with cellular resolution. Determining whether specific cells express a greater density of receptors than others will likely require the resolution attainable through immunohistochemical techniques and receptor-specific antibodies. With regard to the numerous populations of neurons that express CB_1 cannabinoid receptors, additional characterization (co-localization studies) of these cells is needed to further identify the neurons that express the receptor RNA (e.g. neurons in the hypothalamus and horizontal limb of the diagonal band). In some areas, however, the gross morphological characteristics of the neurons or their relative position in the tissues allows accurate identification of the neurons expressing the RNA for cannabinoid receptors (e.g., cerebellar granule cells, hippocampal pyramidal cells).

In addition to tissue-specific expression, the inducibility and temporal dynamics of the expression of the CB_1 receptor gene have been examined. In human immune cell lines, Daaka and Klein (unpublished) found substantial increases in PCR amplification products of cannabinoid receptor RNA in Raji and THP-1 cells treated with *Escherichia coli* 0111:B4 lipopolysaccharide (LPS) or phorbol myristate acetate (PMA), respectively. In addition, a greater than 10-fold increase in cannabinoid receptor RNA was detected in Jurkat (human T cell) and RAW276.4 (mouse macrophage) treated with LPS and phorbol hydroxypalmitate acetate (PHA), respectively. In rat brain, CB_1 receptor RNA levels increase (c. 50%) in the caudate-putamen following adrenalectomy (Mailleux and Vanderhaeghen, 1993); this increase can be reversed by treatment with dexamethasone. In terms of temporal dynamics, CB_1 receptor RNA has been found as early as postnatal day 3 in brain, by PCR amplification (McLaughlin and Abood, 1993) and embryonic day 12 by *in situ* hybridization histochemistry (Herkenham *et al.*, 1991b). Moreover, the receptor RNA in brain appears to decrease by 50% in aged rats (Mailleux and Vanderhaeghen, 1992b).

With regard to the alternative forms of RNAs that arise from different 5′ exons of the CB_1 gene, both forms have been observed on Northern blots that were hybridized with oligonucleotide probes specific for the 5′ leader sequences. Both RNAs are found in total cerebellar RNA whereas only one transcript can be readily observed in total RNA from other brain regions (L.A. Matsuda, unpublished observation). More sensitive techniques for detecting RNA must be used to examine these alternative forms of cannabinoid receptor RNAs before one can conclude that these transcripts are differentially expressed in brain.

Localization of the RNA encoding CB_2 receptors indicates that the gene for this receptor is active in spleen but not in brain (examined by Northern blot analysis or *in situ* hybridization histochemistry; Munro *et al.*, 1993). The RNA for this receptor is not, however, detected in all spleen cells, as demonstrated by the absence of a PCR amplification product from CD5-selected spleenocytes. As reported by Kaminski *et al.* (1992), the RNA encoding the CB_1 cannabinoid receptor can also be detected in spleen, and [^{3}H]CP 55,940 specifically binds to receptors in this tissue. Since this ligand clearly recognizes both CB_1 and CB_2, it is possible that the specific binding of [^{3}H]CP 55,940 in spleen is the sum of the binding to each of the two cannabinoid receptor subtypes. However, the RNA encoding the CB_1 receptor appears to be quite rare and has been detected only as a PCR amplification product or by solution hybridization techniques (RNase protection) (Kaminski *et al.*, 1992). In contrast, CB_2 RNA seems to be considerably more abundant and is clearly observed in total spleen RNA on a Northern blot. If receptor B_{max} values are proportional to RNA levels, the apparent difference in the relative abundance of these receptor RNAs suggests that most of the specifically bound ligand in spleen results from binding to the CB_2 receptor. More information regarding the ligand specificity of the cannabinoid binding site in spleen is needed, however, to verify this. It is also possible that distinct populations of spleen cells will differentially express one or other subtype of cannabinoid receptor.

4.4 Unresolved issues and research implications

4.4.1 Pharmacological issues

The direct interactions between the cannabinoid receptor and other molecules, both intracellular and extracellular, are aspects of this protein that can be examined using a cloned receptor. Similarly, the interaction of a cannabinoid receptor with its membrane environment can also be studied. The functional domains involved in the interaction of receptors with their ligands are characterized in order to design more therapeutically useful drugs. Drugs that are poor therapeutic agents are those that often fail to bind specifically with a single receptor subtype. As described in Chapter 1, a number of potential therapeutic uses for a cannabinoid agonist exist yet few have been realized due to cannabinoid-induced psychoactive effects. Indeed, one of the most promising therapeutic uses for cannabimimetics is to alleviate or minimize pain, however, both the analgesic as well as the psychoactive actions of these compounds appear to be mediated by the same cannabinoid receptor (Compton *et al.*, 1993).

Once a receptor has interacted with an agonist, it communicates this interaction to the cell via its signal transduction system; thus, the interaction with G proteins is another important functional aspect of this receptor superfamily. Indeed, the interactions with distinct second messenger systems is one characteristic that may differ between receptor subtypes that respond to a common ligand. While the functional response of the CB_2 site has not yet been determined, CB_1 appears to activate a variety of different effectors (adenylate cyclase and ion channels). Although some coupling heterogeneity (interactions possibly involving different G-protein subtypes) appears to be involved in cannabinoid-induced decreases in adenylate cyclase activities in brain (Childers *et al.*, 1993; Pacheco *et al.*, 1994), this aspect of cannabinoid receptors has not been extensively addressed. Characterizing the G protein– receptor interactions and the manner in which their respective effectors are activated may also provide additional information needed to design therapeutically useful agents.

Numerous members of the G protein-coupled receptor superfamily have been evaluated in terms of their ligand-binding domains and their interactions with G proteins (Savarese and Fraser, 1992). Commonly, these studies involve the creation of a chimera (exchange of a portion of one receptor protein with that of another) or the mutation of the receptor's gene or cDNA such that the modified protein lacks amino acids (deletion mutation) or contains a single amino acid that varies from the native protein (point mutation). Although studies such as these often give results that are difficult to interpret (changes in a protein can interfere with a particular interaction either directly or indirectly) important information can be obtained. In terms of defining a 'binding pocket' or other functional domain, an alternative approach involves the covalent crosslinking of the receptor with the molecule of interest. This method, however, is not without drawbacks as a ligand's reactive group may interact with the receptor in the vicinity of, yet not actually in, the binding site.

After the discovery of the cDNA encoding the putative CB_2 receptor, the motivation for characterizing the functional domains of cannabinoid receptors has intensified. Although the therapeutic potential of these receptors has not yet been explored thoroughly, the determination of the regions within each receptor that are involved with agonist binding will help in the design of drugs that will activate a single subtype of receptor protein. In addition, as more is learned about the significance of these receptors, it is possible that antagonists may also prove to be useful as therapeutic agents. Although studies that address the structural biology of cannabinoid receptors are only just beginning, the characterization of the functional domains of these proteins remains an important target for future research.

4.4.2 Cellular and molecular genetic issues

As a member of the G-protein-coupled superfamily of receptors, many aspects of the biochemistry and biology of the cannabinoid receptor are assumed to be similar to other superfamily members. Difficulty arises, however, in predicting how similar the cannabinoid receptor is, in reality, to these other proteins. Because of the relatively recent recognition of the receptor-mediated mechanism of cannabinoid action, cannabinoid receptors have not been characterized as extensively as other G protein-coupled receptors. Consequently, there are few data on subcellular localization, post-translational modifications and/or regulatory mechanisms of cannabinoid receptors. Although the primary sequences from the cloned receptors reveal possible sites for various modifications, these biochemical characteristics need to be measured directly before assumptions can be made as to their importance in the function of the receptor.

A ligand capable of interacting with multiple receptors can transmit information in a diverse manner. With regard to the CB_1 receptor, some diversity of function may also result from the manner in which its gene is expressed. The structure of the gene and the expression of its RNA(s) suggest that a number of intriguing regulatory mechanisms may be involved in the expression of this receptor, including alternative splicing or transcription initiated by multiple promoters. Research in this area, again, is only in its early stages and questions are more easily stated than answers. For example, with regard to the expression of notably high levels of receptor RNA by certain neurons, how is this accomplished and to what end? Also, what is the significance of alternative RNAs for this receptor that differ in their 5′ leader sequences or 3′ untranslated regions? Examination of the neurobiological aspects of the cannabinoid receptor may reveal that the manner in which its gene is expressed provides the diversity in function that seems to be required of many chemical messengers and their receptors. However, based on the precedent set by other members of this receptor superfamily and the recent cloning of the CB_2 receptor, diversification via multiple receptors remains highly plausible with regard to cannabinoid actions. Multiple genes for cannabinoid receptors will not only increase the likelihood of a variety of second-messenger transduction systems that respond to cannabinoids, but may also result in receptors with distinct post-translational modifications (phosphorylation and/or acylation) that could further diversify the functionality of these receptors. Indeed, the identification and characterization of cannabinoid receptor subtypes greatly increases the likelihood of obtaining the full therapeutic benefit from cannabimimetic compounds.

4.4.3 Physiological and pathological issues

Based on the effects of cannabinoids in humans, the significance of these receptors is difficult to predict and is perhaps their least understood aspect. In the brain, as well as in peripheral tissues, clarification of the physiological role of cannabinoid receptors will be obtained by the continued study of the effects induced by agonist compounds. The cloned genes for these receptors, however, provide alternative methods to help determine the nature of their role *in vivo*. Recent advances in molecular biology have produced transgenic animals in which a specific gene has been made inoperative. Indeed, these genetically altered animals have provided information relevant to brain functions as complex as learning and memory (Silva *et al.*, 1992a,b). Assuming that this receptor is not critical for survival, successful 'knock out' of the genes for cannabinoid receptors in mice may provide information on the role of these receptors in the physiological processes and developmental programmes involved in central nervous, immune, reproductive and respiratory systems. In a more tissue- or cell-specific manner, the function of these receptors may also be elucidated by interrupting the translation of the proteins at the mRNA level. Antisense oligonucleotides have been used both *in vitro* and *in vivo* to eliminate specific proteins and to study their significance; the function of the neuropeptide Y–Y1 receptor, a G protein-coupled receptor that lacks a pharmacological antagonist, has been examined in this manner (Wahlestedt *et al.*, 1993).

Information obtained from human genes encoding cannabinoid receptors can be used to identify genetic linkages in families affected by hereditary illnesses. In the 'knock out' mouse, the absence of the protein is used to gain clues to its normal function. In humans, determining how a protein is related to a pathological condition may give some indication of its normal role. Since the chromosomal assignment for the CB_1 receptor gene has been made, its linkage with a disease locus can be determined directly by searching for the association of markers for a specific disease with those for the receptor gene (e.g. RFLP). The involvement of this receptor in disease could also be indirectly discovered if a close association exists between a newly identified disease locus and the receptor gene. Should such a physical linkage be found, then the significance of the receptor in the illness can be addressed. However, these approaches are only useful if a genetic anomaly (mutation) occurs within the cannabinoid receptor gene itself. If an aetiology involving the receptor occurs because of problems originating from other proteins that influence the ultimate functionality of the cannabinoid receptor (i.e. proteins that affect its processing or the expression of its gene), the indirect relationship between the receptor and the disease will not be apparent.

An association between a cannabinoid receptor and disease may also be identified from large-scale sequencing of the human population (Hoehe *et al.*, 1993). This approach may detect variations in the coding region of a gene that affect the normal function of a protein. Alternatively, variations within the regulatory regions of a gene may be found that might affect the expression of the gene product. If an anomaly is identified, this change in the receptor or how it is expressed could then be correlated with the condition of that individual and insight might be gained into the relationship of the receptor to disease or to normal physiological functions.

4.5 Summary

Numerous aspects of the molecular biology of cannabinoid receptors have been described and an overview describing future research directions has been provided. Clearly, the behaviour of a cloned receptor in transfected cells has provided a conclusive demonstration of this mechanism of cannabinoid action. Furthermore, the establishment of cell lines that differ only in the presence or absence of a cannabinoid receptor has and will continue to help define the function of this protein and to evaluate candidates for endogenous cannabinoid ligands. The information that can theoretically be accessed by the manipulation of cloned genes for a receptor should not be underestimated. Genetic linkage studies, although in their early stages will, hopefully, shed light on the significance of cannabinoid receptors, as will data gained from transgenic animals. Although the activity of the CB_1 receptor gene is clearly evident in many brain areas, conditions associated with mental health should not be exclusively targeted.

Other research challenges include: the identification and characterization of multiple subtypes of cannabinoid receptors, analysis of the interactions of these receptors with their ligands and G protein(s) and the elucidation of the mechanism and significance of the very high levels of cannabinoid receptor RNAs that are found in some neurons. As the significance of cannabinoid receptors becomes more apparent, the development of novel therapeutic agents that rely on these receptors may again become attractive to the pharmaceutical industry.

References

Abe, T., Sugihara, H., Nawa, H., Shigemoto, R., Mizuno, N. & Nakanishi, S. (1992) Molecular characterization of a novel metabotropic glutamate receptor mGluR5 coupled to inositol phosphate/Ca^{2+} signal transduction. *J. Biol. Chem.* **267**, 13361–13368.

Bidaut-Russell, M., Devane, W.A. & Howlett, A.C. (1990). Cannabinoid receptors and modulation of cyclic AMP accumulation in the rat brain. *J. Neurochem.* **55**, 21–26.

Bouaboula, M., Rinaldi, M., Carayon, P., Carillon, C., Delpech, B., Shire, D., Le Fur G. & Casellas, P. (1993) Cannabinoid-receptor expression in human leukocytes. *Eur. J. Biochem.* **214**, 173–180.

Buckley, N.J., Hulme, E.C. & Birdsall, N.J.M. (1990) Use of clonal cell lines in the analysis of neurotransmitter receptor mechanisms and function. *Biochim. Biophys. Acta* **1055**, 43–53.

Cabral, G.A., Pettit, D.A.D. & Fischer-Stenger, K. (1993) Marijuana inhibition of cytotoxic T lymphocyte activity. *Adv. Biosciences* **86**, 653–662.

Caenazzo, L., Hoehe, M.R., Hsieh, W.-T., Berrettini, W.H., Bonner, T.I. & Gershon, E.S. (1991) HindIII identifies a two allele DNA polymorphism of the human cannabinoid receptor gene (CNR). *Nucl. Acids Res.* **19**, 4798.

Childers, S.R., Roy, M., Stark, S. & Pacheco, M. (1993) G-protein-mediated activities of cannabinoid receptors in cerebellar membranes and cultured cerebellar granule cells. *Int. Cannabis Res. Soc. Abstract.*

Compton, D.R., Johnson, M.R., Melvin, L.S. & Martin, B.R. (1992). Pharmacological profile of a series of bicyclic cannabinoid analogs: classification as cannabimimetic agents. *J. Pharmacol. Exp. Therap.* **260**, 201–209.

Compton, D.R., Rice, K.C., de Costa, B.R., Razdan, R.K., Melvin, L.S., Johnson, M.R. & Martin, B.R. (1993). Cannabinoid structure-activity relationships: correlation of receptor binding and *in vivo* activities. *J. Pharmacol. Exp. Therap.* **265**, 218–226.

Devane, W.A., Spain, J.W., Coscia, C.J. & Howlett, A.C. (1986) An assessment of the role of opioid receptors in the response to cannabimimetic drugs. *J. Neurochem.* **46**, 1929–1935.

Devane, W.A., Dysarz, F.A., Johnson, M.R., Melvin, L.S. & Howlett, A.C. (1988) Determination and characterization of a cannabinoid receptor in rat brain. *Mol. Pharmacol.* **34**, 605–613.

Eidne, K.A., Zabavnik, J., Peters, T., Yoshida, S., Anderson, L. & Taylor, P.L. (1991) Cloning, sequencing and tissue distribution of a candidate G protein-coupled receptor from rat pituitary gland. *FEBS Lett.* **292**, 243–248.

Felder, C.C., Veluz, J.S., Williams, H.L., Briley, E.M. & Matsuda, L.A. (1992) Cannabinoid agonists stimulate both receptor- and non-receptor-mediated signal transduction pathways in cells transfected with and expressing cannabinoid receptor clones. *Mol. Pharmacol.* **42**, 838–845.

Felder, C.C., Briley, E.M., Axelrod, J., Simpson, J.T., Mackie, K. & Devane, W.A. (1993) Anandamide, an endogenous cannabimimetic eicosanoid, binds to the cloned human cannabinoid receptor and stimulates receptor-mediated signal transduction. *Proc. Natl Acad. Sci. USA* **90**, 7656–7660.

Gérard, C., Mollereau, C., Vassart, G. & Parmentier, M. (1990) Nucleotide sequence of a human cannabinoid receptor cDNA. *Nucl. Acids Res.* **18**, 7142.

Gérard, C.M., Mollereau, C., Vassart, G. & Parmentier, M. (1991) Molecular cloning of a human cannabinoid receptor which is also expressed in testis. *Biochem. J.* **279**, 129–134.

Herkenham, M., Lynn, A.B., Little, M.D., Johnson, M.R., Melvin, L.S., de Costa, B.R. & Rice, K.C. (1990) Cannabinoid receptor localization on brain. *Proc. Natl Acad. Sci. USA* **87**, 1932–1936.

Herkenham, M., Lynn, A.B., Johnson, M.R., Melvin, L.S., de Costa, B.R. & Rice, K.C. (1991a) Characterization and localization of cannabinoid receptors in rat brain: a quantitative *in vitro* autoradiographic study. *J. Neurosci.* **11**, 563–583.

Herkenham, M., Fox, R.J., Lynn, A.B., Lolait, S.J. & Matsuda, L.A. (1991b) Development of cannabinoid receptor binding and gene expression in the rat (Neurobehav. Teratol. Soc. Abst.) *Teratol.* **43**, 486.

Hla, T. & Maciag, T. (1990) An abundant transcript induced in differentiating human endothelial cells encodes a polypeptide with structural similarities to G-protein-coupled receptors. *J. Biol. Chem.* **265**, 9308–9313.

Hoehe, M.R., Caenazzo, L., Martinez, M.M., Hsieh, W.-T., Modi, W.S., Gershon, E.S. & Bonner, T.I. (1991) Genetic and physical mapping of the human cannabinoid receptor gene to chromosome 6q14–q15. *New Biologist* **3**, 880–885.

Hoehe, M.R., Goldin, L.R., Martinez, M.M., Caenazzo, L., Hsieh, W.-T., Gershon, E.S., Bonner, T.I. & Church, G.M. (1993) Molecular genetic studies of the human cannabinoid receptor gene. *Int. Cannabis Res. Soc. Abstract.*

Houamed, K.M., Kuijper, J.L., Gilbert, T.L., Haldeman, B.A., O'Hara, P.J., Mulvihill, E.R., Almers, W. & Hagen, F.S. (1991) Cloning, expression, and gene structure of a G protein-coupled glutamate receptor from rat brain. *Science* **252**: 1318–1321.

Howlett, A.C. (1987). Cannabinoid inhibition of adenylate cyclase relative activity of constituents and metabolites of marihuana. *Neuropharmacol.* **26**: 507–512.

Howlett, A.C., & Fleming, R.M. (1984) Cannabinoid inhibition of adenylate cyclase. Pharmacology of the response in neuroblastoma cell membranes. *Mol. Pharmacol.* **26**, 532–538.

Howlett, A.C., Qualy, J.M. & Khachatrian, L.L. (1986) Involvement of G_i in the inhibition of adenylate cyclase by cannabimimetic drugs. *Mol. Pharmacol.* **29**, 307–313.

Howlett, A.C., Johnson, M.R., Melvin, L.S. & Milne, G.M. (1988) Nonclassical cannabinoid analgetics inhibit adenylate cyclase: development of a cannabinoid receptor model. *Mol. Pharmacol.* **33**, 297–302.

Howlett, A.C., Bidaut-Russell, M., Devane, W.A., Melvin, L.S., Johnson, M.R. & Herkenham, M. (1990a) The cannabinoid receptor: biochemical, anatomical and behavioral characterization. *Trends Neurosci.* **13**, 420–423.

Howlett, A.C., Champion, T.M., Wilken, G.H. & Mechoulam, R. (1990b) Stereochemical effects of 11-OH-Δ^8-tetrahydrocannabinol-dimethylheptyl to inhibit adenylate cyclase and bind to the cannabinoid receptor. *Neuropharmacol.* **29**, 161–165.

Ishihara, T., Nakamura, S., Kaziro, Y., Takahashi, T., Takahashi, K. & Nagata, S. (1991) Molecular cloning and expression of a cDNA encoding the secretin receptor. *EMBO J.* **10**, 1635–1641.

Ishihara, T., Shigemoto, R., Mori, K., Takahashi, K. & Nagata, S. (1992) Functional expression and tissue distribution of a novel receptor for vasoactive intestinal polypeptide. *Neuron* **8**, 811–819.

Järbe, T.U.C., Hiltunen, A.J., Mathis, D.A., Hanuš, L., Breuer, A. & Mechoulam, R. (1993) Discriminative stimuls effects and receptor binding of enantiomeric pairs of cannabinoids in rats and pigeons: a comparison. *J. Pharmacol. Exp. Ther.* **264**, 561–569.

Jelinek, L.J., Lok, S., Rosenberg, G.B., Smith, R.A., Grant, F.J., Biggs, S., Bensch, P.A., Kuijper, J.L., Sheppard, P.O., Sprecher, C.A., O'Hara, P.J., Foster, D., Walker, K.M., Chen, L.H.J., McKernan, P.A. & Kindsvogel, W. (1993) Expression cloning and signaling properties of the rat glucagon receptor. *Science* **259**, 1614–1616.

Jüppner, H., Abou-Samra, A.-B., Freeman, M., Kong, X.F., Schipani, E., Richards, J., Kolakowski, L.F., Hock, J., Potts, J.T., Kronenberg, H.M. & Segre, G.V. (1991) A G protein-linked receptor for parathyroid hormone and parathyroid hormone-related peptide. *Science* **254**, 1024–1026.

Kaminski, N.E., Abood, M.E., Kessler, F.K., Martin, B.R. & Schatz, A.R. (1992) Identification of a functionally relevant cannabinoid receptor on mouse spleen cells that is involved in cannabinoid-mediated immune modulation. *Mol. Pharmacol.* **42**, 736–742.

Kaufman, R.J. (1990) Vectors used for expression in mammalian cells. In *Gene Expression Technology, Methods in Enzymology* Vol. 185 (ed. Goeddel, D.V.). pp. 487–511. Academic Press, San Diego, CA.

Lin, H.Y., Harris, T.L., Flannery, M.S., Aruffo, A., Kaji, E.H., Gorn, A., Kolakowski, L.F., Lodish, H.F. & Goldring, S.R. (1991) Expression cloning of an adenylate cyclase-coupled calcitonin receptor. *Science* **254**, 1022–1024.

Lu, R., Hubbard, J.R., Martin, B.R. & Kalimi, M.Y. (1993) Roles of sulfhydryl and disulfide groups in the binding of CP-55,940 to rat brain cannabinoid receptor. *Mol. Cell. Biochem.* **121**, 119–126.

Luckow, V.A. & Summers, M.D. (1988) Trends in the development of baculovirus expression vectors. *Bio/Technol.* **6**, 47–55.

McLaughlin, C.R., & Abood, M.E. (1993) Developmental expression of cannabinoid receptor mRNA. *Develop. Brain Research* **76**, 75–78.

Mailleux, P. & Vanderhaeghen, J.-J. (1992a) Distribution of neuronal cannabinoid receptor in the adult rat brain: a comparative receptor binding radioautography and *in situ* hybridization histochemistry. *Neurosci.* **48**, 655–668.

Mailleux, P. & Vanderhaeghen, J.-J. (1992b) Age-related loss of cannabinoid receptor binding sites and mRNA in the rat striatum. *Neurosci. Lett.* **147**, 179–181.

Mailleux, P. & Vanderhaeghen, J.-J. (1993) Glucocorticoid regulation of cannabinoid receptor messenger RNA levels in the rat caudate-putamen. An *in situ* hybridization study. *Neuroscience Lett.* **156**, 51–53.

Mailleux, P., Parmentier, M. & Vanderhaeghen, J.-J. (1992) Distribution of cannabinoid receptor messenger RNA in the human brain: an *in situ* hybridization histochemistry with oligonucleotides. *Neuroscience Lett.* **143**, 200–204.

Masu, Y., Nakayama, K., Tamaki, H., Harada, Y., Kuno, M. & Nakanishi, S. (1987) cDNA cloning of bovine substance-K receptor through oocyte expression system. *Nature* **329**, 836–838.

Masu, M., Tanabe, Y., Tsuchida, K., Shigemoto, R. & Nakanishi, S. (1991) Sequence and expression of a metabotropic glutamate receptor. *Nature* **349**, 760–765.

Matsuda, L.A., Lolait, S.J., Brownstein, M.J., Young, A.C. & Bonner, T.I. (1990) Structure of a cannabinoid receptor and functional expression of the cloned cDNA. *Nature* **346**, 561–564.

Matsuda, L.A., Bonner, T.I. & Lolait, S.J. (1993a) Localization of cannabinoid receptor mRNA in rat brain. *J. Comp. Neurol.* **327**, 535–550.

Matsuda, L.A., Lolait, S.J., Brownstein, M.J. & Bonner, T.I. (1993b) The THC receptor and its implications. In *Biological Basis of Substance Abuse* (eds. Korenman, S.G. & Barchas, J.D.) pp. 95–106, Oxford University Press, New York.

Mountjoy, K.G., Robbins, L.S., Mortrud, M.T. & Cone, R.D. (1992) The cloning of a family of genes that encode the melanocortin receptors. *Science* **257**, 1248–1251.

Munro, S., Thomas, K.L. & Abu-Shaar, M. (1993) Molecular characterization of a peripheral receptor for cannabinoids. *Nature* **365**, 61–65.

Nakajima, Y., Iwakabe, H., Akazawa, C., Nawa, H., Shigemoto, R., Mizuno, N. & Nakanishi, S. (1993) Molecular characterization of a novel retinal metabotropic glutamate receptor mGluR6 with a high agonist selectivity for L-2-amino-4-phosphonobutyrate. *J. Biol. Chem.* **268**, 11868–11873.

Pacheco, M., Childers, S.R., Arnold, R., Casiano, F. & Ward, S.J. (1991) Aminoalkylindoles: actions on specific G-protein-linked receptors. *J. Pharmacol. Exper. Therap.* **257**, 170–183.

Pacheco, M., Ward, S.J. & Childers, S.R. (1994) Differential requirements of sodium for coupling of cannabinoid receptors to adenylyl cyclase in rat brain membranes. *J. Neurochem.* **62**, 1773–1782.

Probst, W.C., Snyder, L.A., Schuster, D.I., Brosius, J. & Sealfon, S.C. (1992) Sequence alignment of the G-protein coupled receptor superfamily. *DNA Cell Biol.* **11**, 1–20.

Rinaldi-Carmona, M., Barth, F., Héaulme, M., Shire, D., Calandra, B., Congy, C., Martinez,

S., Maruani, J., Néliat, G., Caput, D., Ferrara, P., Soubrié, P., Brelière, J.-C. & Le Fur, G. (1994) SR141716A, a potent and selective antagonist of the brain cannabinoid receptor. *FEBS Letts.* **350**, 240–244.

Ross, P.C., Figler, R.A., Corjay, M.H., Barber, C.M., Adam, N., Harcus, D.R. & Lynch, K.R. (1990) RTA, a candidate G protein-coupled receptor: cloning, sequencing, and tissue distribution. *Proc. Natl Acad. Sci. USA* **87**, 3052–3056.

Savarese, T.M. & Fraser, C.M. (1992) *In vitro* mutagenesis and the search for structure-function relationships among G protein-coupled receptors. *Biochem. J.* **283**, 1–19.

Schibler, U. & Sierra, F. (1987) Alternative promoters in developmental gene expression. *Annu Rev. Gen.* **21**, 237–257.

Silva, A.J., Paylor, R., Wehner, J.M. & Tonegawa, S. (1992a) Impaired spatial learning in α-calcium-calmodulin kinase II mutant mice. *Science* **257**, 206–211.

Silva, A.J., Stevens, C.F., Tonegawa, S. & Wang, Y. (1992b) Deficient hippocampal long-term potentiation in α-calcium-calmodulin kinase II mutant mice. *Science* **257**, 201–206.

Tanabe, Y., Masu, M., Ishii, T., Shigemoto, R. & Nakanishi, S. (1992) A family of metabotropic glutamate receptors. *Neuron* **8**, 169–179.

Thorens, B. (1992) Expression cloning of the pancreatic β cell receptor for the gluco-incretin hormone glucagon-like peptide 1. *Proc. Natl Acad. Sci. USA* **89**, 8641–8645.

Vogel, Z., Barg, J., Levy, R., Saya, D., Heldman, E. & Mechoulam, R. (1993) Anandamide, a brain endogenous compound, interacts specifically with cannabinoid receptors and inhibits adenylate cyclase. *J. Neurochem.* **61**, 352–355.

Wahlestedt, C., Merlo Pich, E., Koob, G.F., Yee, F. & Heilig, M. (1993) Modulation of anxiety and neuropeptide Y–Y1 receptors by antisense oligodeoxynucleotides. *Science* **259**, 528–531.

Watson, S. & Girdlestone, D. (1994) 1994 Receptor nomenclature supplement. *Trends Pharmacol. Sci.*

Young, D., Waitches, G., Birchmeier, C., Fasano, O. & Wigler, M. (1986) Isolation and characterization of a new cellular oncogene encoding a protein with multiple potential transmembrane domains. *Cell* **45**, 711–719.

CHAPTER 5

LOCALIZATION OF CANNABINOID RECEPTORS IN BRAIN AND PERIPHERY

Miles Herkenham

Section on Functional Neuroanatomy, Building 36, Room 2D15, NIMH, Bethesda, MD 20892, USA

Table of Contents

5.1 Introduction

Marijuana (*Cannabis sativa*) is one of the oldest and most widely used drugs in the world (Harris *et al.*, 1977; Abel, 1980; Mechoulam, 1986). The main psychoactive constituent of the marijuana plant is Δ^9-tetrahydrocannabinol (Δ^9-THC) (Mechoulam *et al.*, 1970). Δ^9-THC, Δ^8-THC, their active metabolites and synthetic cannabinoids produce a unique spectrum of CNS-mediated behavioural, physiological and cognitive effects (Dewey, 1986; Hollister, 1986). However, until recently very little was known about the cellular mechanisms through which cannabinoids act.

Early attempts to characterize the cannabinoid receptor using [^{3}H]Δ^9-THC as a ligand (Harris *et al.*, 1978; Roth and Williams, 1979) were unsuccessful because cannabinoids are extremely hydrophobic (causing them to adhere to glassware and

Cannabinoid Receptors
ISBN 0-12-551460-3

filters) and because Δ^9-THC binds to its receptor with relatively low affinity and its tritiated form has low specific activity. An attempt to circumvent these problems with a more soluble form of cannabinoid ([^{3}H]trimethylammonium Δ^8-THC) led to the identification of a high-affinity binding site (Nye *et al.*, 1985) that did not meet the requirements of a cannabinoid receptor (proper structure–activity profile and enantioselectivity of binding) and that later was identified as a myelin-basic protein (Nye *et al.*, 1988).

In the late 1970s, Johnson and Melvin (1986), working at Pfizer, Inc., synthesized a large series of novel cannabinoids for studies of their potential use as analgesics. The Pfizer cannabinoids shared physicochemical properties with the natural cannabinoids and produced many behavioural and physiological effects characteristic of Δ^9-THC, but were 5–1000 times more potent. Each compound was synthesized in pure (–) and (+) forms which showed high enantioselectivity. The availability of the non-classical compounds to researchers revolutionized the study of the biochemical basis of cannabinoid activity and allowed the 'discovery' of the cannabinoid receptor. One of the compounds, (–)-CP 55,940, was tritiated and used by Howlett's group to identify and fully characterize a unique cannabinoid receptor in membranes from rat brain (Devane *et al.*, 1988). The results from the centrifugation assay showed that [^{3}H]CP 55,940 receptor binding is saturable, has high affinity and enantioselectivity and exhibits expected characteristics for a neuromodulator receptor associated with a guanine nucleotide regulatory (G) protein.

We used [^{3}H]CP 55,940 to develop an *in vitro* receptor binding assay using fresh-frozen cryostat-cut slide-mounted brain tissue (Herkenham *et al.*, 1990). The aforementioned technical problems associated with cannabinoid ligands were overcome by the use of this potent highly radiolabelled drug (79 Ci/mmol) in an incubation solution that contained 5% bovine serum albumin to act as a carrier to keep the hydrophobic ligand in solution. Assay conditions were optimized to attain the highest ratio of total to non-specific binding (performed in the presence of 10 μM (–)-CP 55,244, the most potent cannabinoid in the Pfizer series), yielding 90% specific binding; these conditions were used to characterize binding parameters, G-protein coupling, and drug displacement profiles prior to localizing the binding by autoradiography.

The results of association, dissociation, and competition studies showed that section binding was saturable, high-affinity (15 nM), enantioselective and reversible. Profound inhibition of binding by GTP (100% inhibition by 100 μM of the non-hydrolysable GTP analogue, guanylylimidodiphosphate) indicated that the receptor is G-protein linked. The structure–activity profile (rank order of displacement potencies of a series of cannabinoids with known *in vivo* potencies) suggested that the receptor defined by the binding of [^{3}H]CP 55,940 in brain sections was the same as the receptor characterized in the brain-membrane centrifugation assay (Howlett *et al.*, 1990) and was the same receptor that mediates all of the well-characterized behavioural and pharmacological effects of cannabinoids (Abood and Martin, 1992), including the subjective experience termed the human 'high' (Herkenham *et al.*, 1990). All other tested psychoactive drugs, neurotransmitters, steroids, and eicosanoids at 10 μM concentrations failed to bind to this receptor. Anandamide, an endogenous ligand for the cannabinoid receptor, bound with low affinity (K_i=5 μM) even in the presence of 1.0 mM phenylmethylsulfonyl fluoride, an inhibitor of its degradation (A.B. Lynn and M. Herkenham, unpublished data), confirming other studies showing its low affinity at the receptor (Devane *et al.*, 1992). There was no evidence for receptor subtypes from the analysis (Herkenham *et al.*, 1990).

5.2 Localization of cannabinoid binding sites

5.2.1 Localization in brain

The same assay conditions that had been used to validate the binding in brain sections were used for autoradiographic localization of the binding. Sections were incubated, washed, dried, and apposed to ^{3}H-sensitive film for several weeks before developing and viewing. Computer-assisted densitometric analysis of the film autoradiographs showed that in all species examined (rat, guinea-pig, dog, monkey, and human) very dense binding is discretely localized to the globus pallidus, substantia nigra pars reticulata, and the molecular layers of the cerebellum and hippocampal dentate gyrus (Figure 1). Dense binding also occurs in the cerebral cortex, other parts of the hippocampal formation, and the striatum. Moderate levels of binding are found in the hypothalamus, basal amygdala, central grey substance, nucleus of the solitary tract, and laminae I–III and X of the spinal cord. Very sparse and homogeneous binding characterizes the thalamus and most of the brainstem, including all of the monoamine-containing cell groups, reticular formation, primary sensory, visceromotor and cranial motor nuclei, and the area postrema. Interestingly, binding in the hypothalamus, amygdala, and cingulate cortex in human brain (Figure 1a) was relatively greater than in the same areas of rat and monkey tissues (Herkenham *et al.*, 1990), indicating a species difference that could underlie some of the different physiological and behavioural responses to cannabinoids in humans compared with animals.

Quantitative autoradiography confirms the very high numbers of cannabinoid receptors, exceeding 6 pmol/mg protein in densely labelled areas (Herkenham *et al.*, 1991c; Thomas *et al.*, 1992). Whole-brain cannabinoid receptor density is similar to whole-brain densities of receptors for amino acid transmitters such as glutamate (Greenamyre *et al.*, 1984) and GABA (Bowery *et al.*, 1987). When comparing cannabinoid receptors with regionally dense receptor systems, it is found that cannabinoid receptor levels are similar to levels of cortical benzodiazepine (Zezula *et al.*, 1988) and striatal dopamine receptors (Boyson *et al.*, 1986; Richfield *et al.*, 1986). Cannabinoid receptor densities greatly exceed densities for neuropeptide receptors.

Cannabinoid receptors have been mapped using two other selective high-affinity ligands. One ligand is structurally similar to 11-OH-Δ^9-THC, a naturally occurring bioactive metabolite of Δ^9-THC. Compared with the properties of [^{3}H]CP 55,940, this ligand ([^{3}H]11-OH-Δ^9-THC-1,1-dimethylheptyl) shows very similar affinity (K_d=29 nM in sections), capacity (B_{max} per structure ranging from <1 to 13 pmol/mg protein), and distribution pattern in rat brain (Thomas *et al.*, 1992). Another study using a cannabinoid ligand with unique chemistry (an aminoalkylindole named WIN 55,212-2) and similar affinity (K_d=24 nM) also shows similar patterns of distribution, although the study was not an exhaustive one (Jansen *et al.*, 1992). The similarity of patterns produced by structurally distinct cannabinoids provides further evidence that the current set of known cannabinoid ligands all recognize the same receptor and, as a corollary, supports the conclusion that there are no cannabinoid receptor subtypes yet identified in brain.

The overall CNS distribution of cannabinoid receptors is a unique and heterogeneous pattern that does not resemble other receptor maps. Whereas there are some receptor systems that have a propensity for dense forebrain and striatal localization (e.g. monoamine receptors and benzodiazepine receptors), there are none that show a similar overall paucity of receptors throughout the brainstem, medulla, and spinal cord. Interestingly, the cannabinoid receptor distribution does resemble autoradiographic distributions of second messengers mapped by tritiated forskolin (Worley *et al.*, 1986a) or phorbol esters (Worley *et al.*, 1986b), which mark adenylate cyclase/cyclic AMP and protein kinase C/phosphoinositide systems, respectively. These mapping similarities, the very high abundance of the cannabinoid receptor and the profound inhibition of binding by guanine nucleotides suggest that the cannabinoid receptor is closely associated with second messenger systems, although physiological evidence indicates an association only with the cyclic AMP system, specifically via interaction with G_i, the inhibitory guanine nucleotide regulatory protein that links receptors to adenylate cyclase (Howlett *et al.*, 1990; Felder *et al.*, 1992).

Another unique feature of cannabinoid receptors is the conserved pattern of distribution among the species examined. This apparent phylogenetic conservation is in contrast to most other receptor distributions, which show large and seemingly capricious species variations (e.g., McLean *et al.* 1987), and suggests a role for cannabinoid receptors that is basic and conserved in evolution. Perhaps this role is related to the aforementioned close association with the G proteins.

Finally, the cannabinoid receptor distribution pattern serves to 'explain' most of the known actions of cannabinoids (behavioural, physiological, endocrinological). Prior to the identification of the cannabinoid receptor and its localization to specific brain areas, workers had concluded, on the basis of available evidence, that the well-characterized actions of cannabinoids were indeed receptor mediated and, moreover, mediated by receptors in the central nervous system (Martin *et al.*, 1981; Martin, 1986). These predictions were fully borne out, as detailed below.

5.2.1.1 Motor actions

Psychoactive cannabinoids typically inhibit movement in animals in a dose-dependent fashion (Dewey, 1986). At low doses, they decrease spontaneous activity, and at higher doses they elicit catalepsy in mice (Little *et al.*, 1988). Cataleptic mice are conscious and remain for long periods of time in fixed postures with normal muscle tone (Pertwee and Greentree, 1988). In dogs these substances elicit a characteristic static ataxia (Dewey, 1986). Cannabinoid receptors in the basal ganglia and the cerebellum are likely sites for mediation of cannabinoid actions affecting movement. Studies of neuronal localization of receptors using either selective neurotoxins applied to basal ganglia pathways (Herkenham *et al.*, 1991b) or mutant strains of mice (Herkenham *et al.*, 1991a) have revealed that dense binding in the basal ganglia is localized to efferent striatal projections to the pallidum and nigra, whereas dense binding in the cerebellum is localized to granule cell projections to the molecular layer. Thus, in the former location, the neurons containing cannabinoid receptors are GABAergic, and in the latter they are glutamatergic, and in both locations they are 'presynaptically' localized, that is, localized to axons and terminals of the receptor-bearing neurons.

Basal ganglia involvement in the motor effects of cannabinoids is suggested by several research findings. Δ^9-THC injected directly into the caudate nucleus produces catalepsy in rats, although Δ^9-THC injected into the globus pallidus does not (Gough and Olley, 1978). However, Δ^9-THC acts synergistically with GABA agonists injected into the globus pallidus to enhance catalepsy (Pertwee and Wickens, 1991; Wickens and Pertwee, 1993). Δ^9-THC injected into the nucleus accumbens increases spontaneous activity at low doses and has no effect at high doses (Conti and Musty, 1984). Neurochemical blockade of the striatal dopamine system, by reserpine or haloperidol administration, potentiates Δ^9-THC-induced catalepsy (Moss *et al.*, 1984). Other evidence suggests that the motor effects of cannabinoids involve GABAergic systems. Cannabinoids and GABA agonists act synergistically to produce catalepsy in mice and rats (Pertwee and Greentree, 1988; Pertwee and Wickens, 1991; Wickens and Pertwee, 1993). The presence of dense cannabinoid receptors on GABAergic striatal outflow neurons suggests that cannabinoids might modulate GABA-mediated transmission. Cannabinoids do not affect [^{3}H]GABA binding (Leader *et al.*, 1981) nor do they affect GABA turnover in the substantia nigra (Leader *et al.*, 1981; Revuelta *et al.*, 1982). However, cannabinoids stereospecifically attenuate enhanced dopamine accumulation in the rat striatum caused by haloperidol (Leader *et al.*, 1981) as do GABA agonists (Gale and Guidotti, 1976). Together, the results implicate the various components of the basal ganglia in the mediation of cannabinoid effects, but the mechanisms are by no means clear.

Receptors in the cerebellum are also candidates for sites mediating cannabinoid effects on movement. Classic studies of Δ^9-THC effects have been performed in dogs, which show a static ataxia (Dewey, 1986) that resembles human movement disorders caused by cerebellar lesions. Cannabinoid effects on dog ataxia are dose-dependent and show a good structure–activity relationship with receptor binding

(Wilson *et al.*, 1979; Martin *et al.*, 1981; Herkenham *et al.*, 1990). We have previously suggested cerebellar mediation of the ataxia elicited by cannabinoids in dogs on the basis of elevated receptor density in the cerebellum in this species (beagle) relative to other species, notably human (Herkenham *et al.*, 1990).

The localization of cannabinoid receptors in the basal ganglia suggests therapeutic applications. Cannabinoids exacerbate hypokinesia in Parkinson's disease but are beneficial for some forms of dystonia, tremor, and spasticity (Marsden, 1981; Petro and Ellenberger, 1981; Clifford, 1983; Dewey, 1986; Hollister, 1986; Meinck *et al.*, 1989).

5.2.1.2 Reward mechanisms

The marijuana 'high' is described in terms that include 'pleasant,' 'relaxed,' 'dizzy,' 'flight of ideas,' 'hungry,' 'introspective' and 'numbness in body' (Hollister and Gillespie, 1973). Subjects have difficulty in concentrating, and dream-like states are prominent (Hollister, 1986). Levonantradol (CP 50,556) taken orally produces drowsiness, sedation, thought disturbances, and dizziness, with dysphoria more likely than euphoria (Johnson and Melvin, 1986). Dysphoria has been reported also with Δ^9-THC administered orally in clinical settings (Sweet *et al.*, 1981). Thus, clinical data do not support a strong association of cannabinoids with reward mechanisms, a finding consistent with experimental data in animals. Reinforcing properties of cannabinoids have been difficult to demonstrate in studies of self-administration or alterations of self-stimulation thresholds (Harris *et al.*, 1974; Leite and Carlini, 1974; Stark and Dews, 1980; Gardner *et al.*, 1988), and cannabinoids have even been noted to be somewhat aversive (Elsmore and Fletcher, 1972). The abuse potential associated with marijuana smoking nevertheless suggests a reward component.

Mesostriatal dopamine circuits are thought to underlie the rewarding actions of other abused drugs such as cocaine, which blocks uptake sites on dopaminergic nerve terminals to elevate levels of extracellular dopamine (Kornetsky, 1985; Di Chiara and Imperato, 1988). If cannabinoids similarly act on the dopamine system, then it should be possible to show the presence of cannabinoid receptors on dopaminergic nerve terminals in the striatum. To test this, we made selective neurotoxic lesions of the ascending dopamine system by injecting 6-hydroxydopamine (6-OHDA) into the medial forebrain bundle (Herkenham *et al.*, 1991b). The lesions had no effect on levels of cannabinoid receptors in the striatum, indicating that cannabinoid receptors in the basal ganglia are not localized on dopamine neurons.

This result supports other evidence indicating the lack of relationship between cannabinoids and the mesostriatal reward system. Early microdialysis work suggesting that cannabinoids indirectly (through a naloxone-reversible mechanism) elevate extrasynaptic dopamine levels in the Lewis strain of rats (Ng Cheong Ton *et al.*, 1988; Chen *et al.*, 1990) has not been supported by more recent work in freely moving Long–Evans rats given various doses of Δ^9-THC (Castaneda *et al.*, 1991).

5.2.1.3 Cognitive effects

Accounts of cannabis use in humans stress the loosening of associations, fragmentation of thought, and confusion on attempting to remember recent occurrences (Miller, 1984; Hollister, 1986). The most consistent effect of Δ^9-THC on performance is disruption of selective aspects of short-term memory tasks, similar to that found in monkeys and patients with damage to limbic cortical areas (Miller and Branconnier, 1983; Miller, 1984; Aigner, 1988). These cognitive effects may be mediated by receptors in the cerebral neocortex and hippocampus. The hippocampus is pivotally involved with the coding of sensory information (Eichenbaum and Cohen, 1988) and storage of memories (Thompson *et al.*, 1983). In animal studies, Δ^9-THC but not cannabidiol (an inactive cannabinoid) disrupts performance in tasks that require vigilance or estimating passage of time (Schulze *et al.*, 1988; Deadwyler *et al.*, 1990). Δ^9-THC produces deficits in delayed matching-to-sample tasks, as do hippocampal lesions (Deadwyler *et al.*, 1990). The disruption of hippocampal function by cannabinoids (Campbell *et al.*, 1986a,b) may be mediated by actions at receptors localized diffusely throughout the hippocampus but especially densely in the dentate gyrus (see also Chapter 7).

5.2.1.4 Endocrine actions

Cannabinoid effects on endocrine function do not show appropriate structure–activity relationships, suggesting that they are not mediated by the cannabinoid receptor (Pertwee, 1985). Cannabinoid receptors are sparsely and rather homogeneously distributed in the hypothalamus, with lowest densities in the magnocellular nuclei. In the pituitary, very sparse receptors are confined to the neural lobe (Herkenham *et al.*, 1991c). Δ^9-THC elevates plasma corticosterone and ACTH levels in rats (Pertwee, 1985), and this effect disappears when the hypothalamus is deafferented but otherwise left responsive and intact with the pituitary (Puder *et al.*, 1982). A possible extrahypothalamic site of action is the hippocampus, where Δ^9-THC has been shown to inhibit corticosterone uptake (Drew and Slagel, 1973) and compete with moderate affinity (IC_{50}= 250 nM) *in vitro* for [^{3}H]dexamethasone binding to glucocorticoid receptors (Eldridge and Landfield, 1990). A hippocampal glucocorticoid receptor-mediated effect is a plausible explanation for Δ^9-THC effects on hypothalamic–pituitary–adrenal (HPA) axis activity because the hippocampus is thought to relay glucocorticoid negative feedback to the hypothalamus (Sapolsky *et al.*, 1984). Alternatively, the activation of the HPA axis could be secondary to the stress and discomfort produced by Δ^9-THC injections in animals (see also Chapter 6).

5.2.1.5 Autonomic actions

Cannabinoid receptors are rather sparsely distributed throughout the brainstem and spinal cord. However, systemically administered cannabinoids elicit certain autonomic responses and, to the extent that they are receptor mediated, some of the regional variations in density may be relevant. In this regard it should be also noted that absolute levels of cannabinoid receptors are high in the brainstem because of the tremendous abundance of cannabinoid receptors in general.

There are virtually no reports of fatal cannabis overdose in humans (Harris *et al.*, 1977; Hollister, 1986). The safety is likely due to the paucity of receptors in medullary nuclei that mediate vital respiratory (Millhorn and Eldridge, 1986) and cardiovascular functions (Kalia, 1981). Drugs known to depress respiratory and cardiovascular functions (e.g. opiates and benzodiazepines) have high binding levels in the medulla (Atweh and Kuhar, 1977; Zezula *et al.*, 1988).

Cannabinoids produce hypothermia in mice in a dose-dependent, enantioselective fashion and with an appropriate structure–activity relationship (Little *et al.*, 1988). Δ^9-THC reduces body temperature following intraventricular injections or injection directly into the preoptic area (Fitton and Pertwee, 1982; Martin, 1986), suggesting mediation by cannabinoid receptors in this area. Apparently, the sparse and rather homogeneously distributed receptors are localized to key neuronal elements subserving temperature regulation.

Δ^9-THC causes tachycardia in humans and bradycardia in animals (Dewey, 1986); it also causes a fall in blood pressure in both. Cannabinoids may modulate noradrenergic sympathetic activity (Vollmer *et al.*, 1974). Such control could occur in the intermediolateral cell column of the thoracic spinal cord or in rostral centres with afferent connections. Nuclei in this circuitry typically have few receptors. One candidate for control is the caudal part of the nucleus of the solitary tract, which has moderate binding and sends axons to the intermediolateral cell column (Loewy and Burton, 1978). Receptors at more rostral levels of the neuraxis might contribute to the heart rate effects of Δ^9-THC (Vollmer *et al.*, 1974).

The most well-known therapeutic application of cannabinoids is to block emesis (Hollister, 1986), but this action is not easily explained by receptor distributions. The emetic reflex centres for chemical stimuli – the area postrema, parvocellular medullary reticular formation (PRF), and visceromotor nuclei (Borison *et al.*, 1981) – are all low in receptors. Possibly receptors in the nucleus of the solitary tract, which abuts the area postrema, may mediate the effects. Alternatively, the anti-emetic actions of dopamine antagonists may suggest much more remote sites of action, perhaps in the basal ganglia (Borison *et al.*, 1981). Forebrain afferents to the PRF include the substantia nigra pars reticulata and the central nucleus of the amygdala (Mehler, 1983), both elevated in cannabinoid receptors.

Another therapeutic use of cannabinoids is the treatment of glaucoma (Hollister, 1986). Δ^9-THC is known to reduce intra-ocular pressure, and the effect in experimental animals requires intact noradrenergic sympathetic fibres derived from the superior cervical ganglion, with probably a central as well as ganglionic site of action (Green *et al.*, 1977; Green and Kim, 1976). This response may be receptor-mediated because it occurs at relatively low doses i.v. and does not occur when relatively inactive cannabinoids – cannabinol and cannabidiol – are administered (Liu and Dacus, 1987).

The analgesic effects of cannabinoids are, in part at least, mediated by actions

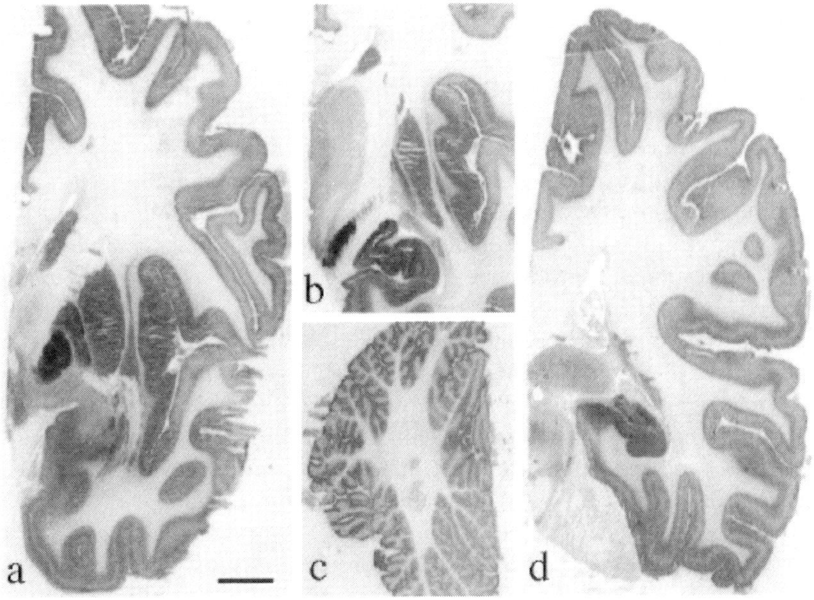

Figure 1 Autoradiography of 10 nM [^{3}H]CP 55,940 binding in coronally cut hemispheres of the human brain. The section in (a) shows dense binding throughout the cerebral cortex, caudate, putamen and globus pallidus, with higher levels in the internal segment than the external segment. There are moderate levels in the amygdala and hypothalamus, and low levels in the thalamus. The white matter is devoid of specific binding. The level in (b) shows putamen, tail of the caudate, substantia nigra, thalamus and posterior hypothalamus. The cerebellum is shown in (c). The hippocampus and brainstem are shown in (d). Binding in the hippocampus is elevated in the dentate gyrus molecular layer. Binding in the pulvinar, tegmentum and pons is low, with variations so that the central grey has somewhat elevated and the pons somewhat reduced relative density. Bar measures 1 cm.

at the level of the spinal cord (Gilbert, 1981; Yaksh, 1981). Several cannabinoid drugs have been shown to inhibit spinal reflexive responses to noxious stimuli given to dogs (Gilbert, 1981) and mice (Smith and Martin, 1992) with spinal transections following intrathecal administration, but data are insufficient to determine a structure–activity profile. In one study, a biodistribution experiment was conducted to show that the intrathecally administered Δ^9-THC did not reach the brain in spinally transected mice (Smith and Martin, 1992), indicating a direct spinal site of action. Cannabinoid receptors in the spinal cord are sparse but relatively elevated in the dorsal horn where primary afferents carrying noxious inputs terminate. The periaqueductal grey is an additional way-station in the pain circuit which has low levels of cannabinoid receptors that are nevertheless elevated over levels in adjacent territories (Figure 1d).

5.2.2. Localization in peripheral tissues

Whereas most of the known physiological effects of cannabinoids have been attributed to actions initiated by binding to CNS receptors, a large body of scientific and historical literature exists suggesting that cannabinoids can also act directly on peripheral target organs, either specifically or non-specifically, possibly through membrane actions (Martin, 1986; see also Chapters 3 and 6). Such 'side-effects' of cannabinoids include cardiovascular alterations, pulmonary effects, immunosupression, reproductive dysfunction, and endocrine modulation (reviewed in Lynn and Herkenham, 1994). To gain insights into whether these actions may be receptor mediated and, if so, whether they are mediated by CNS or peripheral receptors, binding was examined in fresh-frozen rat tissue sections taken from organs or tissues of the immune system, endocrine glands, liver, heart, lung, digestive system, excretory system, female and male reproductive systems, muscle, eye, and secretory structures. Specific (saturable, showing a similar ligand displacement profile as brain) cannabinoid receptor binding was found to be restricted to components of the immune system: the marginal zone of the spleen, cortex of the lymph nodes, and nodular corona of Peyer's patches (Figure 2). All of these sites are B lymphocyte-enriched areas. Specific binding was absent in T lymphocyte-enriched areas, such as the thymus and periarteriolar lymphatic sheaths of the spleen. Although areas showing specific binding also contain specialized forms of macrophages, certain macrophage-enriched areas, i.e. liver and lung, lacked specific binding. Thus, we proposed that the single peripheral cell type that may contain cannabinoid receptors is the B lymphocyte (Lynn and Herkenham, 1994).

No specific binding was found in any of the other peripheral tissues examined, although there were considerable variations in the amount of the non-specific (non-displaceable) binding across the tissues assayed. Thus, it was proposed that other reported effects of cannabinoids on peripheral physiology and function in rats were either mediated by CNS actions or were due to non-specific high-dose effects (Lynn and Herkenham, 1994).

Inhibition of specific binding by GTP analogues in the immune tissues confirmed that cannabinoid receptors are G-protein coupled in peripheral tissues, presumably to an inhibitory subunit. Inhibition of cellular function within the immune system includes *in vitro* reduction of macromolecular synthesis and mitogen-induced lymphocyte proliferation (Klein *et al.*, 1985; Pross *et al.*, 1990), suppression of macrophage burst characteristics in marijuana smokers (Sherman *et al.*, 1991), and suppression of other macrophage extrinsic and intrinsic functional parameters *in vitro* (phagocytosis, cell spreading, susceptibility to infection, and conferring of antiviral activity upon co-cultured cells) (Lopez-Cepero *et al.*, 1986; Barbers *et al.*, 1991; Specter *et al.*, 1991). It is not known which, if any, of these effects is receptor-mediated.

Evidence is emerging for cannabinoid effects on B-cells: *in vitro* lipopolysaccharide-induced B-cell proliferation of mouse splenocytes by Δ^9-THC which was more profound than suppression of T-cell proliferation (Klein and Friedman, 1986).

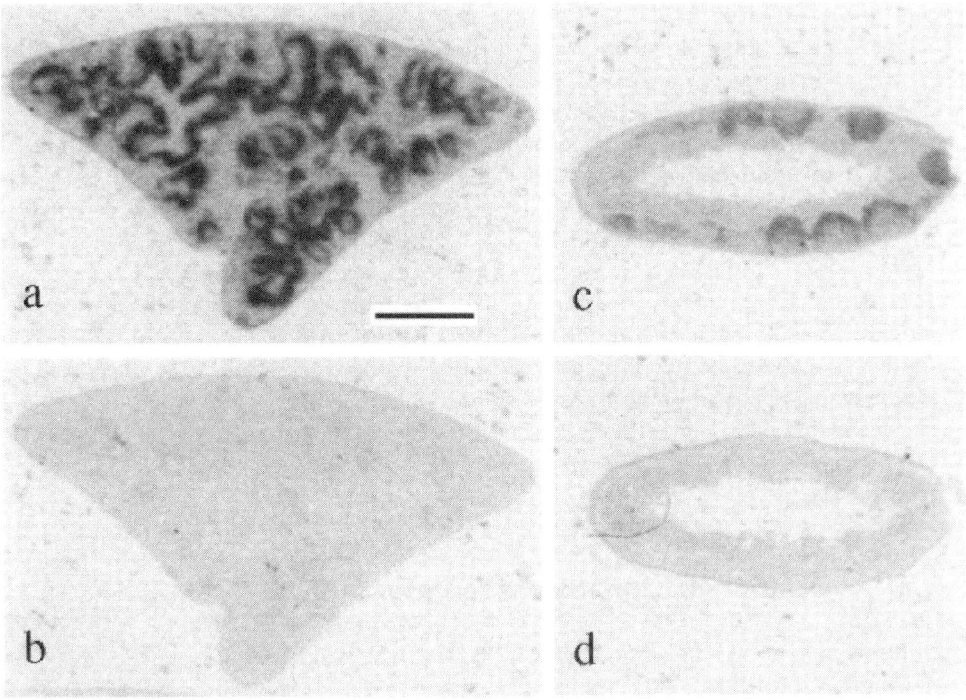

Figure 2 Autoradiography of 10 nM[^{3}H]CP 55,940 binding in peripheral tissues in the rat. Total (a) and non-specific binding (b) in the spleen shows specific binding in the marginal zone. Total (c) and non-specific binding (d) in the intestine shows that specific binding is found only in Peyer's patches, densest in the corona of each nodule and moderate in the internodular areas and germinal centres. Both spleen and Peyer's patches are immune tissues. Bar measures 0.2 mm.

Recently, Kaminski *et al.* (1992) showed localization of cannabinoid receptors in spleen using a membrane-binding assay, and although unable to identify the cell type responsible for binding, they showed profound inhibition of antibody production (a predominantly B-cell task) in splenocytes by CP 55,940.

If cannabinoids do modulate B-cell migration, proliferation, plasma cell morphogenesis, and immunoglobulin production, such effects may have profound implications for the humoral immune response. Whereas suppression of B-cell-mediated function could be detrimental (although there is little evidence that it is *in vivo*), cannabinoid suppression could have a positive therapeutic role in autoimmune disorders. Lyman *et al.* (1989) showed that Δ^9-THC is an effective prophylactic treatment for autoimmune encephalomyelitis (an animal model for multiple sclerosis) in Lewis rats and strain-13 guinea-pigs, and they proposed a mechanism whereby lymphocyte migration and sequestration in the CNS are inhibited by the drug. Only subtle immune suppression has been noted in humans (Nahas *et al.*, 1974; Gross *et al.*, 1991; Sherman *et al.*, 1991), suggesting a species difference in peripheral cannabinoid receptor distributions and densities.

Species differences may also explain the absence of specific receptor binding in two areas of rat that, on the basis of functional assays and structure-activity studies, would appear to have receptors in other species. In excised guinea-pig ileum, the potencies of cannabinoids to inhibit ileal contractions (Rosell and Agurell, 1975; Nye *et al.*, 1985) correlate well with potencies in structure–activity analyses of psychopotency, dog static-ataxia, adenylate cyclase inhibition, mouse analgesia, and brain receptor binding (as reviewed by Herkenham *et al.*, 1990). Similarly, in mouse vas deferens, inhibition of the electrically induced twitch response is a reliable assay for cannabinoid potencies (Devane *et al.*, 1992; Pertwee, 1993). However, there was no detectable receptor binding in rat ileum or vas deferens in our assay. Sections of guinea-pig ileum and mouse vas deferens should therefore be examined for specific cannabinoid receptor binding. The presence of such binding in guinea-pig myenteric plexus has already been reported (Paterson and Pertwee, 1993).

5.3 Localization of cannabinoid receptor mRNA

Matsuda *et al.* (1990) used the adenylate cyclase assay and structure–activity relationships to show that a cloned G protein-coupled receptor is a cannabinoid receptor in transfected cells. Using Northern blot analysis, the mRNA for this gene was found to be expressed in brain but not in the peripheral tissues examined (heart, liver, kidney, spleen, thymus, intestine, testes, and ovary). Subsequently, the human gene was cloned and found to be abundantly expressed in brain and marginally expressed in testes (Gérard *et al.*, 1991).

The similarity of rat brain distribution of the radiolabelled mRNA probe for the cloned receptor with the presently described cannabinoid receptor distribution (Mailleux and Vanderhaeghen, 1992; Matsuda *et al.*, 1993) suggests that they are the same. These similarities extend to the human brain (Figure 3a,b). Noteworthy similarities are the high levels of mRNA expression and binding in the cortex and the neostriatum and low levels in the diencephalon and lower brainstem (Figures 3 and 4). The cortex shows overlapping patterns of high mRNA expression levels and receptor binding densities in superficial (layers I and II) and deep lamina (layers IV, V, or VI, depending on the cortical area and species). The neostriatum (caudate, putamen, and nucleus accumbens) shows a medial-to-lateral gradient of increasing hybridization level that closely parallels the binding gradient. This gradient is more pronounced in the rat than it is in the human (Figure 3).

In both species, the cortical mRNA expression has a characteristically striking feature – that of extremely high levels in rare scattered neurons against a background of low-to-moderate levels in numerous cells. This feature is especially striking in the hippocampus (Figure 4a), where neurons lying outside the domain of the pyramidal and granular layers are those that express at levels several orders of

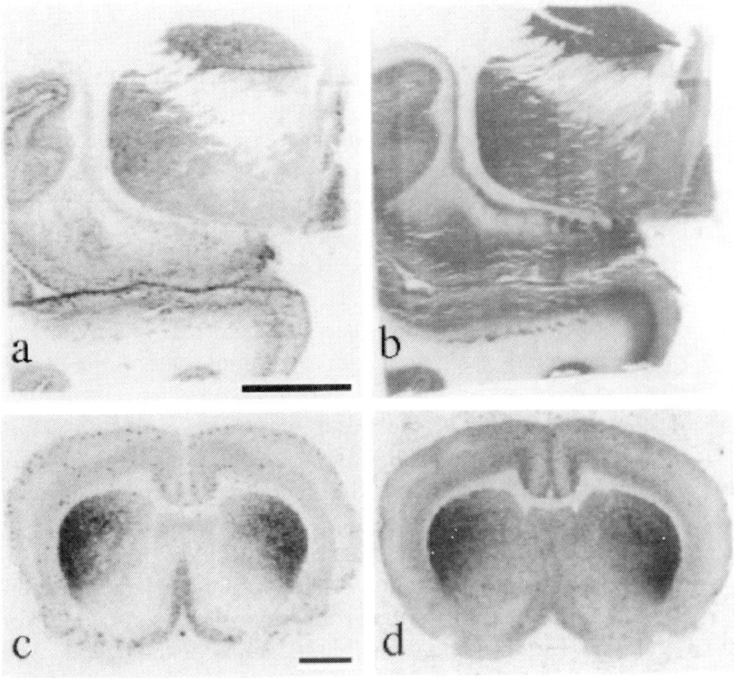

Figure 3 Autoradiography of hSKR6 ribonucleotide probe hybridization (a), rSKR6 oligonucleotide probe hybridization (c) and 10 nM [^{3}H]CP 55,940 binding (b and d) in human (a and b) and rat (c and d) brain at the level of the rostral striatum containing the caudate, putamen, and nucleus accumbens. Bar measures 1 cm in (a) and 0.2 mm in (c).

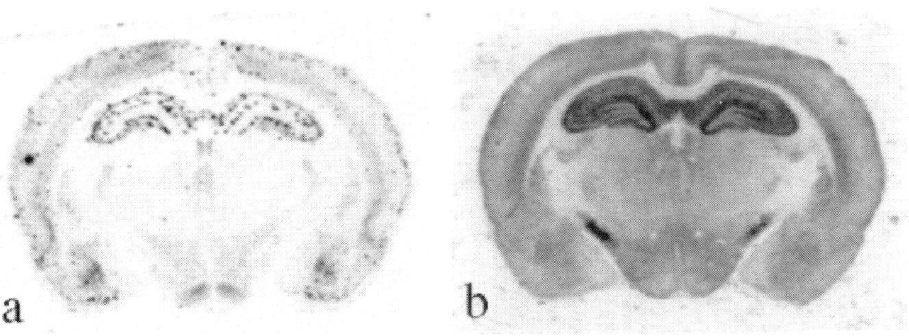

Figure 4 Autoradiography of rSKR6 oligonucleotide probe hybridization (a) and 10 nM [^{3}H]CP 55,940 binding (b) in coronal sections of rat brain at the level of the ventromedial hypothalamic nucleus. See text for details of the distribution.

magnitude higher than the more common cells. It is not known what percentage of receptor binding is to protein made from the abundant mRNA in the highly expressing cells, and an answer to this question would help solve the mystery of the source of dense binding in the dentate gyrus molecular layer. This binding is an enigma because it does not reside on dendrites of granule cells, which have no receptor mRNA expression, and it does not appear to reside on afferent axons because all known inputs have discrete laminar patterns of termination that do not correlate with the pattern of homogenous binding throughout the molecular layer. For example receptor mRNA-expressing neurons in the medial septum and diagonal band (Matsuda *et al.*, 1993) might send receptor-bearing axons into the hippocampus, but the expected termination pattern would be one in which resulting 'presynaptic' receptors would reside mainly in the granule cells of the dentate hilus. Axonally transported receptors arising from mRNA-expressing cells in the entorhinal area would terminate only in the outer one-third of the molecular layer. Commissural connections from the contralateral CA fields would largely avoid the dentate molecular layer, and those from the dentate gyrus would terminate only along the inner margin. A remaining possible source, therefore, is the processes (dendrites and/or local axons) of the intensely mRNA-expressing local circuit neurons of the hippocampus. Identification of the morphology and transmitter phenotype of these cells would aid in the determination of the source and functional role of cannabinoid receptor binding in the hippocampal dentate gyrus. An obvious candidate type would be the local GABA neurons of the hippocampal formation, but the processes of these do not appear to have the requisite patterns of distribution, because the GABA fibre plexus densely surrounds the granule cells themselves and only sparsely innervates the middle portions of the molecular layer (Anderson *et al.*, 1986). Perhaps specialized GABA neurons, or interneurons with other morphologies and transmitters, are the actual sources of the cannabinoid receptor binding in the molecular layer.

Cannabinoid receptor mRNA expression is high in a number of brain areas not enriched in receptor binding, and vice versa. Several examples of each kind of discrepancy are shown in Figure 4. In the first instance, mRNA levels are high and receptor binding is relatively low in the dentate hilus, medial habenula, and the ventromedial hypothalamic nucleus (VMH). In the second instance, mRNA levels are low and receptor binding is high in the dentate gyrus molecular layer, lateral habenula, and entopeduncular nucleus (Figure 4). These discrepancies are predicted by the knowledge that a large number of cannabinoid receptor molecules reside on axons and at nerve terminals distant from the cell bodies that contain the genetic machinery for receptor transcription. Thus, there is no hybridization signal in the globus pallidus, entopeduncular nucleus and substantia nigra, where dense receptor binding is localized to fibres and terminals of projection neurons residing in the striatum (Herkenham *et al.*, 1991b). In the olfactory system, an intense mRNA signal resides in the anterior olfactory nuclei (Mailleux and Vanderhaeghen, 1992; Matsuda *et al.*, 1993), whereas receptor binding is dense in the efferent pathway of the olfactory nuclei (anterior limb of the anterior commissure) and the major target zone (granular and glomerular layers of the main olfactory bulb) – sites where receptor mRNA is not found in abundance (discussed in Herkenham *et al.*, 1991c). The elevated mRNA level in the medial habenula probably results in translated receptor protein in the interpeduncular nucleus, the main target of medial habenular efferents.

The hypothalamic ventromedial nucleus (VMH) contains selectively high mRNA expression and low receptor binding (Figure 4). Axons of VMH neurons distribute to adjacent hypothalamic territory (Saper *et al.*, 1976) where receptor binding is homogeneous and moderately dense. It would be tempting to speculate that the high expression level in the VMH is associated with cannabinoid-induced decreased food intake in animals and increased appetite in humans (Dewey, 1986), but drug effects are mediated through membrane-bound effector-coupled receptors and, therefore, are better understood by knowledge of locations and densities of receptor binding than of mRNA expression.

Wherever cannabinoid receptors are 'presynaptic,' which appears to be a common arrangement, cannabinoids may act through receptor-linked inhibition of cyclic AMP production to affect transmitter release at these sites. In the striatal efferent pathways, the main transmitter is GABA, which is co-localized with enkephalin, dynorphin, and substance P; in the cerebellar granule cell population projecting to the cannabinoid receptor-rich cerebellar molecular layer, the transmitter is glutamate; in the medial habenula it is acetylcholine; in the cerebral cortex it may be glutamate (although the intensely expressing neurons may be interneurons, perhaps GABAergic and peptidergic). Thus, 'presynaptic' cannabinoid actions may be on diverse neurochemical and functional systems (see also Chapter 7).

Recently, a second cannabinoid receptor gene has been cloned from a human promyelocytic leukaemic cell line (Munro *et al.*, 1993). This gene has a 44% homology with the human cannabinoid receptor gene. The rat homologue, isolated by PCR, is expressed in rat spleen but not in rat brain, liver, nasal epithelium, thymus, lung, or kidney. Within the spleen, *in situ* hyridization histochemistry has shown that peripheral cannabinoid receptor mRNA is confined to the same zone that contains specific receptor binding, namely the marginal zone. Cell sorting by antibodies has shown that mRNA expression could be detected by PCR amplification in a macrophage/monocyte population but not in a T lymphocyte population used as a control. Expression of peripheral cannabinoid receptor mRNA in other immune tissues that show specific receptor binding (lymph nodes and Peyer's patches) has not yet been studied, nor has expression in a B lymphocyte cell population. Thus, the suggestion from the binding studies that the B cell is a cannabinoid receptor-bearing cell type remains viable, although not proven.

5.4 Conclusions

Cannabinoids administered to both animals and humans elicit characteristic motor, cognitive, analgesic, and autonomic effects which occur at low doses, suggesting that they are mediated by specific receptors in the central nervous system. Structure–activity relationships generated experimentally show that many of these effects are mediated by the presently identified cannabinoid receptor. Anatomical localization provides likely brain sites of action through which the cannabinoids act. The alterations in motor and cognitive functions elicited by cannabinoids can be understood if one considers the wealth of receptors in the basal ganglia, cerebellum, hippocampus, and cerebral cortex. Therapeutic effects of marijuana smoking on dystonias and hyperkinetic movement disorders may be mediated by receptors in motor areas. Analgesic and autonomic effects of cannabinoids, thought to be promising therapeutically, do not have clear anatomical bases insofar as receptors are relatively quite sparse in the brainstem and spinal areas that mediate the effects. Immunosuppressive effects of cannabinoids may be mediated by receptors localized to macrophages and possibly also B lymphocytes in the spleen and other immune organs.

An understanding of the overall organization of the cannabinoid system is beginning to emerge in the light of the recent explosion of information about cannabinoid binding characteristics, receptor properties, locations, and molecular structure, and features of the endogenous ligand(s). We now know that the cannabinoid receptor is quite ordinary in some ways – it is a typical member of the superfamily of G-protein-coupled receptors containing seven transmembrane-spanning domains, and its ligands bind reversibly and stereoselectively with high affinity. In other ways, however, it is very unusual and in fact unique. First and foremost, it is simple. It is simple because: (1) there are no subtypes in brain (at least to date); (2) it is coupled only to the adenylate cyclase/cyclic AMP second messenger system and not to the phosphoinositide system or to ion channels (effects on ion channels appear to be G-protein mediated, see Chapter 7); (3) its presence is limited to neurons and certain types of immune cells; (4) it has a basic and conserved distribution pattern in brain, and (5) it is simply and profoundly regulated by its ligands (Oviedo *et al.*, 1993). In this sense, therefore, it is the 'aplysia' of receptors and could be used as a model system, especially for G-protein coupling studies. Additional unique (and paradoxical) features include its abundance in brain but apparently minor physiological role. Perhaps related to this are the prevalence of cannabinoid receptors in axons and terminals and the lipophilia of all bioactive ligands, including the endogenous ligand anandamide. Extreme lipophilia may be required for the ligand to interact with its hydrophobic binding domain deeply embedded in one of the transmembrane-spanning sequences of the receptor molecule (Makriyannis and Rapaka, 1990; see also Chapter 3). Endogenous ligands able to interact with such a site are likely to be those that ordinarily reside in membranes. Thus, the cannabinoid receptor may be the first plasmalemma-bound receptor identified that is not involved in intercellular communication. Appreciation of the implications of this will require more work on the anandamide system, but the data suggest that the endogenous ligands are not released into the extracellular spaces but instead 'live' in the cell membranes and act as intracellular messengers that serve to influence local membrane events.

The intimate association with second messenger systems in axons and terminals may be the basis by which cannabinoids in the brain alter presynaptic actions at these locations.

References

Abel, E.L. (1980) *Marihuana, the First Twelve Thousand Years*. Plenum Press, New York.

Abood, M.E. & Martin, B.R. (1992) Neurobiology of marijuana abuse. *Trends Pharmacol. Sci.* **13**, 201–206.

Aigner, T.G. (1988). Delta-9-tetrahydrocannabinol impairs visual recognition memory but not discrimination learning in rhesus monkeys. *Psychopharmacology* **95**, 507–511.

Anderson, K.J., Maley, B.E. & Scheff, S.W. (1986) Immunocytochemical localization of γ-aminobutyric acid in the rat hippocampal formation. *Neurosci. Lett.* **69**, 7–12.

Atweh, S.F. & Kuhar, M.J. (1977) Autoradiographic localization of opiate receptors in rat brain. I. Spinal cord and lower medulla. *Brain Res.* **124**, 53–67.

Barbers, R.G., Evans, M.J., Gong, H.J., & Tashkin, D.P. (1991) Enhanced alveolar monocytic phagocyte (macrophage) proliferation in tobacco and marijuana smokers. *Am. Rev. Resp. Dis.* **143**, 1092–1095.

Borison, H.L., Borison, R., & McCarthy, L.E. (1981) Phylogenic and neurologic aspects of the vomiting process. *J. Clin. Pharmacol.* **21** (Suppl.), 23S–29S.

Bowery, N.G., Hudson, A.L., & Price, G.W. (1987) $GABA_A$ and $GABA_B$ receptor site distribution in the rat central nervous system. *Neuroscience* **20**, 365–383.

Boyson, S.J., McGonigle, P., & Molinoff, P.B. (1986) Quantitative autoradiographic localization of the D_1 and D_2 subtypes of dopamine receptors in rat brain. *J. Neurosci.* **6**, 3177–3188.

Campbell, K.A., Foster, T.C., Hampson, R.E. & Deadwyler, S.A. (1986a) Effects of Δ^9-tetrahydrocannabinol on sensory-evoked discharges of granule cells in the dentate gyrus of behaving rats. *J. Pharmacol. Exp. Therap.* **239**, 941–945.

Campbell, K.A., Foster, T.C., Hampson, R.E. & Deadwyler, S.A. (1986b) Δ^9-tetrahydrocannabinol differentially affects sensory-evoked potentials in the rat dentate gyrus. *J. Pharmacol. Exper. Therap.* **239**, 936–940.

Castaneda, E., Moss, D.E., Oddie, S.D. & Whishaw, I.Q. (1991) THC does not affect striatal dopamine release: microdialysis in freely moving rats. *Pharmacol. Biochem. Behav.* **40**, 587–591.

Chen, J., Paredes, W., Li, J., Smith, D., Lowinson, J. & Gardner, E.L. (1990) Δ^9-Tetrahydrocannabinol produces naloxone-blockable enhancement of presynaptic basal dopamine efflux in nucleus accumbens of conscious, freely-moving rats as measured by intracerebral microdialysis. *Psychopharmacol.* **102**, 156–162.

Clifford, D.B. (1983). Tetrahydrocannabinol for tremor in multiple sclerosis. *Ann. Neurol.* **13**, 669–671.

Conti, L.H. & Musty, R.E. (1984) The effects of delta-9-tetrahydrocannabinol injections to the nucleus accumbens on the locomotor activity of rats. In *The Cannabinoids: Chemical, Pharmacologic, and Therapeutic Aspects* (eds Agurell, S., Dewey, W.L. & Willette, R.E.), pp. 649–655. Academic Press, New York.

Deadwyler, S.A., Heyser, C.J., Michaelis, R.C. & Hampson, R.E. (1990) The effects of delta-9-THC on mechanisms of learning and memory. *NIDA Res. Monogr.* **97**, 79–95.

Devane, W.A., Dysarz, F.A., Johnson, M.R., Melvin, L.S. & Howlett, A.C. (1988) Determination and characterization of a cannabinoid receptor in rat brain. *Mol. Pharmacol.* **34**, 605–613.

Devane, W.A., Hanuš, L., Breuer, A., Pertwee, R.G., Stevenson, L.A., Griffin, G., Gibson, D., Mandelbaum, A., Etinger, A. & Mechoulam, R. (1992). Isolation and structure of a brain constituent that binds to the cannabinoid receptor. *Science* **258**, 1946–1949.

Dewey, W.L. (1986) Cannabinoid pharmacology. *Pharmacol. Rev.* **38**, 151–178.

Di Chiara, G. & Imperato, A. (1988) Drugs abused by humans preferentially increase synaptic dopamine concentrations in the mesolimbic system of freely moving rats. *Proc. Natl Acad. Sci. USA* **85**, 5274–5278.

Drew, W.G. & Slagel, D.E. (1973) Δ^9-THC: selective impairment of corticosterone uptake by limbic structures of the rat. *Neuropharmacol.* **12**, 909–914.

Eichenbaum, H. & Cohen, N.J. (1988) Representation in the hippocampus: what do hippocampal neurons code? *Trends Neurosci.* **11**, 244–248.

Eldridge, J.C. & Landfield, P.W. (1990). Cannabinoid interactions with glucocorticoid receptors in rat hippocampus. *Brain Res.* **534**, 135–141.

Elsmore, T.F. & Fletcher, G.V. (1972) Δ^9-Tetrahydrocannabinol: aversive effects in rat at high doses. *Science* **175**, 911–912.

Felder, C.C., Veluz, J.S., Williams, H.L., Briley, E.M. & Matsuda, L.A. (1992) Cannabinoid agonists stimulate both receptor- and non-receptor-mediated signal transduction pathways in cells transfected with and expressing cannabinoid receptor clones. *Mol. Pharmacol.* **42**, 838–845.

Fitton, A.G. & Pertwee, R.G. (1982) Changes in body temperature and oxygen consumption rate of conscious mice produced by intrahypothalamic and intracerebroventricular injections of Δ^9-tetrahydrocannabinol. *Br. J. Pharmacol.* **75**, 409–414.

Gale, K.N. & Guidotti, A. (1976) GABA-mediated control of rat neostriatal tyrosine hydroxylase revealed by intranigral muscimol. *Nature* **263**, 691–693.

Gardner, E.L., Paredes, W., Smith, D., Donner, A., Milling, C., Cohen, D. & Morrison, D. (1988) Facilitation of brain stimulation reward by Δ^9-tetrahydrocannabinol. *Psychopharmacology* **96**, 142–144.

Gérard, C.M., Mollereau, C., Vassart, G. & Parmentier, M. (1991) Molecular cloning of a human cannabinoid receptor which is also expressed in testis. *Biochem. J.* **279**, 129–134.

Gilbert, P.E. (1981) A comparison of THC, nantradol, nabilone, and morphine in the chronic spinal dog. *J. Clin. Pharmacol.* **21** (Suppl.), 311S–319S.

Gough, A.L. & Olley, J.E. (1978) Catalepsy induced by intrastriatal injections of Δ^9-THC and 11-OH-Δ^9-THC in the rat. *Neuropharmacology* **17**, 137–144.

Green K. & Kim, K. (1976) Mediation of ocular tetrahydrocannabinol effects by adrenergic nervous system. *Exp. Eye Res.* **23**, 443–448.

Green, K., Bigger, J.F., Kim, K. & Bowman, K. (1977) Cannabinoid action on the eye as mediated through the central nervous system and local adrenergic activity. *Exp. Eye Res.* **24**, 189–196.

Greenamyre, J.T., Young, A.B. & Penney, J.B. (1984) Quantitative autoradiographic distribution of L-[^{3}H]glutamate binding sites in rat central nervous system. *J. Neurosci.* **4**, 2133–2144.

Gross, G., Roussaki, A., Ikenberg, H. & Drees, N. (1991) Genital warts do not respond to systemic recombinant interferon alfa-2a treatment during cannabis consumption. *Dermatologica* **183**, 203–207.

Harris, L.S. Dewey, W.L. & Razdan, R.K. (1977) Cannabis. Its chemistry, pharmacology, and toxicology. In *Handbook of Experimental Pharmacology*, 45/II (ed. Martin, W.R.), pp. 371–429. Springer-Verlag, New York.

Harris, L.S., Carchman, R.A. & Martin, B.R. (1978) Evidence for the existence of specific cannabinoid binding sites. *Life Sci.* **22**, 1131–1138.

Harris, R.T., Waters, W. & McLendon, D. (1974) Evaluation of reinforcing capability of Δ^9-tetrahydrocannabinol in rhesus monkeys. *Psychopharmacologia* **37**, 23–29.

Herkenham, M., Lynn, A.B., Little, M.D., Johnson, M.R., Melvin, L.S., de Costa, B.R. &

Rice, K.C. (1990) Cannabinoid receptor localization in brain. *Proc. Natl. Acad. Sci. USA* **87**, 1932–1936.
Herkenham, M., Groen, B.G., Lynn, A.B., de Costa, B.R. & Richfield, E.K. (1991a) Neuronal localization of cannabinoid receptors and second messengers in mutant mouse cerebellum. *Brain Res.* **552**, 301–310.
Herkenham, M., Lynn, A.B., de Costa, B. & Richfield, E.K. (1991b) Neuronal localization of cannabinoid receptors in the basal ganglia of the rat. *Brain Res.* **547**, 267–274.
Herkenham, M., Lynn, A.B., Johnson, M.R., Melvin, L.S., de Costa, B.R. & Rice, K.C. (1991c) Characterization and localization of cannabinoid receptors in rat brain: a quantitative *in vitro* autoradiographic study. *J. Neurosci.* **11**, 563–583.
Hollister, L.E. (1986) Health aspects of cannabis. *Pharmacol. Rev.* **38**, 1–20.
Hollister, L.E. & Gillespie, H.K. (1973) Delta-8- and delta-9-tetrahydrocannabinol. Comparison in man by oral and intravenous administration. *Clin. Pharmacol. Ther.* **14**, 353–357.
Howlett, A.C., Bidaut-Russell, M., Devane, W.A., Melvin, L.S., Johnson, M.R. & Herkenham, M. (1990) The cannabinoid receptor: biochemical, anatomical and behavioral characterization. *Trends Neurosci.* **13**, 420–423.
Jansen, E.M., Haycock, D.A., Ward, S.J. & Seybold, V.S. (1992) Distribution of cannabinoid receptors in rat brain determined with aminoalkylindoles. *Brain Res.* **575**, 93–102.
Johnson, M.R. & Melvin, L.S. (1986) The discovery of nonclassical cannabinoid analgetics. In *Cannabinoids as Therapeutic Agents*, (ed. Mechoulam, R.), pp. 121–145. CRC Press, Boca Raton, FL.
Kalia, M.P. (1981) Anatomical organization of central respiratory neurons. *Annu. Rev. Physiol.* **43**, 105–120.
Kaminski, N.E., Abood, M.E., Kessler, F.K., Martin, B.R. & Schatz, A.R. (1992) Identification of a functionally relevant cannabinoid receptor on mouse spleen cells that is involved in cannabinoid-mediated immune modulation. *Mol. Pharmacol.* **42**, 736–742.
Klein, T.W. & Friedman, H. (1986) Modulation of murine immune cell function by marijuana components. In *Drugs of Abuse and Immune Function* (ed. Watson, R.R.), pp. 87–111. CRC Press, Boca Raton, FL.
Klein, T.W., Newton, C.A., Widen, R. & Friedman, H. (1985) The effect of delta-9-tetrahydrocannabinol and 11-hydroxy-delta-9-tetrahydrocannabinol on T-lymphocyte and B-lymphocyte mitogen responses. *J. Immunopharmacol.* **7**, 451–466.
Kornetsky, C. (1985) Brain-stimulation reward: a model for the neuronal bases for drug-induced euphoria. *NIDA Res. Monogr.* **62**, 30–50.
Leader, J.P., Koe, B.K. & Weissman, A. (1981) GABA-like actions of levonantradol. *J. Clin. Pharmacol.* **21** (Suppl.), 262S–270S.
Leite, J.R. & Carlini, E.A. (1974) Failure to obtain 'cannabis-directed behavior' and abstinence syndrome in rats chronically treated with Cannabis sativa extracts. *Psychopharmacologia* **36**, 133–145.
Little, P.J., Compton, D.R., Johnson, M.R. & Martin, B.R. (1988) Pharmacology and stereoselectivity of structurally novel cannabinoids in mice. *J. Pharmacol. Exp. Ther.* **247**, 1046–1051.
Liu, J.H. & Dacus, A.C. (1987) Central nervous system and peripheral mechanisms in ocular hypotensive effect of cannabinoids. *Arch. Opthalmol.* **105**, 245–248.
Loewy, A.D. & Burton, H. (1978) Nuclei of the solitary tract: efferent projections to the lower brain stem and spinal cord of the cat. *J. Comp. Neurol.* **181**, 421–449.
Lopez-Cepero, M., Freidman, M., Klein, T.W. & Friedman, H. (1986) Tetrahydrocannabinol-induced suppression of macrophage spreading and phagocytic activity *in vitro*. *J. Leukocyte Biol.* **39**, 679–686.
Lyman, W.D., Sonett, J.R., Brosnan, C.F., Elkin, R. & Bornstein, M.B. (1989) Delta 9-

tetrahydrocannabinol: a novel treatment for experimental autoimmune encephalomyelitis. *J. Neuroimmunol.* **23**, 73–81.

Lynn, A.B. & Herkenham, M. (1994) Localization of cannabinoid receptors and nonsaturable high-density cannabinoid binding sites in peripheral tissues of the rat: implications for receptor-mediated immune modulation by cannabinoids. *J. Pharmacol. Exp. Therap.* **268**, 1612–1623.

Mailleux, P. & Vanderhaeghen, J.-J. (1992) Distribution of neuronal cannabinoid receptor in the adult rat brain: a comparative receptor binding radioautography and *in situ* hybridization histochemistry. *Neuroscience* **48**, 655–668.

Makriyannis, A. & Rapaka, R.S. (1990) The molecular basis of cannabinoid activity. *Life Sci.* **47**, 2173–2184.

Marsden, C.D. (1981) Treatment of torsion dystonia. In *Disorders of Movement, Current Status of Modern Therapy, Vol. 8,* (ed. Barbeau, A.), pp. 81–104. Lippincott, Philadelphia, PA.

Martin, B.R. (1986) Cellular effects of cannabinoids. *Pharmacol. Rev.* **38**, 45–74.

Martin, B.R., Balster, R.L., Razdan, R.K., Harris, L.S. & Dewey, W.L. (1981) Behavioral comparisons of the stereoisomers of tetrahydrocannabinols. *Life Sci.* **29**, 565–574.

Matsuda, L.A., Bonner, T.I. & Lolait, S.J. (1993) Localization of cannabinoid receptor mRNA in rat brain. *J. Comp. Neurol.* **327**, 535–550.

Matsuda, L.A., Lolait, S.J., Brownstein, M.J., Young, A.C. & Bonner, T.I. (1990) Structure of a cannabinoid receptor and functional expression of the cloned cDNA. *Nature* **346**, 561–564.

McLean, S., Rothman, R.B., Jacobsen, A.E., Rice, K.C. & Herkenham, M. (1987) Distribution of opiate receptor subtypes and enkephalin and dynorphin immunoreactivity in the hippocampus of squirrel, guinea pig, rat and hamster. *J. Comp. Neurol.* **255**, 497–510.

Mechoulam, R. (1986) The pharmacohistory of *Cannabis sativa.* In *Cannabinoids as Therapeutic Agents* (ed. Mechoulam, R.), pp. 1–19. CRC Press, Boca Raton, FL.

Mechoulam, R., Shani, A., Edery, H. & Grunfeld, Y. (1970) Chemical basis of hashish activity. *Science* **169**, 611–612.

Mehler, W.R. (1983) Observations on the connectivity of the parvicellular reticular formation with respect to a vomiting center. *Brain Behav. Evol.* **23**, 63–80.

Meinck, H.-M., Schonle, P.W. & Conrad, B. (1989) Effect of cannabinoids on spasticity and ataxia in multiple sclerosis. *J. Neurol.* **236**, 120–122.

Miller, L.L. (1984) Marijuana: acute effects on human memory. In *The Cannabinoids: Chemical, Pharmacologic, and Therapeutic Aspects* (eds Agurell, S., Dewey, W.L. & Willette, R.E.), pp. 21–46. Academic Press, New York.

Miller, L.L. & Branconnier, R.J. (1983) Cannabis: effects on memory and the cholinergic limbic system. *Psych. Bull.* **93**, 441–456.

Millhorn, D.E. & Eldridge, F.L. (1986) Role of ventrolateral medulla in regulation of respiratory and cardiovascular systems. *J. Appl. Physiol.* **61**, 1249–1263.

Moss, D.E., Montgomery, S.P., Salo, A.A. & Steger, R.W. (1984) Tetrahydrocannabinol effects on extrapyramidal motor behaviors in an animal model of Parkinson's disease. In *The Cannabinoids: Chemical, Pharmacologic, and Therapeutic Aspects* (eds Agurell, S., Dewey, W.L. & Willette, R.E.), pp. 815–828. Academic Press, New York.

Munro, S., Thomas, K.L. & Abu-Shaar, M. (1993) Molecular characterization of a peripheral receptor for cannabinoids. *Nature* **365**, 61–65.

Nahas, G.G., Suciu-Foca, N., Armand, J.-P. & Morishima, A. (1974) Inhibition of cellular mediated immunity in marihuana smokers. *Science* **183**, 419–420.

Ng Cheong Ton, J.M., Gerhardt, G.A., Friedemann, M., Etgen, A.M., Rose, G.M., Sharpless, N.S. & Gardner, E.L. (1988) The effects of Δ^9-tetrahydrocannabinol on potassium-evoked release of dopamine in the rat caudate nucleus: an *in vivo* electrochemical and *in vivo* microdialysis study. *Brain Res.* **451**, 59–68.

Nye, J.S., Seltzman, H.H., Pitt, C.G. & Snyder, S.H. (1985) High-affinity cannabinoid binding sites in brain membranes labeled with [^{3}H]-5′-trimethylammonium Δ^8-tetrahydrocannabinol. *J. Pharmacol. Exp. Therap.* **234**, 784–791.

Nye, J.S., Voglmaier, S., Martenson, R.E. & Snyder, S.H. (1988) Myelin basic protein is an endogenous inhibitor of the high-affinity cannabinoid binding site in brain. *J. Neurochem.* **50**, 1170–1178.

Oviedo, A., Glowa, J. & Herkenham, M. (1993) Chronic cannabinoid administration alters cannabinoid receptor binding in rat brain: a quantitative autoradiographic study. *Brain Res.* **616**, 293–302.

Paterson, S.J. & Pertwee, R.G. (1993) Characterization of the cannabinoid binding site in the guinea-pig. *Br. J. Pharmacol.* **109**, 82P.

Pertwee, R.G. (1985) Cannabis. In *Psychopharmacology 2, Part 1: Preclinical Psychopharmacology* (eds Grahame-Smith, D.G. & Cowan, P.J.), pp. 364–391. Elsevier, Amsterdam.

Pertwee, R.G. (1993) The evidence for the existence of cannabinoid receptors. *Gen. Pharmacol.* **24**, 811–824.

Pertwee, R.G. & Greentree, S.G. (1988) Delta-9-tetrahydrocannabinol-induced catalepsy in mice is enhanced by pretreatment with flurazepam or chlordiazepoxide. *Neuropharmacology* **27**, 485–491.

Pertwee, R.G. & Wickens, A.P. (1991) Enhancement by chlordiazepoxide of catalepsy induced in rats by intravenous or intrapallidal injections of enantiomeric cannabinoids. *Neuropharmacology* **30**, 237–244.

Petro, D.J. & Ellenberger, C.E. (1981) Treatment of human spasticity with delta-9-tetrahydrocannabinol. *J. Clin. Pharmacol.* **21** (Suppl.), 413S–416S.

Pross, S.H., Klein, T.W., Newton, C.A., Smith, J., Widen, R. & Friedman, H. (1990) Differential suppression of T-cell subpopulations by THC (delta-9-tetrahydrocannabinol). *Int. J. Immunopharmacol.* **12**, 539–544.

Puder, M., Weidenfield, J., Chowers, I., Nir, I., Conforti, N. & Siegel, R.A. (1982) Corticotrophin and corticosterone secretion following Δ^1-tetrahydrocannabinol, in intact and in hypothalamic deafferented male rats. *Exp. Brain Res.* **46**, 85–88.

Revuelta, A.V., Cheney, D.L. & Costa, E. (1982) The dimethylheptyl derivative of (−)-Δ^8-tetrahydrocannabinol reduces the turnover rate of gamma-aminobutyric acid in the septum and nucleus accumbens. *Life Sci.* **30**, 1841–1846.

Richfield, E.K., Young, A.B. & Penney, J.B. (1986) Properties of D_2 dopamine receptor autoradiography: high percentage of high-affinity agonist sites and increased nucleotide sensitivity in tissue sections. *Brain Res.* **383**, 121–128.

Rosell, S. & Agurell, S. (1975) Effects of 7-hydroxy-Δ^6-tetrahydrocannabinol and some related cannabinoids on the guinea pig isolated ileum. *Acta Physiol. Scand.* **94**, 142–144.

Roth, S.H. & Williams, P.J. (1979) The non-specific membrane binding properties of Δ^9-tetrahydrocannabinol and the effects of various solubilizers. *J. Pharm. Pharmacol.* **31**, 224–230.

Saper, C.B., Swanson, L.W. & Cowan, W.M. (1976) The efferent connections of the ventromedial nucleus of the hypothalamus of the rat. *J. Comp. Neurol.* **169**, 409–442.

Sapolsky, R.M., Krey, L.C. & McEwen, B.S. (1984) Glucocorticoid-sensitive hippocampal neurons are involved in terminating the adrenocortical stress response. *Proc. Natl Acad. Sci. USA* **81**, 6174–6177.

Schulze, G.E., McMillan, D.E., Bailey, J.R., Scallet, A., Ali, S.F., Slikker, W.J. & Paule, M.G. (1988) Acute effects of Δ^9-tetrahydrocannabinol in rhesus monkeys as measured by performance in a battery of complex operant tests. *J. Pharmacol. Exp. Ther.* **245**, 178–186.

Sherman, M.P., Campbell, L.A., Gong, H.J., Roth, M.D. & Tashkin, D.P. (1991) Antimicrobial and respiratory burst characteristics of pulmonary alveolar macrophages recovered from smokers of marijuana alone, smokers of tobacco alone, smokers of marijuana and tobacco, and nonsmokers. *Amer. Rev. Respir. Dis.* **144**, 1351–1356.

Smith, P.B. & Martin, B.R. (1992) Spinal mechanisms of Δ^9-tetrahydrocannabinol-induced analgesia. *Brain Res.* **578**, 8–12.

Specter, S., Lancz, G. & Goodfellow, D. (1991) Suppression of human macrophage function in vitro by Δ^9-tetrahydrocannabinol. *J. Leuk. Biol.* **50**, 423–426.

Stark, P. & Dews, P.B. (1980) Cannabinoids. I. Behavioral effects. *J. Pharmacol. Exp. Ther.* **214**, 124–130.

Sweet, D.L., Miller, N.J., Weddington, W., Senay, E. & Sushelsky, L. (1981) Δ^9-tetrahydrocannabinol as an antiemetic for patients receiving cancer chemotherapy. A pilot study. *J. Clin. Pharmacol.* **21** (Suppl.), 70S–75S.

Thomas, B.F., Wei, X. & Martin, B.R. (1992) Characterization and autoradiographic localization of the cannabinoid binding site in rat brain using [^{3}H]11-OH-Δ^9-THC-DMH. *J. Pharmacol. Exper. Ther.* **263**, 1383–1390.

Thompson, R.F., Berger, T.W. & Madden, J. (1983) Cellular processes of learning and memory in the mammalian CNS. *Annu. Rev. Neurosci.* **6**, 447–491.

Vollmer, R.R., Cavero, I., Ertel, R.J., Soloman, T.A. & Buckley, J.P. (1974) Role of the central autonomic nervous system in the hypotension and bradycardia induced by (–)-Δ^9-*trans*-tetrahydrocannabinol. *J. Pharm. Pharmacol.* **26**, 186–192.

Westlake, T.M., Howlett, A.C., Bonner, T.I., Matsuda, L.A. & Herkenham, M. (1994) Cannabinoid receptor binding and messenger RNA expression in human brain: an *in vitro* receptor autoradiography and *in situ* hybridization histochemistry study of normal aged and Alzheimer's brains. *Neuroscience* **63**, 637–652.

Wickens, A.P. & Pertwee, R.G. (1993) Delta-9-tetrahydrocannabinol and anandamide enhance the ability of muscimol to induce catalepsy in the globus pallidus of rats. *Eur. J. Pharmacol.* **250**, 205–208.

Wilson, R.S., May, E.L. & Dewey, W.L. (1979) Some 9-hydroxycannabinoid-like compounds. Synthesis and evaluation of analgesic and behavioral properties. *J. Med. Chem.* **22**, 886–888.

Worley, P.F., Baraban, J.M., de Souza, E.B. & Snyder, S.H. (1986a) Mapping second messenger systems in the brain: differential localization of adenylate cyclase and protein kinase. *C. Proc. Natl Acad. Sci. USA* **83**, 4053–4057.

Worley, P.F., Baraban, J.M. & Snyder, S.H. (1986b) Heterogeneous localization of protein kinase C in rat brain: autoradiographic analysis of phorbol ester receptor binding. *J. Neurosci.* **6**, 199–207.

Yaksh, T.L. (1981) The antinociceptive effects of intrathecally administered levonantradol and desacetyllevonantradol in the rat. *J. Clin. Pharmacol.* **21** (Suppl.), 334S–340S.

Zezula, J., Cortés, R., Probst, A. & Palacios, J.M. (1988) Benzodiazepine receptor sites in the human brain: autoradiographic mapping. *Neuroscience* **25**, 771–795.

CHAPTER 6

CANNABINOID COMPOUNDS AND SIGNAL TRANSDUCTION MECHANISMS

Allyn C. Howlett

Department of Pharmacological and Physiological Science, Saint Louis University School of Medicine, St. Louis, MO 63104, USA

Table of Contents

Cannabinoid Receptors
ISBN 0-12-551460-3

6.1 Cellular mechanisms for cannabinoid compounds

6.1.1 CB_1 receptors

The cannabinoid receptor (CB_1) has been described as the membrane-associated receptor that transduces signals from CNS-active constituents of *Cannabis sativa*, as well as synthetic bicyclic and tricyclic cannabinoid analogues and aminoalkylindole cannabimimetic compounds. This receptor is coupled to G_I and is able to inhibit adenylate cyclase activity and decrease intracellular cyclic AMP content. The CB_1 receptor has been cloned, and the amino acid sequence is consistent with a tertiary structure resembling that of the seven-transmembrane spanning proteins that couple to G proteins (Gérard *et al.*, 1990, 1991; Matsuda *et al.*, 1990). Several recent reviews have described the pharmacology and biochemistry of this receptor and its distribution in the central nervous system (CNS) (Howlett, 1990; Howlett *et al.*, 1990b, 1992; Abood and Martin, 1992; Pertwee, 1993). The present review will describe and evaluate the evidence for the interaction of cannabimimetic compounds with various signal transduction pathways, either as a direct interaction with the CB_1 receptor, or by alternative nonreceptor-mediated mechanisms.

6.1.2 CB_2 receptors

Homology cloning techniques have revealed the existence of an alternative cannabinoid-binding seven-transmembrane spanning receptor (Munro *et al.*, 1993). The clone (hCX5.36) was derived from a human promyelocytic leukaemia cell HL60 cDNA library. The rat or human CX5 clones were able to hybridize with mRNA in undifferentiated HL60 cells and in HL60 cells that had been differentiated into granulocytes or macrophages. Probes also hybridized with a splenic macrophage/ monocyte preparation but not with splenic T cells. A signal was observed in Northern blots of RNA from spleen, but not from mature blood neutrophils, thymus, liver, brain, lung, or kidney. Thus, the CX5 pattern of distribution is distinctly different from that of the CB_1 receptor. Probes specific for CB_1 receptors have been used for reverse transcription – polymerase chain reaction (PCR) amplification, and a band has been observed by Northern analysis from splenocytes (Kaminski *et al.*, 1992) and various mononuclear cell types (Bouaboula *et al.*, 1993). Although these investigators interpreted their findings to mean that the CB_1 receptor was expressed in those cells, it is possible that the observed Northern bands, for which sequence information was not obtained, may correspond to a signal produced by the same gene as that reported by Munro *et al.* (1993).

The CX5 protein, proposed by Munro *et al.* (1993) to be called the CB_2 receptor, shows 68% identity to the CB_1 receptor within the helical regions, and 44% identity to the total protein. When expressed in COS cells, the protein binds [^{3}H]CP 55,940 and [^{3}H]WIN 55,212–2 with K_d values of 1.6 nM and 3.7 nM, respectively. The pharmacological distinction between the *hCX5.36* gene product binding profile and that of the CB_1 receptor is that cannabinol appears to bind with about the same affinity as Δ^9-tetrahydrocannabinol (Δ^9-THC) to the CX5 receptor protein. An extensive study of binding displacement profiles was not provided in this initial report, and the K_i values for the five compounds tested were calculated from displacement of either of two different radioligands (Munro *et al.*, 1993). Kaminski *et al.* (1992) reported some degree of enantioselectivity for the two enantiomeric pairs, CP 55,940 and CP 56,667 and the 1,1-dimethylheptyl homologues of (+)- and (−)-11-hydroxy-Δ^8-THC (HU-210 and HU-211, respectively). However, this was not as great a selectivity as expected for CB_1 receptor binding. Bouaboula *et al.* (1993) reported potency orders for Δ^9-THC, CP 55,940 and WIN 55,225 that were not what would have been expected for the CB_1 receptor. Thus, a thorough pharmacological description of this receptor awaits further investigation.

No studies that describe potential signal transduction pathways for the CX5 gene product were reported by Munro *et al.* (1993). Δ^9-THC at concentrations greater than 10 μM were able to inhibit forskolin-stimulated cyclic AMP production in spleen cells (Schatz *et al.*, 1992). This concentration of Δ^9-THC is greater than what would be expected to inhibit cyclic AMP production via CB_1 receptors in intact neuroblastoma cells (Devane *et al.*, 1986; Howlett *et al.*, 1986). However, the difference may represent the pharmacological distinction between the CB_1 and CB_2 subtypes. In interpreting these findings, it should not necessarily be expected that the CB_2 receptor would couple to the same G proteins or regulate the same signal transduction pathways as does the CB_1 receptor. The finding of an inhibition of cyclic AMP accumulation by Δ^9-THC might have been due to the presence of CB_1 receptors in the splenocyte cell population that was examined or may have been a non-receptor mediated event (Schatz *et al.*, 1992).

6.1.3 Trimethylammonium-Δ^8-THC cation binding sites

In an attempt to overcome the problems of poor aqueous solubility of Δ^9-THC, Nye *et al.* (1985) developed the charged quaternary ammonium derivative, trimethylammonium-Δ^8-THC (TMA-Δ^8-THC) as a ligand for labelling cannabinoid binding sites. [^{3}H]-TMA-Δ^8-THC bound reversibly with a K_d of about 90 nM to a component of membranes from rat brain and also to membranes from kidney, liver, testis, spleen, ileum, heart, lung and muscle (Nye *et al.*, 1985). It is not possible to define this particular binding site as a cannabinoid receptor because its pharmacological characterization, using over 40 constituents of *Cannabis* extract and metabolites, failed to correlate with cannabimimetic activity in animal models or the *in vitro* model of attenuation of guinea-pig ileum contractions (Nye *et al.*, 1985). When injected by a peripheral route, TMA-Δ^8-THC fails to behave as a cannabimimetic agonist in antinociception or ring-immobility tests, but does appear to produce hypothermia and hypoactivity in mice (Compton and Martin, 1990). However, this may be due to differences in the pharmacokinetic behaviour; it is unclear how much charged analogue passes across the blood–brain barrier to reach the areas of the brain that may be important for the analgesia or catalepsy. Thirty minutes after intravenous injection, the ratio of TMA-Δ^8-THC to Δ^8-THC in the brain (which had not been perfused with saline to remove plasma) was smaller than the ratio in the plasma (Compton and Martin, 1990). This might suggest that the charged compound remains in the plasma volume and does not efficiently cross the blood–brain endothelial barrier. Results of intracerebroventricular administration of Δ^8-THC or TMA-Δ^8-THC were not reported for antinociceptive or immobility tests (Compton and Martin, 1990).

Accumulated data suggest that the charged portion of the molecule may be dominant in defining the binding site rather than the cannabinoid nucleus. The evidence for this is that divalent cations, polycations, and myelin basic protein were all effective displacers of binding (Nye *et al.*, 1988, 1989). No signal transduction pathways were reported to be coupled to these binding sites. However, a description of effects of ions and nucleotides on the binding properties is not consistent with this binding site serving as a G protein-coupled receptor (Nye *et al.*, 1989).

6.1.4 Cannabinoid compounds and steroid receptors

Considerable evidence exists to link cannabinoid drugs to endocrine physiology and, in particular, to reproductive physiology (for review see Murphy *et al.*, 1990). Clearly, many effects might be attributable to cannabinoceptive neurons in brain pathways that ultimately regulate pituitary function. However, because both cannabinoid and steroid molecules are multi-ringed structures possessing hydroxyl groups, it is not surprising that there might exist some affinity of one class of compounds for the other class of receptors. When steroid ligands, including oestradiol, progesterone, pregnenolone sulfate, androsterone, cortisone, and corticosterone, were tested for their ability to displace [^{3}H]CP 55,940 or [^{3}H]WIN 55,212–2 from binding to CB_1 receptors in rat brain membranes, no interaction was noted at concentrations as high as 1 μM or 10 μM (Bidaut-Russell *et al.*, 1990; Howlett *et al.*, 1992; Kuster *et al.*, 1993). Thus, it is unlikely that steroid compounds present at their physiological free concentrations in biological fluids would participate in the regulation of cannabinoid receptors in the brain.

The interaction of cannabinoid compounds with oestrogen receptors has been studied and has provided conflicting results. Rawitch *et al.* (1977) performing heterologous competition assays for [^{3}H]oestradiol binding to oestrogen receptors in uterine cytosol from Fisher-344 rats, found that 5 μM to 50 μM Δ^9-THC or 11-OH-Δ^9-THC could displace specific binding to a maximum of about 22%. In that same study, the [^{3}H]oestradiol–receptor complex cosedimented on a sucrose density gradient with [^{14}C]Δ^9-THC-labelled peaks at 4S and 10.4S values. In contrast, Okey and Bondy (1977) failed to observe any displacement of [^{3}H]oestradiol-17β from 4S or 8S peaks of oestrogen receptors by either 10 μM Δ^9-THC or *Cannabis* resin at an amount that would contribute 10 μM Δ^9-THC, plus other cannabinoid compounds. Tests were performed using uterine cytosol fractions obtained from either sexually mature mice or 21-day old Wistar rats. As a control, 100 nM diethylstilbestrol completely displaced radioligand binding. To further address this issue, Sauer *et al.* (1983) examined the competition of either crude marijuana extract (containing approx. 20 μM Δ^9-THC), 6 μM cannabidiol, 70 μM Δ^9-THC or 20 μM concentrations of 10 hydroxylated metabolites of Δ^9-THC, for binding of [^{3}H]oestradiol to the estrogen receptor from rat uterine cytosol. The only competition observed was in response to cannabidiol or the crude extract of marijuana smoke. It was shown that the agent responsible in the extract was apigenin, a flavinoid phyto-oestrogen found in *Cannabis* leaves. In that same study (Sauer *et al.*, 1983), administration of the crude marijuana extract to immature rats behaved neither as an oestrogen to promote uterine development nor as an antagonist to the effects of oestradiol. Thus, the evidence in support of a physiologically important interaction of cannabinoid compounds with oestrogen receptors is, at best, weak.

The interaction between cannabinoid compounds and glucocorticoid receptors was initially investigated *in vivo* using the model of glucocorticoid-mediated induction of the hepatic enzyme tyrosine aminotransferase (Friedman and Wrenn, 1977; Wrenn and Friedman, 1978). Exceptionally high doses (100 to 200 mg/kg i.p.) of Δ^9-THC to mice attenuated enzyme induction in response to hydrocortisone but not to glucagon or to tryptophan in adrenalectomized mice. These findings led to testing if [^{3}H]Δ^8-THC would be able to bind to corticoid receptors in crude nuclear fractions from hepatoma tissue culture (HTC) cells (Harris *et al.*, 1978). Because only a small fraction of the bound [^{3}H]Δ^8-THC could be displaced by dexamethasone, it is not clear whether this represented binding to corticoid receptors.

Studies of the corticoid receptors in rat hippocampal cytosol have suggested that some affinity for cannabinoid compounds exists in the brain (Eldridge and Landfield, 1990; Eldridge *et al.*, 1991). Displacement of [^{3}H]dexamethasone by 250 nM Δ^9-THC or cannabidiol reached 30% and 10%, respectively, compared with displacement by 250 nM RU 28,362 of 80% (denoting type II receptors) and displacement by 250 nM dexamethasone of 100% (denoting both type I plus type II receptors). Displacement by RU 28,362 plus Δ^9-THC or cannabidiol was not additive, suggesting that the cannabinoid compounds displaced binding to some fraction of the type II corticoid receptors. Analysis of the labelled fraction by sucrose density gradient revealed that the 8S to 9S receptor peak was sensitive to cannabinoid compounds. Δ^9-THC and cannabidiol displaced [^{3}H]dexamethasone with an IC_{50} of about 250 nM. Only about 50% of the specific binding to total corticoid receptors could be displaced by Δ^9-THC, and less than 25% could be displaced by cannabidiol, even at concentrations that exceeded 10 μM.

The functional significance of the binding of cannabinoid compounds to glucocorticoid receptors has yet to be defined. Only inferences can be made from studies addressing whether Δ^9-THC behaves as an agonist or an antagonist at these receptors. Chronic administration of Δ^9-THC to rats mimics the effects of chronic stress or glucocorticoid administration to cause degenerative changes in hippocampal neurons. This finding suggests that Δ^9-THC may have behaved as an agonist (Eldridge *et al.*, 1991). Alternatively, the results could have been due to glucocorticoid release in response to the stress of Δ^9-THC administration. It is of note that Δ^9-THC is able to stimulate glucocorticoid release via stimulation of corticotrophin release (Kubena *et al.*, 1971; Pertwee, 1974; Miczek and Dixit, 1980; Puder *et al.*, 1982; Pertwee, 1985). A second argument supporting the hypothesis that Δ^9-THC behaves as a corticoid receptor agonist is that Δ^9-THC can prevent the upregulation of glucocorticoid receptors occurring after adrenalectomy (Eldridge *et al.*, 1991).

The true test of whether cannabinoid compounds exhibit agonist ability at brain corticoid receptors will be the direct demonstration of effects on transcription. The Δ^9-THC–corticoid receptor complex would be expected to regulate the glucocorticoid response element controlling gene expression if Δ^9-THC behaves as an agonist. Such a demonstration of agonist activity, combined with a demonstration that Δ^9-THC can mimic limbic system and neuroendocrine regulation by corticoids, would be substantial confirmation that cannabinoid compounds act via corticoid receptors in the brain.

6.1.5 Cannabinoid compounds and membrane interactions

Δ^9-THC is poorly soluble in physiological solutions at concentrations above 1 μM to 10 μM, depending upon temperature, ion strength and other conditions (Garrett and Hunt, 1974; Roth and Williams, 1979). It is therefore not surprising that this compound and some of its metabolites might partition to a great extent into physiological membranes. In studies of synaptosomal membranes in 10 mM phosphate buffer, the membrane/buffer partition coefficient was reported to be 12 500 when Δ^9-THC was present at 0.01 μM to 1 μM (Roth and Williams, 1979). Other investigators have reported membrane/buffer partition coefficients for synaptosomes or brain mitochondrial fractions to be 500 to 800 at Δ^9-THC concentrations of 1 μM to 10 μM (Seeman *et al.*, 1972; Colburn *et al.*, 1974). Roth and Williams (1979) calculated that the maximum binding to the membrane was about 17 mmol/kg dry membrane, whereas Seeman *et al.* (1972) calculated the value to be about 4 mmol/kg dry membrane. Both values fall below the Meyer–Overton rule for anaesthesia, which demands that the anaesthetizing membrane concentration be 30 mmol/kg. Thus, Δ^9-THC has been categorized as a partial rather than a full anaesthetic (Seeman *et al.*, 1972). This ‘non-specific’ binding within membranes rather than to high-affinity protein receptors cannot qualify as the stimulator of a signal transduction pathway. However, membrane perturbations can ultimately alter the behaviour of membrane proteins and thereby influence various signal transduction mechanisms in what is termed a non-specific (i.e. non-receptor-mediated) manner.

One manifestation of the membrane actions of cannabinoid drugs has been the apparent alteration of neuromodulator receptor–ligand binding parameters in *in vitro* assays. This was first noted by Hillard and Bloom (1982, 1984), who observed that Δ^9-THC or 11-OH-Δ^9-THC at concentrations of greater than 3 μM increased, and concentrations of greater than 30 μM decreased, the binding of [^{3}H]dihydroalprenolol to the β-adrenergic receptor in mouse cerebral cortex membranes. High μM concentrations of cannabinoid drugs also reduced binding of [^{3}H]spiperone or [^{3}H]ADTN to dopamine receptors in mouse striatal membranes (Bloom, 1984; Bloom and Hillard, 1985) and [^{3}H]naloxone binding to opioid receptors in mouse brain membranes (Bloom and Hillard, 1985). Effects of other ligands on radioligand binding were complex in these studies. Vaysse *et al.* (1987) showed that inhibition of the binding of [^{3}H]dihydromorphine and several other radioligands to rat brain μ- and δ- opioid receptors occurred at concentrations between 3 μM and 100 μM Δ^9-THC. A variety of other CNS-active or inactive cannabinoid compounds also interfered with binding in the high μM range. In that study, the concentrations of Δ^9-THC that effectively inhibited binding depended upon the concentration of membranes used in the assay, suggesting a drug interaction with the membranes.

To explain the effects of cannabinoid ligands on properties of ligand binding to various neuromodulator receptors, Hillard and Bloom proposed that the cannabinoid compounds perturbed membrane properties and thereby altered ligand binding to the integral membrane receptors (Hillard and Bloom, 1982, 1984; Bloom, 1984; Bloom and Hillard, 1985). However, an alternative explanation could be that above aqueous-soluble concentrations, the cannabinoid compounds may have complexed with the lipophilic radioligands and reduced the effective concentration available for interaction with the receptor. In support of the former, Hillard *et al.* (1985) performed a careful analysis of the effects of various cannabinoid compounds on synaptosomal plasma membrane lipid order using diphenylhexatriene as a fluorescence polarization probe. They observed that Δ^9-THC and 11-OH-Δ^9-THC at concentrations of about 3 μM increased the mobility of the probe, an indication of increased disorder within the membrane core. At concentrations of 30 μM, Δ^9-THC, cannabinol and cannabidiol, but not 11-OH-Δ^9-THC, increased polarization of the probe (i.e. decreased disorder of the terminal region of the acyl chains). Using several probes, differing effects were observed in model phospholipid membranes that were dependent upon the cannabinoid compound, the initial bilayer fluidity, and the nature of the headgroup of the phospholipids. This would mean that:

(1) The influence of cannabinoid drugs on integral membrane proteins could differ depending upon the type of cell and the composition of its plasma membrane (or other organelle membranes);
(2) The response could vary between gel and liquid crystal domains within the same biological membrane, and thus, the domain within which the pertinent membrane proteins reside is an important factor;
(3) The effects of cannabinoid drugs on membrane proteins could vary depending upon the interaction of the protein with the membrane (integral membrane positioning or association in a more superficial manner);
(4) A change in the functioning of the protein would have to depend upon the vulnerability of region(s) of the protein to conformational changes resulting from alterations in the specific regions of the membrane (e.g., along the acyl chains, or at the region of the phospholipid headgroups).

The ramifications for signal transduction mechanisms reside with the fact that receptors for many hormones and neuromodulators are transmembrane proteins, and transducer and effector proteins may also be associated with the membrane for at least part, if not all, of their activation cycle. Thus, the potential exists for modification of signal transduction mechanisms as the result of membrane perturbations. Whether findings observed *in vitro* can be readily extrapolated to cellular biology *in vivo* is a persistent question. One aspect of this question that is particularly pertinent to cannabinoid pharmacology is the comparison between drug concentrations required to observe the effects *in vitro* versus drug concentrations that may actually be present in the cellular environment *in vivo* (for review and discussion see Pertwee, 1988; see also Chapter 3).

A role has been proposed for the neuronal plasma membrane in orienting the active cannabinoid compounds for an orderly approach to cannabinoid receptors prior to ligand–receptor interaction (Makriyannis and Rapaka, 1990; Yang *et al.*, 1991). The model proposing that a permissible orientation within the membrane be attained for ligand–receptor interaction has not been directly tested for cannabinoid receptor interactions. The model is based upon determinations of the orientation of cannabinoid compounds within model membranes using small-angle X-ray diffraction and solid-state ^{2}H-NMR studies (Makriyannis *et al.*, 1989; Mavromoustakos *et al.*, 1990, 1991; see also Chapter 3). Those studies have demonstrated that Δ^9-THC and other centrally active cannabinoid compounds are positioned in the membrane such that the tricyclic nucleus is anchored in the region of the phospholipid head groups, and is stabilized in that orientation by the phenolic hydroxyl. The hydrophobic C3 side-chain can then be inserted parallel to the hydrocarbon chains of the fatty acids. Evidence against the achievement of a proper membrane orientation as a prerequisite for interaction with CB_1 receptors is that the detergent-solubilized receptor, which is entirely devoid of a normal membrane environment, is able to recognize cannabinoid compounds with the same affinities and order of potency as does the receptor within the membrane (Houston and Howlett, 1993).

6.2 Signal transduction via the CB_1 receptor as a G protein-coupled receptor

6.2.1 Overview of G protein-coupled receptors

A very large family of receptors exists that mediate the regulation of effector proteins by coupling to G proteins. These include rhodopsin, chemoattractant and odorant receptors, as well as receptors for peptide hormones, eicosanoids and neurotransmitters (see Birnbaumer *et al.*, 1990, for review). As a broad generalization, these receptors are structurally characterized by: (1) being a single peptide chain; (2) having the *N*-terminal extended extracellularly, which may have one or several glycosylation sites; (3) possessing seven regions of 15 to 25 amino acids with predominant hydrophobicity consistent with α-helix formation; (4) the transmembrane passage of these seven helices, creating three extracellular and three intracellular loops of amino acid sequences of varying length; and (5) a *C*-terminal that is extended intracellularly, which may have sites for phosphorylation by various protein kinases. The analysis of the folding of this class of receptors and the localization of binding sites for ligands is a subject of great current interest, and has been reviewed recently by several laboratories (Strader *et al.*, 1989; Savarese and Fraser, 1992; Baldwin, 1993; Hibert *et al.*, 1993). Much of our understanding of ligand binding to these receptors is based upon the binding of aminergic neurotransmitters (noradrenaline, dopamine, serotonin, acetylcholine) to their various pharmacological receptors. For these, the receptor helices are believed to associate within the membrane to form a barrel such that the ligand-binding site is within the interior pore of the barrel. The binding pocket comprises amino acid residues from several different helices that are able to bind to the ligand molecule through hydrogen bonding, ionic, and hydrophobic interactions. Agonist ligands are believed to promote a conformation that is able to interact with the G protein. The third intracellular loop and the *C*-terminal extension are believed to subserve the function of G-protein regulation. It is believed that an amphiphilic sequence of amino acids may form an intracellular α-helix that is able to interact with G proteins by virtue of the clustering of charged residues. The same domain of the receptor may be important for desensitization and receptor sequestration. Phosphorylation, interactions with proteins of the arrestin family, and membrane internalization mechanisms are important in regulating the function of this class of receptors.

The G proteins with which these receptors interact are heterotrimers comprising α subunits, which bind GTP and divalent cations, and β and γ subunits, which are strongly associated with each other as a dimer. The functional mechanisms by which G proteins act, and interactions of G proteins with receptor and effector proteins have been reviewed by several laboratories (Casey and Gilman, 1988; Freissmuth *et al.*, 1989; Birnbaumer *et al.*, 1990; Lamb and Pugh, 1992; Conklin and Bourne, 1993). The G protein superfamily can be categorized into families according to function as well as sequence homology within the α subunits (see Simon *et al.*, 1991). The G_s family was termed such because the α subunits interact with adenylate cyclase to stimulate its activity. By homology, the $G_{I/O}$ family includes $G_{I1,2,3}$, $G_{OA,B}$, $G_{T1,2}$ (transducin) and G_Z. The original designation of G_I referred to the ability of this class of G proteins to inhibit adenylate cyclase. However, it should be noted that diverse regulatory mechanisms involving α as well as βγ subunits of G_S and G_I influence the various subtypes of catalytic proteins of adenylate cyclase in different ways. For this reason, the simple notion that G_S stimulates and G_I inhibits adenylate cyclase may be a generalization that is limited by the cellular stoichiometry of the various G protein α and βγ subunits and their effector molecules. The $G_{I/O}$ family of proteins is also involved in the regulation of ion channels. The transducins regulate a specific cyclic GMP phosphodiesterase in rods and cones of the visual system. One important commonality for distinguishing the α subunit members of the $G_{I/O}$ family is that each possesses a cys residue in the

C-terminal which, when ADP-ribosylated by pertussis toxin, prevents the interaction with its respective receptor. Thus, pertussis toxin functionally uncouples G_I-mediated pathways. The G_Q family consists of at least five proteins, several of which regulate G protein-coupled phospholipase Cs. Several other G proteins that have been discovered by homology cloning ($G_{12,13}$) are presently without receptor or effector complements.

6.2.2 Cannabinoid compounds and adenylate cyclase

The first evidence that cannabinoid compounds interact with the adenylate cyclase second messenger pathway in an inhibitory fashion required an analysis of the mechanism. Criteria expected for a receptor-mediated mechanism transduced by G_I would minimally include the physiologically relevant cannabimimetic pharmacology expected for the receptor, the regulation by GTP and cations expected for the G protein and a non-essential role for cannabinoid–membrane interactions. These criteria were first fulfilled using a simple model system of neuroblastoma cells and have since been investigated in brain slice and membrane preparations, in primary cultures, and in transfected cell models.

6.2.2.1 Pharmacology of cannabimimetic compounds

The criteria for a physiologically relevant cannabimimetic pharmacology were first developed and tested using the neuroblastoma N18TG2 cell line as a model system. This established that a neuronal line possesses an adenylate cyclase system that is stimulated by prostacyclin/PGE_1 receptors, secretin/vasoactive intestinal peptide receptors, and adenosine A_2 receptors. Inhibition of adenylate cyclase in this cell line is regulated by α_2-adrenergic, M_4-muscarinic, δ-opioid, and CB_1 cannabinoid receptors. The evidence that the inhibition due to cannabinoid compounds resulted from an agonist–cannabinoid receptor interaction was initially indirect: it was demonstrated that 1 μM concentrations of either Δ^9-THC, Δ^8-THC, or levonantradol or its desacetylated derivative produced a non-competitive inhibition irrespective of the receptor type used to stimulate adenylate cyclase (Howlett and Fleming, 1984). This would indicate that the cannabinoid drug was not competing for these receptors and, by inference, would be acting distally to the agonist-stimulatory receptor interaction. The possibility of a direct interaction of the cannabinoid compounds with adrenergic, muscarinic or opioid receptors was eliminated by the failure of antagonists for those receptors to block the cannabinoid response in either the N18TG2 neuroblastoma or the NG108–15 neuroblastoma–glioma hybrid cell lines (Howlett and Fleming, 1984; Devane *et al.*, 1986; Howlett *et al.*, 1986). A failure of cannabimimetic agents to interact directly with a variety of neuromodulator receptors in the brain has subsequently been demonstrated. Using brain slice preparations from striatum, the D_2-dopaminergic antagonist, spiperone, was able to block the inhibition of cyclic AMP production by D_2 agonists, but not by the cannabimimetic desacetyllevonantradol (Bidaut-Russell and Howlett, 1991).

In that same preparation, the opioid antagonist, naloxone, was able to block the response to morphine but not to the cannabinoid agonist. Pacheco *et al.* (1991) used a potent aminoalkylindole agonist to demonstrate cannabimimetic inhibition of basal adenylate cyclase activity in rat cerebellar membranes. A variety of opioid, adrenergic, dopaminergic and serotonergic antagonist ligands failed to block this inhibition.

It should be noted that prostaglandin production in response to cannabinoid compounds may influence cyclic AMP accumulation. As discussed below, cannabinoid compounds at μM concentrations increase free arachidonic acid concentrations in a number of cell systems, and this may result in increased production of prostaglandins. A role for prostaglandins in the stimulation of adenylate cyclase activity in mouse brain homogenates was proposed by Hillard and Bloom (1983). These investigators found that 10 μM to 30 μM concentrations of Δ^9-THC, 11-OH-Δ^9-THC, cannabidiol, cannabinol, or levonantradol (but not dextronantradol) stimulated adenylate cyclase above basal activity. It was shown that the response to Δ^9-THC could be blocked by quinacrine, indomethacin, or aspirin, implicating the involvement of phospholipase A_2 and cyclooxygenase in the response. Such a mechanism may also have been responsible for cannabinoid effects in rat cerebellar brain slice preparations, in which 1 μM desacetyllevonantradol stimulated cyclic AMP accumulation, whereas 10 nM did not, and this stimulation could be blocked by indomethacin (Bidaut-Russell *et al.*, 1990). However, in rat cerebellar membranes, an aminoalkylindole agonist at 2 μM showed clear inhibition of adenylate cyclase, and no effects of indomethacin were observed (Pacheco *et al.*, 1991). Perhaps the reason for the difference is that an intact cell or crude homogenate system is required in order for all the necessary enzymes to be present for both prostaglandin production and regulation of adenylate cyclase via prostaglandin receptors. Alternatively, the aminoalkylindole class of compounds may not initiate accumulation of free arachidonic acid leading to prostaglandin production, whereas the cannabinoid class of compounds do.

The cannabimimetic pharmacology of the inhibition of adenylate cyclase has been demonstrated in a number of experimental systems (Table 1). The order of potency for a variety of *Cannabis* products and metabolites of Δ^9-THC or Δ^8-THC to inhibit adenylate cyclase activity in N18TG2 membranes or intact cells paralleled that observed for subjective and cardiovascular effects in humans, and cannabimimetic effects in a variety of animal model systems (Howlett and Fleming, 1984; Howlett *et al.*, 1986; Howlett, 1987). Enantioselectivity within the classical cannabinoid structure was demonstrated in this system using the dimethylheptyl derivative of 11-OH-Δ^8-THC (Howlett *et al.*, 1990c). The (−) isomer of 11-OH-Δ^8-THC-dimethylheptyl (HU-210) was several orders of magnitude more potent in inhibiting cellular cyclic AMP accumulation and membrane adenylate cyclase activity than was the (+) isomer (HU-211). Pharmacological specificity and enantioselectivity within the non-classical bicyclic and tricyclic series of cannabinoid derivatives were demonstrated, and the order of potency correlated with the antinociceptive properties of these compounds (Howlett and Fleming, 1984;

Table 1 Pharmacological potencies of certain cannabinoid analogues

Compound	Adenylate cyclase inhibition IC$_{50}$ (nM)			*In vivo* activities ED$_{50}$ or MPE$_{50}$ (mg/kg)			Antinociception	
	N18TG2 membranes[a]	Rat cerebellar membranes[b]	CB_1- transfected CHO cells[c]	Locomotor inhibition[d]	Hypothermia[d]	Immobility (catalepsy)[d]	Tail flick[d]	PBQ stretching[e]
Classical cannabinoid structures								
Δ^9-THC	360	—	13.5;13	3.1	24	1.6	1.3	5.9
Δ^8-THC	560	—	27.4;82	1.9	15	5.2	1.5	—
11-OH-Δ^9-THC	100	—	8.9	—	—	—	—	—
11-OH-Δ^8-THC	260	—	—	—	—	—	—	—
Cannabinol	1400 (60% max)	—	>100	—	—	—	—	—
Cannabidiol	>10 000	—	>1000	34	18	>100	>100	—
Synthetic tricyclic and bicyclic structures								
HU210	7.2	—	0.02	0.004	0.021	0.019	0.009	—
HU211	>10 000	—	191	>30	>30	>30	>30	—
Levonantradol	100	200	—	0.123	0.769	1.5	0.287	0.07
Desacetyllevonantradol	7	—	—	—	—	—	—	0.062
Dextronantradol	>5000	—	—	>10	>10	>10	>10	6.5
CP 55,940	25	—	0.87;0.99	0.040	0.349	0.347	0.087	0.059
CP 56,667	>5000	—	96;346	3	>10	>10	6	14.6
CP 55,244	5	—	0.128	0.004	0.050	0.085	0.010	0.018
CP 55,243	>10 000	—	>10 000	8	>10	>10	>10	>100
Aminoalkylindole structures								
Pravadoline	—	3600	—	—	—	—	—	—
WIN 55,212—2	—	320	24	0.25	23	2.1	0.82	—
WIN 55,212—3	—	>10 000	>2000	>57	>57	>57	>57	—

[a]Inhibition of secretin-stimulated production of [^{32}P]cyclic AMP from [α-^{32}P]ATP (Howlett, 1987; Howlett *et al.*, 1988, 1990b)
[b]Inhibition of basal production of [^{3}H]cyclic AMP from [^{3}H]ATP in low pH-treated rat brain membranes (Pacheco *et al.*, 1991, 1993)
[c]Inhibition of forskolin-stimulated cyclic AMP accumulation in intact CB_1-transfected CHO cells (Matsuda *et al.*, 1990; Gérard *et al.*, 1991; Felder *et al.*, 1992)
[d]Activities in mice after intravenous administration of drugs: photocell activity monitor (10 min); rectal temperature (60 min); ring-stand immobility test (90 min) tail flick reaction time (20 min) (Little *et al.*, 1988, 1989; Compton *et al.*, 1992a,b)
[e]Phenylbenzylquinone-induced stretching in mice 2 h after subcutaneous administration of drugs (Howlett *et al.*, 1988)

Howlett *et al.*, 1988). Using primary cultures of cerebellar granule cells, Pacheco *et al.* (1993) demonstrated that levonantradol and CP 55,940, members of the latter series of non-classical cannabinoid compounds, were more potent in inhibiting cyclic AMP accumulation than was Δ^9-THC, which is consistent with observations in the neuroblastoma cell. Several potent classical and bicyclic and tricyclic cannabinoid derivatives were able to inhibit forskolin-activated adenylate cyclase in mouse brain synaptosomes (Little and Martin, 1991). However, in those experiments, some problems in demonstrating the effects of other cannabinoid compounds may have resulted from the relatively high concentrations of the dimethylsulfoxide vehicle used in the assays (Little and Martin, 1991). Cannabinoid pharmacological and enantiomeric specificity was observed for the inhibition of forskolin-stimulated cyclic AMP production in SKR6-transfected CHO-K1 cells expressing the rat CB_1 receptor (Matsuda *et al.*, 1990; Felder *et al.*, 1992) and in pSVL/BS08-transfected CHO-K1 cells expressing the human CB_1 receptor (Gérard *et al.*, 1991).

Within the aminoalkylindole series of cannabimimetic agonists, Pacheco *et al.* (1991) developed structure–activity profiles for the inhibition of adenylate cyclase activity in membrane preparations from rat cerebellum and striatum. The data showed pharmacological and enantiomeric selectivity, and provided an excellent correlation with the ability of these aminoalkylindole analogues to inhibit electrically stimulated contractions of mouse vas deferens preparations. Cyclic AMP accumulation in cultured rat cerebellar granule cells was also shown to be enantioselective for the potent aminoalkylindole agonist (+)-WIN 55,212–2 as opposed to its (−) enantiomer, WIN 55,212–3 (Pacheco *et al.*, 1993). It is of particular interest that one of the members of this series, WIN 56,098, behaved as a competitive antagonist to the aminoalkylindole response in mouse vas deferens preparations (Pacheco *et al.*, 1991). WIN 56,098 at 10 μM was also able to shift the log dose–response curve for WIN 55,212–2-induced inhibition of adenylate cyclase, which it did by eight-fold. This rightward shift in the log dose–response curve is consistent with this analogue being a competitive antagonist for the second messenger system.

The above studies, which are summarized in Table 1, have described a receptor coupled to the signal transduction mechanism of inhibition of adenylate cyclase. The receptor described for this cellular mechanism exhibits the same pharmacological profile as has been described for the *in vivo* cannabinoid responses of analgesia, hypothermia, decreased spontaneous activity and immobility in mice, and ataxia in dogs (Little *et al.*, 1988, 1989; Martin *et al.*, 1991; Compton *et al.*, 1992ab, 1993). The *in vitro* actions of cannabimimetic compounds to inhibit neuronally mediated smooth muscle contraction (Pacheco *et al.*, 1991; Pertwee *et al.*, 1992) can also be described by the same pharmacological profile as exhibited for the inhibition of adenylate cyclase. Thus, it must be concluded that this receptor, termed CB_1, is capable of decreasing cyclic AMP content within neurons, and that this may be a potential mechanism whereby the activity of these neurons modifies critical synaptic pathways involved in these activities.

6.2.2.2 G protein involvement in the effects of cannabimimetic compounds

The hypothesis was tested that a pharmacologically distinct cannabinoid receptor (i.e. CB_1) produces its effect by the same mechanism as receptors known to couple to a discrete pool of G proteins that interact with adenylate cyclase. Biochemical studies to demonstrate a role for G_I in transducing the cannabinoid inhibition of adenylate cyclase came from studies in density gradient-purified membrane preparations from N18TG2 cells (Howlett, 1985). The addition of desacetyllevonantradol or Δ^9-THC produced a reduction in adenylate cyclase activity at low mM concentrations of Mg^{2+}, which could be overcome by increasing the divalent cation concentration. The inhibition was particularly sensitive to reversal in the presence of Mn^{2+} as the divalent cation. Typically, the GTP cycle of G_I requires a lower concentration of divalent cation than that of G_S. The optimal GTP concentration for inhibition by cannabimimetic compounds was 10 μM or greater. Similarly high concentrations of GTP were also required for optimal inhibition of adenylate cyclase by aminoalkylindole analogues in rat cerebellar membranes (Pacheco *et al.*, 1991). Typically, G_I exhibits a lower affinity for GTP than does G_S. Pacheco *et al.* (1991) further showed that the inhibition of adenylate cyclase was promoted by treatment of the membranes at pH 4.5. These data are consistent with the cannabinoid effects on the regulation of adenylate cyclase occurring at the level of G_I, as opposed to G_S. No cannabinoid inhibition in the presence of the GTP analogues Gpp(NH)p or GTP-γ-S was observed (Howlett, 1985; Pacheco *et al.*, 1991), except under conditions that suppressed the effects of Gpp (NH)p on G_S (Howlett, 1985). This finding also is consistent with the effects of cannabimimetic compounds being due to an interaction with G_I.

Additional evidence supporting the requirement for the presence of G_I came from experiments using pertussis toxin as a tool to ADP-ribosylate a critical cys residue found on $G_{I/O}$ subtypes of G proteins, thereby precluding the interaction of $G_{I/O}$ with receptors. Pertussis toxin was shown to deplete functional $G_{I/O}$ proteins in parallel with its ability to prevent the cannabinoid inhibition of cyclic AMP production in N18TG2 cells or membranes (Howlett *et al.*, 1986). Pertussis toxin-treatment of cerebellar granule cells, under conditions that promoted maximal ADP-ribosylation of $G_{I/O}$ proteins, blocked the inhibition of cyclic AMP accumulation in response to both cannabinoid and aminoalkylindole agonists (Pacheco *et al.*, 1993).

Applying the pertussis toxin technique to deplete functional $G_{I/O}$ within heterologous CNS systems can yield results that are difficult to interpret. For example, pertussis toxin was bilaterally implanted at multiple sites within the rat striatum, and time was allowed for diffusion, cellular uptake and processing of the toxin, and enzymatic modification of cellular G proteins by the processed toxin. The inhibition of cyclic AMP production by desacetyllevonantradol was only partly blocked (Bidaut-Russell *et al.*, 1990), and the ability of cannabinoid compounds to produce catalepsy was not affected (Howlett *et al.*, 1990a). This latter result could mean that regulation of cellular cyclic AMP content in neurons in the striatum is not a factor in the catalepsy behaviour. On the other hand, the failure to observe a change could also have resulted from the failure to inactivate a predominance (close to 100%) of the $G_{I/O}$ proteins in each neuron, or alternatively, the failure to inactivate the $G_{I/O}$ proteins in most (close to 100%) of the neurons. Another equally good explanation might be that inactivation of $G_{I/O}$ in neurons that respond to cannabinoid agonists is counteracted by the effects of inactivating $G_{I/O}$ in other populations of cells in that same region. Yet another explanation could be that the striatum is not dominant in the cannabinoid-promoted catalepsy behaviour. One can appreciate the dilemma engendered by such a protocol *in vivo*.

If the inhibitory regulation of adenylate cyclase by CB_1 receptors is mediated by $G_{I/O}$ as is the response to certain other neuroreceptors, and if there is an excess of receptors and a limited number either of G proteins or of effector molecules, then one would predict a lack of additivity for the actions of the two maximally stimulated receptors when they are combined. This would be the case because both receptor types would be acting via the same saturable process. The alternative would be that the response produced by one mechanism would be additive to, or synergistic with, the response generated by the second mechanism. Presumptive evidence supporting the prediction that the CB_1 receptor acts by the same mechanism as other receptors coupled to $G_{I/O}$ came from additivity studies. These studies demonstrated that the response to stimulation of two receptor populations was no greater than the maximal response to either one alone. The maximal inhibition of cyclic AMP production obtained with cannabimimetic drugs in the homogeneous population of N18TG2 cells, or membranes derived from these, was not increased by addition of maximally effective concentrations of muscarinic, adrenergic or opioid agonists (Howlett and Fleming, 1984; Devane *et al.*, 1986). Using the (predominantly) granule cell population of cultured cerebellar neurons, Pacheco *et al.* (1993) observed the lack of an additive inhibition of cyclic AMP accumulation when either the cannabinoid levonantradol or the aminoalkylindole WIN 55,212–2 were combined with the $GABA_B$ agonist baclofen.

This same non-additivity for inhibition of cyclic AMP production was also observed using CNS preparations. Membranes prepared from rat cerebella exhibited a marginally greater maximal inhibition of adenylate cyclase when cannabinoid or aminoalkylindole compounds were combined with the $GABA_B$ agonist (Pacheco *et al.*, 1993). Using rat striatal slices, the interaction of cannabimimetic drugs with opioid systems to inhibit cyclic AMP accumulation was neither additive nor synergistic (Bidaut-Russell and Howlett, 1991). These studies in CNS preparations suggest that these pharmacological receptor types must be co-localized on at least some of the same neurons and, furthermore, that they share a common coupling mechanism (see also section 7.6). As mentioned previously, this interpretation of the data assumes a stoichiometry described by an excess of receptors and limited numbers of either G proteins or effector molecules.

Multiple $G_{I/O}$ subtypes exist, and it is currently not known which is/are capable of coupling the CB_1 receptor to adenylate cyclase. There is probably more rigid selectivity of the effector than of the receptor for a specific G protein subtype . The ability of the agonist–receptor complex to directly interact with G proteins can be measured as the ability to stimulate GTP hydrolysis by the α subunit. Low K_m GTPase activity stimulated by cannabimimetic drugs has been demonstrated in cerebellar and dentate gyrus membranes (Deadwyler *et al.*, 1993; Pacheco *et al.*, 1993). However, it is not possible to ascertain, using this method, which G protein subtype(s) were coupled to the CB_1 receptor, or to which effector(s) each was coupled. The complex behaviour of GTP analogues in disrupting the interaction between G proteins and CB_1 receptors in detergent solution suggests that this receptor may be associated with more that one G protein subtype having differing affinities for the GTP analogues (Houston and Howlett, 1993). In studies of rat cerebellar membranes, the ability of cannabimimetic agonists to stimulate low K_m GTPase activity was additive with that stimulated by the $GABA_B$ agonist baclofen (Pacheco *et al.*, 1993). This finding could be interpreted to mean that multiple pools of G proteins exist, and that the pool that interacts with the CB_1 receptors is distinct from that which interacts with the $GABA_B$ receptors. An alternative interpretation is that the stoichiometry of receptors, $G_{I/O}$ proteins and effector molecules is such that the $G_{I/O}$ proteins are in relative excess compared with the other components of the system. The finding that additivity between pharmacologically distinct receptors is pronounced for regulation of GTPase activity but is minimal or non-existent for inhibition of adenylate cyclase might suggest that some pool(s) of G proteins is/are stimulated by the receptors but not coupled to adenylate cyclase. Coupling to alternative effector systems is an explanation for this finding, highlighting the notion that the coupling of CB_1 receptors to specific subtypes of $G_{I/O}$ may serve a regulatory function. It may be that the relative stoichiometry of receptors to various G protein α and βγ subunits of various subtypes may ultimately be shown to determine how the response to cannabinoid drugs differs in different cell types.

6.2.2.3 Implications for protein kinase A regulatory mechanisms

Increases in cellular cyclic AMP content can be expected to stimulate protein kinase A activity, leading to the phosphorylation of enzymes and proteins that represent key steps regulating differentiated functions of the cell. Conversely, when the cyclic AMP content of the cell is reduced, or increases are attenuated by the influence of an inhibitory neurotransmitter, then one would expect that phosphorylation processes would be curtailed. The state of phosphorylation of those key regulatory proteins would revert to the 'dephospho' state as the result of protein phosphatases within the cell. In order to demonstrate that such a schema is possible, Fleming *et al.* (1992) were able to define *in vitro* conditions such that adenylate cyclase could utilize the substrate App(NH)p, an ATP analogue that does not act as a substrate for protein kinase A. Rat brain membranes were incubated first with 0.5 mM App(NH)p to yield cyclic AMP, and then with 1 μM [γ-^{32}P]ATP to monitor protein phosphorylation stimulated by protein kinase A. Using this protocol, it is possible to monitor phosphorylation states that are directly coupled to cyclic AMP increases or decreases. Fleming *et al.* (1992) were able to discern a decrement in the forskolin-stimulated phosphorylation of rat brain membrane proteins by the addition of D-Ala-enkephalin, an opioid agonist that inhibits adenylate cyclase activity via G_I. Two phosphorylated bands found on sodium dodecylsulfate–polyacrylamide gel electrophoresis were of particular interest. These bands corresponded in relative molecular weight to synapsins I and II, proteins important for neurotransmitter release. This study demonstrates a clear role for inhibition of neuronal cyclic AMP content in the behaviour of the cell. The results of comparable experiments using cannabimimetic ligands will be particularly exciting because it could be predicted that a homologous pathway should exist for CB_1 receptors.

6.2.2.4 Membrane perturbations and adenylate cyclase

The influence of cannabinoid drugs on membrane enzymes is a factor that must be considered, particularly in *in vitro* assays utilizing concentrations that exceed 1 μM of the drug. It was noted in studies of adenylate cyclase in N18TG2 membranes that inhibition was reversed as the concentrations of Δ^9-THC, Δ^8-THC, or the more potent levonantradol or desacetyllevonantradol approached and exceeded 10 μM (Howlett and Fleming, 1984). In those experiments, the effects of the ethanol vehicle may have been additive with those of the cannabinoid compounds. Stimulation of adenylate cyclase by ethanol (>0.2%) has been well documented (see Rabin *et al.*, 1986 and references therein). However, in subsequent studies, ethanol was eliminated from the reaction mixture by evaporation from the stocks under N_2 gas, initially forming a suspension as 0.1 mM or 1 mM stocks of cannabinoid compounds in 50 mg/ml fatty acid-deficient bovine serum albumin, and subsequently diluting the stocks such that the greatest concentration of serum albumin in the reaction mixtures would be 0.5 mg/ml. Under these conditions, the experimental points for cannabinoid compounds at concentrations above 3 μM fall below the maxima expected from the computer-generated Langmuir adsorption isotherms based on non-linear regression on the data points (for example, see Figures 1 and 3 in Howlett, 1987).

The deviation from the expected behaviour at >3 μM concentrations of cannabinoid compounds led to testing the hypothesis that, at such concentrations, the inhibitory effect of cannabinoid compounds on adenylate cyclase mediated by the CB_1 receptor is opposed by a stimulatory effect on the same enzyme, which stems from the ability of cannabinoid compounds to increase the fluidity of the membrane lipid environment (Howlett *et al.*, 1989). It was known that agents that increase membrane bilayer fluidity, as shown by fluorescence polarization or electron spin resonance probe measurements, are able to augment hormone-stimulated adenylate cyclase activity (for review see Houslay and Gordon, 1983; Salesse and Garnier, 1984). It was also known that cannabinoid compounds can increase the fluidity of membrane lipids (see Section 6.1.5).

In order to study the influence of membrane fluidity on adenylate cyclase activity, the enzymatic activity is measured as a function of temperature. Within the lipid bilayer, there are domains that exist in a fluid state and domains that exist in the more rigid liquid crystalline structure. The temperature at which the liquid crystalline state 'melts' to become a fluid state is called the transition temperature. The influence of the physical state of the lipids in the immediate environment of the protein on enzymatic activity is dependent upon temperature. The activation energy is smaller at temperatures greater than the phase transition temperature, above which the membrane lipids exist in the fluid state. Some agents that fluidize the rigid membrane domains do so by intercalating within the head-group region of the phospholipid bilayer, thereby spreading the head-groups apart and allowing increased mobility of the fatty acids extending into the core of the bilayer. Such membrane fluidizing agents will lower the thermal energy required to melt the liquid crystalline domains, i.e. they will decrease the phase transition temperature. If enzymatic activity is affected by the state of the lipid domains of the membrane, then the decrease in the phase transition temperature will be reflected in the enzymatic activity as a function of temperature. The effect of temperature on enzyme activity is analysed by transforming the data into an Arrhenius plot, by which the log of the specific activity of adenylate cyclase is plotted as a function of the inverse of the temperature in degrees Kelvin. The transition temperature occurs at the position on the curve met by the two straight lines of differing slopes.

In order to understand the influence of cannabinoid compounds on membrane lipid domains and how this influence may alter adenylate cyclase activity, Howlett *et al.* (1989) studied the temperature-dependence of hormone-stimulated adenylate cyclase in N18TG2 membranes. The break-point in the Arrhenius plot line, indicative of the lipid phase transition temperature was 20°C (Howlett *et al.*, 1989). A prototypical membrane fluidizing agent, benzyl alcohol, partitions into lipid bilayers and disrupts phospholipid interactions, resulting in a decrease in the temperature required for the lipid phase transition. For the N18TG2 membranes, the temperature was reduced to 14°C by optimal concentrations of benzyl alcohol, without altering the activation energies above or below the thermal transition temperature (Howlett *et al.*, 1989). When cannabinoid compounds were examined, it was observed that 30 μM cannabidiol significantly reduced the transition temperature to 17.5°C without altering the activation energies. This behaviour resembled the effects of benzyl alcohol. In contrast, Δ^9-THC failed to alter the transition temperature in this system, whether tested at 1 μM or at 30 μM. These data support the hypothesis that 30 μM cannabidiol is able to alter membrane fluidity in domains that are critical for adenylate cyclase activity in this experimental system. However, the data also suggest that Δ^9-THC does not act in this way, at least at concentrations of 1 μM or 30 μM.

The thermodynamic studies (Howlett *et al.*, 1989) indicated that for either carbachol, acting through the muscarinic receptor, or Δ^9-THC, acting through the cannabinoid receptor, a small decrease in the enthalpy change, and a smaller decrease in the entropy change resulted in a net change in free energy that was increased by 0.2 kcal/mol in the presence of inhibitory agonists. This change in net free energy is opposite to that evoked by the presence of stimulatory hormones. These effects are not related to changes in membrane fluidity, but rather, reflect the net changes in free energy by which the agonist–receptor complex promotes catalytic activity. When the stimulatory agonist interacts with its receptor, the dissociation of the G_S heterotrimer represents an increased disorder of the system. The evidence suggests that the presence of the inhibitory agonist–receptor–G_I interaction opposes the increased disorder that was promoted by the stimulatory agonist–receptor–G_S interaction.

Thermodynamic analyses of the effects of cannabinoid compounds on hormone-stimulated adenylate cyclase in rat liver and heart membranes have been performed by Hillard *et al.* (1986, 1990). For both systems, Δ^9-THC enhanced hormone-stimulated activity by 20% to 40%. Liver and heart tissue do not express mRNA for CB_1 (Gérard *et al.*, 1991), and so it would not be expected that a G_I-coupled mechanism for the inhibition of adenylate cyclase would be present in these tissues. In the case of the cardiac membranes, enhancement by 3 μM Δ^9-THC or 11-OH-Δ^9-THC was accompanied by a decrease in the phase transition temperature, denoted as a break in the Arrhenius plots of β-adrenergically stimulated adenylate cyclase activity and diphenylhexatriene fluorescence polarization temperature curves (Hillard *et al.*, 1990). In these experiments, neither cannabinol nor cannabidiol mimicked the effects of Δ^9-THC. However, these agents did eliminate the break in the Arrhenius plot for enzymatic activity, suggesting that a heterogeneity was set up by the compounds such that a clear phase transition temperature could no longer be discerned.

A different influence of cannabinoid compounds was observed for glucagon-stimulated adenylate cyclase activity in hepatic membranes (Hillard *et al.*, 1986). In that system, 3 μM Δ^9-THC had no effect, but 30 μM increased hormonal stimulation of the enzyme by shifting the glucagon log concentration–response curve to the left. No decrease in the phase transition temperature denoted by the Arrhenius plot of activity resulted from the addition of cannabinoid compounds. Rather, a decrease in activation energy at higher temperatures was observed, suggesting that the presence of the cannabinoid compounds may have somehow facilitated the stimulatory agonist–receptor interaction.

These thermodynamic studies have not presented us with a cohesive description of the effects of membrane-perturbing cannabinoid compounds on the regulation of adenylate cyclase. The differences in sensitivity to particular cannabinoid compounds may be attributable to the differences in lipid composition between diverse tissues and cells. The enhancement produced by membrane fluidization agents (usually less than 50% above control) is minor when contrasted with the physiologically relevant stimulation of activity mediated by hormonal receptors (two-fold to 20-fold in many cell types). One must also consider whether the high concentrations of cannabinoid compounds required to produce membrane pertubation in *in vitro* assay systems such as these would be likely to be encountered in the cellular environment in animals.

6.2.3 Ion channels coupled to the cannabinoid receptor

6.2.3.1 Ca^{2+} currents

Inhibition of a Ca^{2+} current coupled directly to CB_1 receptors was demonstrated using the differentiated NG108–15 neuroblastoma-glioma hybrid cell model system (Caulfield and Brown, 1992; Mackie and Hille, 1992). A number of receptor types act via the $G_{I/O}$ family of G proteins to both inhibit adenylate cyclase and inhibit a voltage-gated Ca^{2+} current (see Birnbaumer *et al.* (1990), Sternweiss and Pang (1990), and Brown (1991) for review and discussion of G protein-regulated ion channels plus original references). Thus, it was hypothesized that CB_1 receptors would also regulate Ca^{2+} current in an analogous manner. Using whole-cell voltage-clamp experiments, Caulfield and Brown (1992) showed that 30 μM Δ^9-THC or 1 μM CP 55,940 were able to inhibit both peak and end-of-pulse Ca^{2+} current by about 40%. Mackie and Hille (1992) demonstrated, using a pluronic-nystatin perforated-patch technique, that 0.1 μM WIN 55,212–2 or 0.1 μM CP 55,940 caused a 40% inhibition of the Ca^{2+} current. Mackie *et al.* (1993) also demonstrated using whole-cell voltage-clamp of the N18 neuronal cell, that 0.3 μM anandamide could maximally inhibit the Ca^{2+} current (30%), but that this response was less efficacious than that observed for WIN 55,212–2 (55%). Furthermore, anandamide partially antagonized the response to WIN 55,212–2, behaving as a partial agonist.

The signal transduction mechanism for this response was demonstrated to result from a direct interaction of the CB_1 receptor with a G protein. The cannabinoid inhibition of the Ca^{2+} current was slow in onset (minutes) and recovery (minutes to hours). Regulation of adenylate cyclase by cannabimimetic compounds was apparently not a factor in the regulation of the Ca^{2+} current, because inhibition of the Ca^{2+} current was not reversed by addition of the analogues dibutyryl-cyclic AMP or 8-chlorophenylthio-cyclic AMP in the presence of the cyclic nucleotide phosphodiesterase inhibitor isobutylmethylxanthine (Mackie and Hille, 1992). Coupling via $G_{I/O}$ was determined by the demonstration that treatment of the cells with pertussis toxin reduced the response to maximally effective concentrations of Δ^9-THC or CP 55,940 (Caulfield and Brown, 1992) and of WIN 55,212–2 (Mackie and Hille, 1992) from 40% to less than 10%. There is some likelihood that G_O may be the G protein associated with neuronal Ca^{2+} channels (McFadzean *et al.*, 1989; Sternweis and Pang, 1990); however, coupling via G_O has not yet been demonstrated for CB_1 receptors. The Ca^{2+} channel was sensitive to ω-conotoxin, which reduced the cannabinoid inhibition to less than 10%, implicating the N-type Ca^{2+} channel as the effector in this response. Despite this strong evidence for the coupling of the CB_1 receptor to a G protein that regulates an N-type Ca^{2+} ion channel, the study by Caulfield and Brown (1992) suggested that cannabinoid compounds may also exert influences on L- or T-type Ca^{2+} currents that cannot be explained by this mechanism alone.

Biochemical studies have demonstrated the ability of cannabinoid compounds to inhibit $^{45}Ca^{2+}$ uptake by rat brain synaptosomes when this is triggered by K^+-induced depolarization (Harris and Stokes, 1982). Δ^9-THC at 10 nM to 100 nM attenuated depolarization-dependent (but not resting) uptake of $^{45}Ca^{2+}$ in synaptosomes prepared from the cerebellum or brainstem, but not the cortex or striatum. Although 100 nM cannabidiol evoked a small response in brainstem synaptosomes, cannabinol did not affect $^{45}Ca^{2+}$ uptake in any region. More recent investigations have used fura-2-loaded synaptosomes from rat cortex to demonstrate an effect of Δ^9-THC, 11-OH-Δ^9-THC, and cannabidiol on depolarization-induced Ca^{2+} influx (Hillard and Pounds, 1994). In that study, Δ^9-THC inhibited both the transient spike and the sustained plateau of Ca^{2+} uptake evoked by K^+-induced depolarization.

In contrast to the inhibition of depolarization-dependent Ca^{2+} levels, Felder *et al.* (1992) found a cannabinoid-induced increase in intracellular Ca^{2+}. This study, using the fura-2 technique with CHO cells, demonstrated that the cannabinoid agonists HU-210 (11-OH-Δ^8-THC-dimethylheptyl), Δ^9-THC or CP 55,940, at concentrations greater than 1 μM, were able to increase intracellular Ca^{2+}. This response was not mediated via CB_1 receptors, however, because identical responses were observed in control CHO cells as well as CHO cells expressing the rat CB_1 receptor. Furthermore, the poorly active (+) isomer, HU-211, produced responses comparable to those of HU-210. The source of the Ca^{2+} was considered to be intracellular because a similarly robust response was observed in the absence of extracellular Ca^{2+}. The nature of the mechanism is unresolved, but is not considered to involve inositol phosphate metabolism (Felder *et al.*, 1992). Perhaps a compromise in the intracellular membranous organelles controlling intracellular Ca^{2+} fluxes might be considered.

Harris and Stokes (1982) demonstrated that at 10 μM concentrations, all cannabinoid drugs tested (Δ^9-THC, 11-OH-Δ^9-THC, cannabinol, cannabidiol) profoundly inhibited rat or mouse brain synaptosomal $^{45}Ca^{2+}$ uptake. This response probably represented damage to the synaptosomes, because 10 μM Δ^9-THC (but not 1 μM) markedly reduced the membrane potential determined using the tetraphenylphosphonium accumulation method, and allowed considerable Ca^{2+} efflux from preloaded synaptosomes (Harris and Stokes, 1982). Neurochemical and morphological disruption of synaptosomal membranes by these concentrations of cannabinoid drugs has been previously reported (Hershkowitz *et al.*, 1977).

6.2.3.2 K^+ currents

Receptors that regulate adenylate cyclase via $G_{I/O}$ frequently also open a K^+ channel (see Birnbaumer *et al.*, 1990; Brown, 1991; North, 1989). Deadwyler *et al.* (1993) presented evidence for a cannabinoid receptor utilization of this signal transduction mechanism using whole-cell patch clamp recording from primary cultures of hippocampal cells (see Chapter 7). A voltage-sensitive K^+ channel that produces A-current (I_A), is characterized by its activation during membrane depolarization, and its susceptibility to blockade by mM concentrations of 4-aminopyridine. The application of CP 55,244 decreased the inactivation by producing a positive shift in the voltage dependence, and increased the residual A-current, again by a voltage shift. Other K^+ channels were not affected by the cannabimimetic drug. CP 55,244 exhibited an EC_{50} value approximating 1 μM in the application pipette. At concentrations of 2 μM, WIN 55,212–2 was equally as effective as CP 55,244, and levonantradol and CP 55,940 were nearly as effective. Inactive enantiomers of these compounds produced no response. The effect of CP 55,244 disappeared within minutes of wash-out.

The involvement of G proteins in the cannabinoid regulation of I_A has been demonstrated (Deadwyler *et al.*, 1993). First, pretreatment of the cells with pertussis toxin attenuated the effects of cannabimimetic compounds on the depolarizing voltage shift in the inactivation of I_A and the negative voltage shift in relative activation of I_A. Second, dialysis of cells with the non-hydrolysable GTP analogue GTP-γ-S, mimicked the effects of cannabinoid drugs on I_A, as would be expected for a direct activation of the G protein circumventing the requirement for stimulation by the agonist–receptor complex. Furthermore, the agonist–receptor complex created by the addition of WIN 55,212–2 failed to produce greater effects than those produced by intracellular GTP-γ-S.

These data provide evidence in support of a CB_1 receptor and $G_{I/O}$-regulated K^+ A-current (Deadwyler *et al.*, 1993). However, whether this results from a direct coupling of the G protein to the ion channel or whether the mechanism is secondary to a change in intracellular cyclic AMP and the resulting phosphorylation state of the ion channel protein cannot be discerned by this initial characterization. Further studies have addressed this by demonstrating a negative voltage-dependent shift upon bath-application of the cyclic AMP analogue, 8-bromo-cyclic AMP, or the adenylate cyclase activator, forskolin (Mu *et al.*, 1993). The response to forskolin was reversed by pressure-injection of WIN 55,212–2. This would be consistent with a mechanism by which the synthesis of cyclic AMP in response to forskolin was inhibited by agonist-stimulated cannabinoid receptors. However, a direct coupling of the G protein(s) to the ion channels as a concurrent mechanism cannot be ruled out by this set of experiments.

6.2.3.3 Implications of signal transduction via ion channels

A reduction in Ca^{2+} current may affect the functioning of neurons by decreasing the probability of neurotransmitter release at the presynaptic terminal. Thus, this signal transduction pathway for cannabimimetic compounds may predominate in regulation of transmitter release from neurons in the CNS as well as in the periphery. In the latter case, it is believed that cannabimimetic compounds attenuate electrically stimulated smooth muscle contraction at the presynaptic level by causing a decrement in the release of the contracting neurotransmitter (Rosell *et al.*, 1976; Roth, 1978). Electrophysiological studies describing the effects of cannabinoid compounds with nerve-muscle preparations have been reviewed previously (Pertwee, 1988).

The physiological significance of the CB_1 receptor-regulated K^+ conductance mechanism would relate to the resultant hyperpolarization that would attenuate repetitive spiking in particular neurons that utilize this pathway. The voltage shift would also act to neutralize any depolarizing events when a neuron is at its resting potential. Thus, this signal transduction mechanism could not only alter the endogenous firing pattern of neurons, but also alter the response of the cell to synaptic inputs.

6.3 Phospholipase C

Evidence for an involvement of cannabinoid compounds in the regulation of the polyphosphoinositide second messenger system has been sought but not found. White and Tansik (1980) failed to find increases in phosphatidic acid from phospholipids prelabelled with [^{14}C]arachidonic acid after incubation of platelets with Δ^9-THC or cannabinol in the 10 μM to 80 μM range. Chang *et al.* (1991) were unable to detect [^{14}C]arachidonic acid in the diglyceride fraction after incubation of radiolabelled phosphatidylcholine with sea-urchin sperm homogenates and cannabinoid drugs. These products would have been expected if phospholipase C had acted on the phospholipid fraction in response to the cannabinoid compounds. Testing the activity of phosphoinositide phospholipase C activity directly, Felder *et al.* (1992) were unable to obtain release of IP, IP_2 or IP_3 in either CHO cells expressing rat CB_1 receptors or L cells expressing human CB_1 receptors in response to concentrations of HU-210 as high as 100 μM.

Of relevance to the central nervous system, it was found that 8 μM Δ^9-THC elicited a loss of [^{14}C]arachidonic acid-labelled phosphatidylinositol, concomitant with an increase in unesterified [^{14}C]arachidonic acid in guinea-pig brain slice preparations (Reichman *et al.*, 1988). However, in contrast to what was seen with the agonist acetylcholine acting through the muscarinic receptor to regulate phosphoinositide phospholipase C activity, Δ^9-THC failed to increase the content of labelled diglycerides or phosphatidic acid (Reichman *et al.*, 1991). Using [^{3}H]glycerol as the label, Δ^9-THC also failed to increase the content of labelled diglycerides or phosphatidic acid, again in contrast to what was observed with the acetylcholine control. Finally, using [^{3}H]inositol as the label, Δ^9-THC failed to elicit increases in IP, IP_2, IP_3, or a variety of other inositol phosphate metabolites, in contrast to the marked increases observed using the control acetylcholine via the muscarinic system. Furthermore, cannabinoid compounds were unable to release arachidonic acid from the prelabelled diglyceride pool in guinea-pig brain slices (Reichman *et al.*, 1988). This suggests that the pathway of phospholipase C production of diacylglycerides as the substrate for diglyceride lipase is not a viable mechanism for cannabinoid-induced arachidonic acid release (see below). The accumulated evidence fails to provide any indication that the CB_1 receptor regulates phosphoinositide metabolism as a signal transduction pathway in the brain.

6.4 Cannabinoid compounds and eicosanoids

6.4.1 Cannabinoid-induced release of arachidonic acid

Signal transduction via the release of arachidonic acid and subsequent production of biologically active metabolites provides a second layer of second messenger systems, because a number of eicosanoid metabolites regulate adenylate cyclase or Ca^{2+} release mechanisms. Cannabinoid compounds appear to promote the accumulation of free arachidonic acid in a number of different experimental model systems. The mechanism may involve several steps along the pathway of arachidonic acid metabolism. In addition, cannabinoid compounds influence the production of metabolites, including prostaglandins and leukotrienes. These cannabinoid actions generally require relatively high concentrations *in vitro*, and appear not to be related to CB_1-stimulated events.

Cultured HeLa cells, pretreated with [^{14}C]arachidonic acid for 1 h to incorporate this fatty acid into phospholipids, could be stimulated to release free [^{14}C]arachidonic acid by cannabinoid compounds (Burstein and Hunter, 1978). Accumulation doubled over control during 1 h incubations with 16 μM Δ^9-THC, 11-OH-Δ^9-THC, cannabinol, and cannabidiol, and achieved five- to nine-fold stimulation over control at 160 μM. A variety of other cannabinoid compounds also promoted free arachidonic acid accumulation at 160 μM. Similar mobilization of free [^{14}C]arachidonic acid from isolated Leydig cells and WI-38 fibroblasts was evoked by 3 μM to 16 μM concentrations of Δ^9-THC or cannabichromene (Burstein and Hunter, 1981). These studies have been replicated using mouse peritoneal cells (macrophages), in which 3.2 μM to 32 μM concentrations of Δ^9-THC, 11-OH-Δ^9-THC, 6α- and 6β-OH-Δ^9-THC, but not Δ^9-THC-11-oic acid promoted production of unesterified [^{14}C]arachidonic acid (Burstein *et al.*, 1984; Audette *et al.*, 1991). For intact human platelets, concentrations of 30 μM Δ^9-THC or cannabidiol were required to release pre-esterified [^{14}C]arachidonic acid (White and Tansik, 1980). In S49 lymphoma cells, 16 μM Δ^9-THC could stimulate the appearance of radioactivity from cells pretreated with [^{14}C]arachidonic acid (Audette *et al.*, 1991). The addition of 20 μM Δ^9-THC to islet cells that had been pretreated with [^{14}C]arachidonic acid enhanced the loss of radiolabelled phospholipids in response to glucose (Laychock *et al.*, 1986). Thus, in a variety of diverse cell types, cannabinoid compounds in μM concentrations promote the accumulation of unesterified arachidonic acid.

Cannabinoid compounds can also increase unesterified arachidonic acid in brain preparations. Δ^9-THC was able to release [^{14}C]arachidonic acid from phospholipids in prelabelled synaptosomes and other membrane fractions from mouse brain (Hunter *et al.*, 1986). In guinea-pig brain cerebral cortex slice preparations that had been prelabelled with [^{14}C]arachidonic acid, Δ^9-THC increased free [^{14}C]arachidonic acid over four-fold at 32 μM, with an EC_{50} of about 8 μM (Reichman *et al.*, 1988). Although (+)-Δ^9-THC was equipotent in promoting this response, cannabinol, Δ^8-THC or cannabidiol were required in greater concentrations. The origin of over 80% of the released arachidonic acid was from the phospholipid fraction, and much of the remainder was from the triglyceride fraction. Cannabinoid compounds were unable to release arachidonic acid from the diglyceride pool, suggesting that the pathway of phospholipase C production of diacylglycerides does not produce a major source of substrate for cannabinoid-induced arachidonic acid release. Over half of the loss in the labelled phospholipid fraction appeared to be derived from phosphatidylinositol, the major store of arachidonic acid in the phospholipid pool, and about 20% was derived from phosphatidylethanolamine. It is notable that the release from phosphatidylcholine was minor, considering that this phospholipid is a major source of arachidonic acid in the 2 position in the brain (Reichman *et al.*, 1988), and that release of arachidonic acid from prelabelled phosphatidylcholine has previously been demonstrated in synaptosomal and other preparations (Hunter *et al.*, 1986; Laychock *et al.*, 1986).

6.4.2 Cannabinoid-induced release of eicosanoids

In some cell systems, the accumulation of arachidonic acid has been associated with release of prostaglandins (PGs) or other eicosanoids. The arachidonic acid mobilization in HeLa cells was accompanied by PGE_2 (but not $PGF_{2\alpha}$) synthesis only at 160 μM Δ^9-THC. Release of PGE_2 from isolated human synovial cells was stimulated two- to three-fold by 10 μM to 30 μM concentrations of Δ^9-THC, cannabinol or cannabidiol (Barrett *et al.*, 1985). Δ^9-THC, cannabidiol or cannabinol in the range of 3.2 μM to 32 μM increased PGE_2 production in WI-38 human lung fibroblasts (Burstein *et al.*, 1982). Because this response occurred in the same concentration range as that which could promote free [^{14}C]arachidonic acid accumulation, and the phospholipase inhibitor mepacrine blocked the prostaglandin production, the cannabinoid effects could be attributed to increased free arachidonic acid as the mechanism for enhanced eicosanoid accumulation. This was further supported by the demonstration that a number of cannabinoid compounds increased PGE production in WI-38 cells with an order of potency that correlated well with the order of potency for [^{14}C]arachidonic acid accumulation (Burstein *et al.*, 1983).

Products of arachidonic acid metabolism from mouse peritoneal macrophages were distributed on thin-layer chromatography at positions that correspond to HETEs as well as to prostaglandins (Burstein *et al.*, 1984). [^{3}H]arachidonic acid that is released from phospholipids in islet cells in response to Δ^9-THC plus high glucose is converted predominantly to 12-HETE, but also to PGE_2 (Laychock *et al.*, 1986). The distribution of [^{14}C]arachidonic acid metabolites in human platelets indicates a predominance of lipoxygenase products, particularly HETE and HHT, compared with the small amount of the cyclooxygenase product thromboxane (White and Tansik, 1980).

6.4.3 Cannabinoid-induced inhibition of cyclooxygenase

It should be noted, with respect to the synthesis of prostaglandins, that cannabinoid compounds have been demonstrated to inhibit cyclooxygenase activity in a variety of model systems. Δ^9-THC (IC_{50} about 320 μM), as well as cannabichromene, cannabidiol, Δ^8-THC, cannabidiolic acid, and cannabinol (IC_{50}s between 70 μM and 260 μM) inhibited cyclooxygenase activity in bovine seminal vesicle microsomes (Burstein *et al.*, 1973). Cyclooxygenase inhibition in rat striatal synaptosomes was reported in response to concentrations of 0.1 μM to 10 μM Δ^9-THC (Howes and Osgood, 1976). WI-38 fibroblast and mouse brain cyclooxygenase appeared to be particularly sensitive to Δ^9-THC-7-oic acid (IC_{50} around 16 μM) (Burstein *et al.*, 1986). Because olivetol (3-pentylresorcinol) reproduced this response with an IC_{50} of 100 μM, and this structure is common to all the cannabinoid compounds shown to inhibit cyclooxygenase activity, it has been suggested that the phenolic ring plus the pentyl chain is the portion of the cannabinoid structure critical for this effect (Burstein *et al.*, 1973). These findings were also noted in human synovial cells, in which release of PGE_2 by Δ^9-THC, cannabinol and cannabidiol was reversed at higher concentrations (Barrett *et al.*, 1985). Olivetol inhibited 12-*O*-tetradecanoylphorbol-13-acetate-induced prostaglandin production with no stimulation phase at lower concentrations.

Thus, although cannabinoid compounds promote the accumulation of free arachidonic acid, the production of cyclooxygenase products might be curtailed because of a concurrent inhibition of this enzyme. This appears to be the case in human platelets, where prostaglandin and thromboxane synthesis is blocked at the same concentration range of Δ^9-THC and cannabidiol as is the stimulation of phospholipase A_2, and yet lipoxygenase activity is not affected (White and Tansik, 1980). The mixed action of cannabinoid compounds may also explain the findings in guinea-pig brain slice preparations that Δ^9-THC could have different effects on unstimulated or acetylcholine-stimulated prostaglandin efflux, and different effects depending upon the region of the brain (Reichman *et al.*, 1987).

6.4.4 Possible mechanisms by which cannabinoid compounds induce arachidonic acid release

The mechanism of action for the stimulation of arachidonic acid accumulation by cannabinoid compounds has been addressed in several ways. Audette *et al.* (1991) treated S49 cells or mouse peritoneal macrophages with either pertussis toxin or cholera toxin in an effort to determine if a G protein might be involved. Those experiments were inconclusive because the toxins had similar or greater effects on the unstimulated release as they did on the Δ^9-THC-stimulated release. The decrement in Δ^9-THC-stimulated release incurred after treatment with either cholera or pertussis toxin was less than 10% in the S49 cells, although the magnitude of the change was greater for the peritoneal macrophages. One interpretation of an inhibition of the response by both cholera and pertussis toxins could relate to the fact that both toxins are expected to increase cellular cyclic AMP content. However, cyclic AMP has not been implicated in cannabinoid-induced arachidonic acid accumulation. Thus, it is not clear how to interpret these findings.

Felder *et al.* (1992) determined that the arachidonic acid mobilization could not be attributed to an effect mediated by CB_1 receptors. An identical response occurred in Chinese hamster ovarian fibroblasts (CHO control cells) as occurred is CHO cells transfected with the rat CB_1 DNA, indicating that the cannabinoid receptor is not necessary for the response. In those studies, CP 55,940 and HU-210, as well as the inactive isomer HU-211, stimulated [^{3}H]arachidonic acid release at concentrations of 0.1 mM and 1 mM, indicating that the strict enantioselectivity expected for a cannabinoid receptor-mediated process is lacking. The possibility was tested that stimulation of the CB_1 receptor might act synergistically with ATP-stimulated arachidonic acid release as does carbachol via the m_2 muscarinic receptor and G_I (Felder *et al.*, 1992). However, neither CP 55,940 at concentrations ranging from 1 pM to 1 μM, nor HU-210 or WIN 55,212–2 at concentrations up to 100 μM, exhibited any ability to augment ATP-stimulated arachidonic acid release. This would indicate that this physiological regulatory mechanism is not activated by CB_1 receptors. The polyphosphatidylinositol pathway was not activated by cannabinoid compounds in transfected or untransfected CHO cells, as it is for the carbachol activation in m_5-transfected CHO cells (Felder *et al.*, 1992). One mechanism whereby phospholipase A_2 might be activated is via the increased intracellular free Ca^{2+} concentrations evoked by μM concentrations of cannabinoid compounds (Felder *et al.*, 1992). However, alteration of this single step may not fully explain the enhanced accumulation of free arachidonic acid.

The hypothesis that cannabinoid compounds increase arachidonic acid by directly stimulating some isoform of phospholipase A_2 has been tested in broken-cell preparations. The ability of phospholipase from a 10 000×*g* supernatant fraction from rat Leydig cells to cleave [^{14}C]arachidonic acid from prelabelled phospholipids was enhanced two-fold by 3.2 μM Δ^9-THC, and five-fold by 16 μM Δ^9-THC, 11-OH-Δ^9-THC, or cannabinol (Burstein and Hunter, 1981). In islet homogenates, 20 μM Δ^9-THC increased arachidonic acid release from ^{14}C-labelled phosphatidylcholine by nearly two-fold (Laychock *et al.*, 1986). In invertebrate systems, Δ^9-THC or cannabinol, in the range of 10 μM to 100 μM, stimulated the release of [^{14}C]arachidonic acid from synthetic 1-stearoyl-2-arachidonyl phosphatidylcholine by sea urchin sperm homogenates (Chang *et al.*, 1991). Cleavage of the substrate dicapryloyllecithin by phospholipase A_2 from *Naja naja* venom was stimulated up to three-fold by Δ^9-THC, cannabidiol, cannabinol or cannabigerol in the 10 μM to 20 μM range (Evans *et al.*, 1987). However, greater concentrations, up to 150 μM, completely reversed the enhancement.

A second mechanism proposed for the increased arachidonic acid accumulation in cells treated with cannabinoid compounds is the failure to re-esterify free arachidonic acid in the phospholipid synthetic pathway. Acyltransferase (acyl CoA:lysophosphatidylcholine acyltransferase; EC2.3.1.23) activity was determined directly by transfer of oleoyl CoA into [^{32}P]lysophosphatidylcholine in a preparation of splenic lymphocytes (Greenberg *et al.*, 1977; Greenberg and Mellors, 1978). After pre-incubation of the cells with 0.2 μM to 2 μM Δ^9-THC, or between 20 μM and 1 mM concentrations of Δ^8-THC, cannabinol, cannabigerol, or olivetol, acyltransferase activity was inhibited. Acyltransferase activity in rat pancreatic islet homogenates, measured by [^{14}C]arachidonic acid incorporation into lysophosphatidylcholine, was reduced by 15% and 20% after 30 min incubation with 2 μM and 20 μM Δ^9-THC respectively (Laychock *et al.*, 1986). After 60 min, activity was reduced by 40% with as little as 0.2 μM Δ^9-THC. Felder *et al.* (1992) showed that 100 μM HU-210 could totally eliminate [^{3}H]arachidonic acid uptake not only into CHO cells that had been transfected with rat CB_1 cDNA but also into control CHO cells that lacked this receptor, demonstrating that this response was not cannabinoid-receptor mediated. The results with cannabinoid compounds contrast with what would be expected for receptor-mediated arachidonic acid accumulation. For example, carbachol triggers arachidonic acid release from m_5-CHO-transfected cells, an effect that is m_5 receptor-mediated. Conversely, it fails to affect uptake of [^{3}H]arachidonic acid into these cells, a process that is not mediated by m_5 receptors (Felder *et al.*, 1992).

Reichman *et al.* (1991) have provided evidence, using guinea-pig brain slice preparations, that phospholipase A_2 activation is not involved in the cannabinoid-evoked increase in arachidonic acid accumulation, but that inhibition of acyltransferase activity can account for observed effects. Using [^{3}H]glycerol as the labelling substrate, 8 μM Δ^9-THC produced no significant increase in lysophosphatidylcholine or lysophosphatidylethanolamine, and only a minor appearance of lysophosphatidylinositol compared with the total labelled phosphatidylinositol pool. These products would have been expected if arachidonic acid had been cleaved from the 2 position by phospholipase A_2 activity, as demonstrated by the control stimulation of that activity by melittin. In these brain slice preparations, acetylcholine acts via a receptor-mediated regulation of phosphoinositide phospholipase C to produce diacylglycerol, which can subsequently be acted upon by diacylglycerol lipase to release fatty acids and glycerol. The effects of combining Δ^9-THC with acetylcholine on the decrease in phosphatidylinositol and the production of diglyceride and glycerol were greater than additive. This finding, plus other evidence (see above), suggests that two different mechanisms are involved in the arachidonic acid accumulation evoked by the two different classes of compounds and, furthermore, that the phospholipase C pathway is not responsible for the effects of cannabinoid compounds (Reichman *et al.*, 1991).

Reichman *et al.* (1991) have proposed that the action of cannabinoid compounds of increasing arachidonic acid accumulation can be entirely attributed to the ability of cannabinoid compounds to block lipid re-acylation subsequent to arachidonic acid cleavage resulting from basal or endogenous stimuli in the brain slice preparations. This proposal is supported by studies showing that Δ^9-THC significantly reduced the amount of [^{3}H]arachidonic acid that could be incorporated into phosphatidylinositol, phosphatidylethanolamine and triglycerides. The concentration–response curve for Δ^9-THC to block acylation was identical to that observed for radiolabelled arachidonic acid accumulation from prelabelled phospholipids. Nevertheless, the response was much more robust when acylation of each of the phospholipid pools was measured than when the accumulation of unesterified arachidonic acid released from each of the prelabelled phospholipid pools was measured.

The physiological relevance of cannabinoid-mediated inhibition of acyltransferase activity was addressed by Huszar *et al.* (1977), who injected mice with increasing doses of Δ^9-THC, and measured the acyltransferase activity in tissues. Doses of greater than 10 mg/kg and 25 mg/kg were required to produce a 10% decrement in splenic lymphocyte and synaptosomal acyltransferase activity, respectively. By extrapolation, it was found that concentrations of greater than 50 to 70 mg/kg would be required to produce a 50% inhibition of the enzyme activities. These estimates of IC_{50} doses exceed by 50-fold the IC_{50} values required to produce typical cannabinoid responses in mice (see Abood and Martin, 1992). Reichman *et al.* (1991) performed log dose–response curves for both (–) and (+) stereoisomers of Δ^9-THC, and Δ^8-THC, cannabinol and cannabidiol for inhibition of esterification of [^{3}H]arachidonic acid using brain cortex slices. The order of potency was not related to that expected for cannabimimetic compounds in behavioural tests. Thus, although inhibition of acyltransferase may not be correlated with analgesia and other activities, one could speculate that this effect may relate to the CNS depression elicited by high doses of cannabinoid compounds.

6.5 Protein kinase C

Δ^9-THC has been shown to activate protein kinase C by a mechanism which has been hypothesized to be related to the effects of cannabinoid compounds to alter the physical state of the membrane lipid (Hillard and Auchampach, 1994). In that study, a semi-purified preparation of protein kinase C from rat forebrain was tested for its ability to phosphorylate histone. It was shown that Δ^9-THC, cannabinol, cannabidiol, and 11-OH-Δ^9-THC at concentrations of 3 μM to 30 μM were able to activate phosphatidylserine- and Ca^{2+}-dependent isoforms of protein kinase C. Δ^9-THC at 10 μM had no effect on the concentration requirements for phosphatidylserine, but reduced the K_{act} for Ca^{2+} two-fold. Δ^9-THC in the same concentration range also promoted binding of [^{3}H]phorbol dibutyrate to the diacylglycerol site of protein kinase C. These same concentrations of Δ^9-THC increased the fluorescence anisotropy of diphenylhexatriene (i.e. increased lipid ordering) in fluorescence polarization studies of the phosphatidylserine vesicles used in the protein kinase C studies. Hillard and Auchampach (1994) proposed that the cannabinoid compounds facilitated insertion of protein kinase C into the hydrophobic domain of the vesicle, and thereby reduced the requirement for Ca^{2+} to promote association of the enzyme with the membrane and its activation. This explanation is logical because activation of protein kinase C occurs concomitantly with its membrane insertion, and Ca^{2+} may promote this process by a mechanism that reduces the energy barrier for protein insertion. The authors point out that other membrane-acting agents including cholesterol also activate protein kinase C, suggesting that this mechanism is perhaps relevant to the physiological regulation of protein kinase C. The significance of protein kinase C activation to the pharmacological actions of cannabinoid drugs *in vivo* has not yet been determined.

6.6 Implications and future directions

In agreement with the conclusions of Martin (1986) in his review of the cellular effects of cannabinoid compounds, there appear to be numerous actions that Δ^9-THC and other cannabinoid compounds exert *in vitro*. Many of these actions have been analysed in the present discussion of the literature describing the influences of cannabinoid compounds on recognized signal transduction pathways. What has not been clear from studies in the literature is which of these actions observed in *in vitro* model systems may apply to the regulation of cellular function in the intact organism.

One difficulty in approaching the problem of relevance to cannabinoid behaviours in animals has been the lack of availability of a potent and efficacious cannabinoid receptor antagonist. The aminoalkylindole antagonist, WIN 56,098, is not a useful antagonist. Although it blocks responses to aminoalkylindole agonists in the *in vitro* assays of mouse vas deferens relaxation and inhibition of adenylate cyclase, this compound has not been shown to antagonize cannabinoid behaviours *in vivo* (Pacheco *et al.*, 1991). Perhaps this first-generation antagonist does not have the potency necessary to achieve concentrations at the receptor sites that would be sufficient to result in a decrement in overt behaviours. An important advance, reported during the proof stage of this book, has been the development of a cannabinoid receptor antagonist that is both potent and selective (Rinaldi-Carmona *et al.*, 1994). This antagonist compound should be useful for blocking specific behaviours in order to assess the role played by the antagonized receptor. It should also be invaluable in determining synaptic pathways involved in cannabinoid actions at important brain regions including the basal ganglia and the hippocampal–dentate gyrus structures. At the biochemical level, the antagonist will be particularly useful for determining the role of membrane perturbations on the effects of less potent cannabinoid compounds, including Δ^9-THC, that may be produced by actions in addition to stimulation of cannabinoid receptors.

In an effort to overcome the problem of the lack of an antagonist, the approach of developing irreversibly binding cannabinoid agonist ligands has been undertaken (Compton *et al.*, 1990; Richardson *et al.*, 1991; Charalambous *et al.*, 1992). These ligands are expected to attain equilibrium binding with the cannabinoid receptor as a function of their affinity and then covalently interact with vulnerable chemical groups on the protein as a function of their reactivity. These ligands could be useful tools *in vitro* to study second messenger responses if covalent occupancy of the receptor is able to block the ability of agonist ligands to enter the binding pocket and initiate a response. In theory, these ligands should be able to block receptor-mediated responses both *in vitro* and *in vivo*. In practice, utilization of covalently binding ligands in intact animals is dependent upon their reaching the receptors in sufficient quantities to sustain an observable decrement in the response. This could be a problem if the compound reacts with the infinite numbers of potential interaction sites during drug distribution. In addition, the minimum number of receptors required to be occupied in order to produce various responses in intact animals is not known. It is possible that inactivation of a major fraction (e.g. >99%) of receptor sites may be necessary before a decrement in response is observed. Given these problems, interpretation of negative results will be difficult when using covalently binding ligands.

Another impediment to interpreting the relevance of *in vitro* studies to cannabinoid behaviours in animals is our inability to directly determine what concentrations of cannabinoid compounds are achieved in the extracellular neuronal environment after administration of drugs in sufficient doses to evoke behavioural responses. Although the brain concentrations of Δ^9-THC and its metabolites have been determined at the peak time of behavioural effects (Ohlsson *et al.*, 1980), it is not clear what concentrations of these compounds are available to interact with cannabinoid receptors. The problem is related to the pharmacokinetics of the cannabinoid compounds. These drugs are rapidly metabolized to both active and inactive products. Both parent and biotransformed compounds may be sequestered in fat stores and also potentially interact with membrane lipids in virtually all cells in the body. Thus, for events that may be related to membrane perturbations, we do not have an appreciation of how much total cannabinoid compound is present in the membranes of the relevant brain cells after distribution *in vivo*. For events that require interactions with cannabinoid receptors, we do not know how much is available in the extracellular fluids to interact with receptors on the neuronal surface. Thus, extrapolating the results of responses observed in model systems to possible responses in whole animals is difficult at best. This conundrum has been discussed in a review of the central neuropharmacology of cannabinoid compounds by Pertwee (1988).

In conclusion, cannabinoid compounds, including both CNS-active and CNS-inactive compounds, are able to influence a variety of signal transduction systems. A cannabinoid receptor, CB_1, is found in great abundance in the brain, and is coupled to G proteins of the $G_{I/O}$ class to regulate intracellular cyclic AMP content and influence ion channel opening. Future investigations of alternative cannabinoid receptor subtype(s) may ultimately describe other signal transduction pathways that can be regulated by these compounds. Cannabinoid compounds can also interact with membranes to influence receptors and enzymes involved in signal transduction in *in vitro* model systems. Cannabinoid compounds are able to interact with enzymes that regulate arachidonic acid metabolism. These actions may be related to the relative structural similarities that the cannabinoid compounds have to eicosanoids (see Milne and Johnson, 1981). Those non-receptor-mediated actions of cannabinoid compounds are readily observable in *in vitro* model systems. However, the role these mechanisms may play in the actions of cannabinoid compounds in intact animals is not understood.

References

Abood, M.E. & Martin, B.R. (1992) Neurobiology of marijuana abuse. *Trends Pharmacol. Sci.* **13**, 201–206.

Audette, C.A., Burstein, S.H., Doyle, S.A. & Hunter, S.A. (1991) G-protein mediation of cannabinoid-induced phospholipase activation. *Pharmacol. Biochem. Behav.* **40**, 559–563.

Baldwin, J.M. (1993) The probable arrangement of the helices in G protein-coupled receptors. *EMBO J.* **12**, 1693–1703.

Barrett, M.L., Gordon, D. & Evans, F.J. (1985) Isolation from *Cannabis sativa* L. of cannflavin – a novel inhibitor of prostaglandin production. *Biochem. Pharmacol.* **34**, 2019–2024.

Bidaut-Russell, M. & Howlett, A.C. (1991) Cannabinoid receptor-regulated cyclic AMP accumulation in the rat striatum. *J. Neurochem.* **57**, 1769–1773.

Bidaut-Russell, M., Devane, W.A. & Howlett, A.C. (1990) Cannabinoid receptors and modulation of cyclic AMP accumulation in the rat brain. *J. Neurochem.* **55**, 21–26.

Birnbaumer, L., Abramowitz, J. & Brown A.M. (1990) Receptor-effector coupling by G proteins. *Biochim Biophys. Acta* **1031**, 163–224.

Bloom, A.S. (1984) Effects of cannabinoids on neurotransmitter receptors in the brain. In *The Cannabinoids: Chemical, Pharmacologic, and Therapeutic Aspects* (eds Agurell, S., Dewey, W. & Willette, R.), pp. 575–589. Academic Press, Orlando, FL.

Bloom, A.S. & Hillard, C.J. (1985) Cannabinoids, neurotransmitter receptors and brain membranes. In *Marihuana '84. Proceedings of the Oxford Symposium on Cannabis* (eds Harvey, D.J., Paton, W. & Nahas, G.G.), pp. 217–231. IRL Press, Oxford, UK.

Bouaboula, M., Rinaldi, M., Carayon, P., Carillon, C., Delpech, B., Shire, D., Le Fur, G. & Casellas, P. (1993) Cannabinoid-receptor expression in human leukocytes. *Eur. J. Biochem.* **214**, 173–180.

Brown, A.M. (1991) A cellular logic for G protein-coupled ion channel pathways. *FASEB J.* **5**, 2175–2179.

Burstein, S. & Hunter, S.A. (1978) Prostaglandins and cannabis-VI. Release of arachidonic acid from HeLa cells by Δ^1-tetrahydrocannabinol and other cannabinoids. *Biochem. Pharmacol.* **27**, 1275–1280.

Burstein, S. & Hunter, S.A. (1981) Prostaglandins and cannabis-VIII. Elevation of phospholipase A_2 activity by cannabinoids in whole cells and subcellular preparations. *J. Clin. Pharmacol.* **21** (Suppl.), 240S–248S.

Burstein, S., Levin, E. & Varanelli, C. (1973) Prostaglandins and cannabis–II. Inhibition of biosynthesis by the naturally occurring cannabinoids. *Biochem. Pharmacol.* **22**, 2905–2910.

Burstein, S., Hunter, S.A., Sedor, C. & Shulman, S. (1982) Prostaglandins and cannabis-IX. Stimulation of prostaglandin E_2 synthesis in human lung fibroblasts by Δ^1-tetrahydrocannabinol. *Biochem. Pharmacol.* **31**, 2361–2365.

Burstein, S., Hunter, S.A. & Ozman, K. (1983) Prostaglandins and cannabis. XII. The effect of cannabinoid structure on the synthesis of prostaglandins by human lung fibroblasts. *Mol. Pharmacol.* **23**, 121–126.

Burstein, S., Hunter, S.A., Ozman, K. & Renzulli, L. (1984) Prostaglandins and cannabis–XIII. Cannabinoid-induced elevation of lipoxygenase products in mouse peritoneal macrophages. *Biochem. Pharmacol.* **33**, 2653–2656.

Burstein, S., Hunter, S.A., Latham, V. & Renzulli, L. (1986) Prostaglandins and cannabis–XVI. Antagonism of Δ^1-tetrahydrocannabinol action by its metabolites. *Biochem. Pharmacol.* **35**, 2553–2558.

Casey, P.J. & Gilman, A.G. (1988) G protein involvement in receptor-effector coupling. *J. Biol. Chem.* **263**, 2577–2580.

Caulfield, M.P. & Brown, D.A. (1992) Cannabinoid receptor agonists inhibit Ca current in NG108–15 neuroblastoma cells via a pertussis toxin-sensitive mechanism. *Brit. J. Pharmacol.* **106**, 231–232.

Chang, M.C., Berkery, D., Laychock, S.G. & Schuel, H. (1991) Reduction of the fertilizing capacity of sea urchin sperm by cannabinoids derived from marihuana. III. Activation of phospholipase A_2 in sperm homogenate by Δ^9-tetrahydrocannabinol. *Biochem. Pharmacol.* **42**, 899–904.

Charalambous, A., Yan, G., Houston, D.B., Howlett, A.C., Compton, D.R., Martin, B.R.

& Makriyannis, A. (1992) 5′-Azido-Δ^8-THC: a novel photoaffinity label for the cannabinoid receptor. *J. Med. Chem.* **35**, 3076–3079.

Colburn, R.W., Ng, L.K.Y., Lemberger, L. & Kopin, I.J. (1974) Subcellular distribution of Δ^9-tetrahydrocannabinol in rat brain. *Biochem. Pharmacol.* **23**, 873–877.

Compton, D.R. & Martin, B.R. (1990) Pharmacological evaluation of water soluble cannabinoids and related analogs. *Life Sci.* **46**, 1575–1585.

Compton, D.R. & Little, P.J., Martin, B.R., Gilman, J.W., Saha, J.K., Jorapur, V.S., Sard, H.P. & Razdan, R.K. (1990) Synthesis and pharmacological evaluation of amino, azido, and nitrogen mustard analogues of 10-substituted cannabidiol and 11- or 12-substituted Δ^8-tetrahydrocannabinol. *J. Med. Chem.* **33**, 1437–1443.

Compton, D.R., Gold, L.H., Ward, S.J., Balster, R.L. & Martin, B.R. (1992a) Aminoalkylindole analogs: cannabimimetic activity of a class of compounds structurally distinct from Δ^9-tetrahydrocannabinol. *J. Pharmacol. Exp. Therap.* **263**, 1118–1126.

Compton, D.R., Johnson, M.R., Melvin, L.S. & Martin, B.R. (1992b) Pharmacological profile of a series of bicyclic cannabinoid analogs: classification as cannabimimetic agents. *J. Pharmacol. Exp. Therap.* **260**, 201–209.

Compton, D.R., Rice, K.C., de Costa, B.R., Razdan, R.K., Melvin, L.S., Johnson, M.R. & Martin, B.R. (1993) Cannabinoid structure–activity relationships: correlation of receptor binding and *in vivo* activities. *J. Pharmacol. Exp. Therap.* **265**, 218–226.

Conklin, B.R. & Bourne, H.R. (1993) Structural elements of Gα subunits that interact with Gβτ, receptors, and effectors. *Cell* **73**, 631–641.

Deadwyler, S.A., Hampson, R.E., Bennett, B.A., Edwards, T.A., Mu, J., Pacheco, M.A., Ward, S.J. & Childers, S.R. (1993) Cannabinoids modulate potassium current in cultured hippocampal neurons. *Recep. Chan.* **1**, 121–134.

Devane, W.A., Spain, J.W., Coscia, C.J. & Howlett, A.C. (1986) An assessment of the role of opioid receptors in the response to cannabimimetic drugs. *J. Neurochem.* **46**, 1929–1935.

Eldridge, J.C. & Landfield, P.W. (1990) Cannabinoid interactions with glucocorticoid receptors in rat hippocampus. *Brain Res.* **534**, 135–141.

Eldridge, J.C., Murphy, L.L. & Landfield, P.W. (1991) Cannabinoids and the hippocampal glucocorticoid receptor: recent findings and possible significance. *Steroids* **56**, 226–231.

Evans, A.T., Formukong, E. & Evans, F.J. (1987) Activation of phospholipase A_2 by cannabinoids. Lack of correlation with CNS effects. *FEBS Lett.* **211**, 119–122.

Felder, C.C., Veluz, J.S., Williams, H.L., Briley, E.M. & Matsuda, L.A. (1992) Cannabinoid agonists stimulate both receptor- and non-receptor-mediated signal transduction pathways in cells transfected with and expressing cannabinoid receptor clones. *Mol. Pharmacol.* **42**, 838–845.

Fleming, L.M., Ponjee, G. & Childers, S.R. (1992) Inhibition of protein phosphorylation by opioid-inhibited adenylyl cyclase in rat brain membranes. *J. Pharmacol. Exp. Therap.* **260**, 1416–1424.

Freissmuth, M., Cassey, P.J. & Gilman, A.G. (1989) G proteins control diverse pathways of transmembrane signaling. *FASEB J.* **3**, 2125–2131.

Friedman, M.A. & Wrenn, J.M. (1977) Suppression by Δ^9-tetrahydrocannabinol of induction of hepatic tyrosine aminotransferase and tryptophan oxygenase. *Toxicol. Appl. Pharmacol.* **41**, 345–352.

Garrett, E.R. & Hunt, C.A. (1974) Physiochemical properties, solubility, and protein binding of Δ^9-tetrahydrocannabinol. *J. Pharm. Sci.* **63**, 1056–1064.

Gérard, C., Mollereau, C., Vassart, G. & Parmentier, M. (1990) Nucleotide sequence of a human cannabinoid receptor cDNA. *Nucl. Acids Res.* **18**, 7142.

Gérard, C.M., Mollereau, C., Vassart, G. & Parmentier, M. (1991) Molecular cloning of a human cannabinoid receptor which is also expressed in testis. *Biochem J.* **279**, 129–134.

Greenberg, J.H. & Mellors, A. (1978) Specific inhibition of an acyltransferase by Δ^9-tetrahydrocannabinol. *Biochem. Pharmacol.* **27**, 329–333.

Greenberg, J.H., Saunders, M.E. & Mellors, A. (1977) Inhibition of a lymphocyte membrane enzyme by Δ^9-tetrahydrocannabinol *in vitro*. *Science* **197**, 475–477.

Harris, L.S., Carchman, R.A. & Martin, B.R. (1978) Evidence for the existence of specific cannabinoid binding sites. *Life Sci.* **22**, 1131–1137.

Harris, R.A. & Stokes, J.A. (1982) Cannabinoids inhibit calcium uptake by brain synaptosomes. *J. Neurosci.* **2**, 443–447.

Hershkowitz, M., Goldman, R. & Raz, A. (1977) Effect of cannabinoids on neurotransmitter uptake, ATPase activity and morphology of mouse brain synaptosomes. *Biochem. Pharmacol.* **26**, 1327–1331.

Hibert, M.F., Trumpp-Kallmeyer, S., Hoflack, J. & Bruinvels, A. (1993) This is not a G protein-coupled receptor. *Trends Pharmacol. Sci.* **14**, 7–12.

Hillard, C.J. & Auchampach, J.A. (1994) *In vitro* activation of brain protein kinase C by the cannabinoids. *Biochim. Biophys. Acta.* **1120**, 163–170.

Hillard, C.J. & Bloom, A.S. (1982) Δ^9-Tetrahydrocannabinol-induced changes in β-adrenergic receptor binding in mouse cerebral cortex. *Brain Res.* **235**, 370–377.

Hillard, C.J. & Bloom, A.S. (1983) Possible role of prostaglandins in the effects of the cannabinoids on adenylate cyclase activity. *Eur. J. Pharmacol.* **91**, 21–27.

Hillard, C.J. & Bloom, A.S. (1984) Further studies of the interactions of delta-9-tetrahydrocannabinol with the beta-adrenergic receptor. In *The Cannabinoids: Chemical, Pharmacologic, and Therapeutic Aspects* (eds Agurell, S., Dewey, W. & Willette, R.), pp. 591–602. Academic Press, Orlando, FL.

Hillard, C.J. & Pounds, J.J. (1993) Cannabinoid effects on depolarization-induced calcium influx in cerebral cortical synaptosomes.

Hillard, C.J., Harris, R.A. & Bloom, A.S. (1985) Effects of the cannabinoids on physical properties of brain membranes and phospholipid vesicles: fluorescence studies. *J. Pharmacol. Exp. Therap.* **232**, 579–588.

Hillard, C.J., Bloom, A.S. & Houslay, M.D. (1986) Effects of Δ^9-tetrahydrocannabinol on glucagon receptor coupling to adenylate cyclase in rat liver plasma membranes. *Biochem. Pharmacol.* **35**, 2797–2803.

Hillard, C.J., Pounds, J.J., Boyer, D.R. & Bloom, A.S. (1990) Studies of the role of membrane lipid order in the effects of Δ^9-tetrahydrocannabinol on adenylate cyclase activation in heart. *J. Pharmacol. Expt. Ther.* **252**, 1075–1082.

Houslay, M.D. & Gordon, L.M. (1983) The activity of adenylate cyclase is regulated by the nature of its lipid environment. *Curr. Top. Membr. Transp.* **18**, 179–231.

Houston, D.B. & Howlett, A.C. (1993) Solubilization of the cannabinoid receptor from rat brain and its functional interaction with guanine nucleotide-binding proteins. *Mol. Pharmacol.* **43**, 17–22.

Howes, J.F. & Osgood, P.F. (1976) Cannabinoids and the inhibition of prostaglandin synthesis. In *Marijuana: Chemistry, Biochemistry and Cellular Effects* (ed. Nahas, G.G.), pp. 415–424. Springer-Verlag, New York.

Howlett, A.C. (1985) Cannabinoid inhibition of adenylate cyclase. Biochemistry of the response in neuroblastoma cell membranes. *Mol. Pharmacol.* **27**, 429–436.

Howlett, A.C. (1987) Cannabinoid inhibition of adenylate cyclase: relative activity of constituents and metabolites of marijuana. *Neuropharmacology* **26**, 507–512.

Howlett, A.C. (1990) Reverse pharmacology applied to the cannabinoid receptor. *Trends Pharmacol. Sci.* **11**, 395–397.

Howlett, A.C. & Fleming, R.M. (1984) Cannabinoid inhibition of adenylate cyclase. Pharmacology of the response in neuroblastoma cell membranes. *Mol. Pharmacol.* **26**, 532–538.

Howlett, A.C., Qualy, J.M. & Khachatrian, L.L. (1986) Involvement of G_i in the inhibition of adenylate cyclase by cannabimimetic drugs. *Mol. Pharmacol.* **29**, 307–313.

Howlett, A.C., Johnson, M.R., Melvin, L.S. & Milne, G.M. (1988) Nonclassical cannabinoid analgetics inhibit adenylate cyclase: development of a cannabinoid receptor model. *Mol. Pharmacol.* **33**, 297–302.

Howlett, A.C., Scott, D.K. & Wilken, G.H. (1989) Regulation of adenylate cyclase by cannabinoid drugs. Insights based on thermodynamic studies. *Biochem. Pharmacol.* **38**, 3297–3304.

Howlett, A.C., Bidaut-Russell, M., Champion-Dorow, T., Devane, W.A., Westlake, T.M., Wilken, G.H., Melvin, L.S. & Johnson, M.R. (1990a) Cannabinoid receptors in the brain and cultured neuroblastoma cells. *Eur. J. Pharmacol.* **183**, 2149–2150.

Howlett, A.C., Bidaut-Russell, M., Devane, W.A., Melvin, L.S., Johnson, M.R. & Herkenham, M. (1990b) The cannabinoid receptor: biochemical, anatomical and behavioural characterization. *Trends Neurosci.* **13**, 420–423.

Howlett, A.C., Champion, T.M., Wilken, G.H. & Mechoulam, R. (1990c) Stereochemical effects of 11-OH-Δ^8-tetrahydrocannabinol-dimethylheptyl to inhibit adenylate cyclase and bind to the cannabinoid receptor. *Neuropharmacology* **29**, 161–165.

Howlett, A.C., Evans, D.M. & Houston, D.B. (1992) The cannabinoid receptor. In *Marijuana/Cannabinoids: Neurobiology and Neurophysiology* (eds Murphy, L. & Bartke, A.), pp. 35–72. CRC Press, Boca Raton, FL.

Hunter, S.A., Burstein, S. & Renzulli, L. (1986) Effects of cannabinoids on the activities of mouse brain lipases. *Neurochem. Res.* **11**, 1273–1288.

Huszar, L.A., Greenberg, J.H. & Mellors, A. (1977) Effects of Δ^9-tetrahydrocannabinol on lymphocyte and synaptosomal lysophosphatidylcholine acyltransferases *in vivo*. *Mol. Pharmacol.* **13**, 1086–1091.

Kaminski, N.E., Abood, M.E., Kessler, F.K., Martin, B.R. & Schatz, A.R. (1992) Identification of a functionally relevant cannabinoid receptor on mouse spleen cells that is involved in cannabinoid-mediated immune modulation. *Mol. Pharmacol.* **42**, 736–742.

Kubena, R.K., Perhach, J.L. & Barry, H. (1971) Corticosterone elevation mediated centrally by Δ^1-tetrahydrocannabinol in rats. *Eur. J. Pharmacol.* **14**, 89–92.

Kuster, J.E., Stevenson, J.I., Ward, S.J., D'Ambra, T.E. & Haycock, D.A. (1993) Aminoalkylindole binding in rat cerebellum: selective displacement by natural and synthetic cannabinoids. *J. Pharmacol. Exp. Therap.* **264**, 1352–1363.

Lamb, T.D. & Pugh, E.N. (1992) G-protein cascades: gain and kinetics. *Trends Neurosci.* **15**, 291–297.

Laychock, S.G., Hoffman, J.M., Meisel, E. and Bilgin, S. (1986) Pancreatic islet arachidonic acid turnover and metabolism and insulin release in response to delta-9-tetrahydrocannabinol. *Biochem. Pharmacol.* **35**, 2003–2008.

Little, P.J. & Martin, B.R. (1991) The effects of Δ^9-tetrahydrocannabinol and other cannabinoids on cAMP accumulation in synaptosomes. *Life Sci.* **48**, 1133–1141.

Little, P.J., Compton, D.R., Johnson, M.R., Melvin, L.S. & Martin, B.R. (1988) Pharmacology and stereoselectivity of structurally novel cannabinoids in mice. *J. Pharmacol. Exp. Therap.* **247**, 1046–1051.

Little, P.J., Compton, D.R., Mechoulam, R. & Martin, B.R. (1989) Stereochemical effects of 11-OH-Δ^8-THC-dimethylheptyl in mice and dogs. *Pharmacol. Biochem. Behav.* **32**, 661–666.

Mackie, K. & Hille, B. (1992) Cannabinoids inhibit N-type calcium channels in neuroblastoma-glioma cells. *Proc. Natl Acad. Sci. USA* **89**, 3825–3829.

Mackie, K., Devane, W.A. & Hille, B. (1993) Anandamide, an endogenous cannabinoid, inhibits calcium currents as a partial agonist in N18 neuroblastoma cells. *Mol. Pharmacol.* **44**, 498–503.

Makriyannis, A. & Rapaka, R.S. (1990) The molecular basis of cannabinoid activity. *Life Sci.* **47**, 2173–2184.

Makriyannis, A., Banijamali, A., Jarrell, H.C. & Yang, D.-P. (1989) The orientation of (−)-Δ^9-tetrahydrocannabinol in DPPC bilayers as determined from solid-state ^{2}H-NMR. *Biochim. Biophys. Acta* **986**, 141–145.

Martin, B.R. (1986) Cellular effects of cannabinoids. *Pharmacol. Rev.* **38**, 45–74.

Martin, B.R., Compton, D.R., Thomas, B.F., Prescott, W.R., Little P.J., Razdan, R.K., Johnson, M.R., Melvin, L.S., Mechoulam, R. & Ward, S.J. (1991) Behavioral, biochemical, and molecular modeling evaluations of cannabinoid analogs. *Pharmacol. Biochem. Behav.* **40**, 471–478.

Matsuda, L.A., Lolait, S.J., Brownstein, M.J., Young, A.C. & Bonner, T.I. (1990) Structure of a cannabinoid receptor and functional expression of the cloned cDNA. *Nature* **346**, 561–564.

Mavromoustakos, T., Yang, D.P., Charalambous, A., Herbette, L.G. & Makriyannis, A. (1990) Study of the topography of cannabinoids in model membranes using X-ray diffraction. *Biochim. Biophys. Acta* **1024**, 336–344.

Mavromoustakos, T., Yang, D.P., Broderick, W., Fournier, D. & Makriyannis, A. (1991) Small angle X-ray diffraction studies on the topography of cannabinoids in synaptic plasma membranes. *Pharmacol. Biochem. Behav.* **40**, 547–552.

McFadzean, I., Mullaney, I., Brown, D.A. & Milligan, G. (1989) Antibodies to the GTP binding protein, Go, antagonize noradrenaline-induced calcium current inhibition in NG108-15 hybrid cells. *Neuron* **3**, 177–182.

Miczek, K.A. & Dixit, B.N. (1980) Behavioural and biochemical effects of chronic Δ^9-tetrahydrocannabinol in rats. *Psychopharmacology* **67**, 195–202.

Milne, G.M., Jr. & Johnson, M.R. (1981) Levonantradol: a role for central prostanoid mechanisms? *J. Clin. Pharmacol.* **21** (Suppl.), 367S–374S.

Mu, J., Kirby, M.T., King, V.C., Bennett, B.A., Hampson, R.E., Childers, S.R. & Deadwyler, S.A. (1993) Cannabinoids increase potassium A-current and decrease synaptic transmission via cAMP dependent processes in hippocampus. *Soc. Neurosci. Abs* **19**, 280.

Munro, S., Thomas, K.L. & Abu-Shaar, M. (1993) Molecular characterization of a peripheral receptor for cannabinoids. *Nature* **365**, 61–65.

Murphy, L.L., Steger, R.W. & Bartke, A. (1990) Psychoactive and nonpsychoactive cannabinoids and their effects on reproductive neuroendocrine parameters. In *Biochemistry and physiology of substance abuse, Vol. 2* (ed Watson, R.R.), pp. 73–93. CRC Press, Boca Raton, FL.

North, R.A. (1989) Twelfth Gaddum memorial lecture. Drug receptors and the inhibition of nerve cells. *Brit. J. Pharmacol.* **98**, 13–28.

Nye, J.S., Seltzman, H.H., Pitt, C.G. & Snyder, S.H. (1985) High-affinity cannabinoid binding sites in brain membranes labeled with $[^3H]$-5′-trimethylammonium Δ^8-tetrahydrocannabinol. *J. Pharmacol. Exp. Ther.* **234**, 784–791.

Nye, J.S., Voglmaier, S., Martenson, R.E. & Snyder, S.H. (1988) Myelin basic protein is an endogenous inhibitor of the high-affinity cannabinoid binding site in brain. *J. Neurochem.* **50**, 1170–1178.

Nye, J.S., Snowman, A.M., Voglmaier, S. & Snyder, S.H. (1989) High-affinity cannabinoid binding site: regulation by ions, ascorbic acid, and nucleotides. *J. Neurochem.* **52**, 1892–1897.

Ohlsson, A., Widman, M., Carlsson, S., Ryman, T. & Strid, C. (1980) Plasma and brain levels of Δ^6-THC and seven monooxygenated metabolites correlated to the cataleptic effect in the mouse. *Acta Pharmacol. Toxicol.* **47**, 308–317.

Okey, A.B. & Bondy, G.P. (1977) Is delta-9-tetrahydrocannabinol estrogenic? *Science* **195**, 904–906.

Pacheco, M., Childers, S.R., Arnold, R., Casiano, F. & Ward, S.J. (1991) Aminoalkylindoles: actions on specific G-protein-linked receptors. *J. Pharmacol. Exp. Ther.* **257**, 170–183.

Pacheco, M.A., Ward, S.J. & Childers, S.R. (1993) Identification of cannabinoid receptors in cultures of rat cerebellar granule cells. *Brain Res*, **603**, 102–110.

Pertwee, R.G. (1974) Tolerance to the effect of Δ^1-tetrahydrocannabinol on corticosterone levels in mouse plasma produced by repeated administration of cannabis extract or Δ^1-tetrahydrocannabinol. *Brit. J. Pharmacol.* **51**, 391–397.

Pertwee, R.G. (1985) Cannabis. In *Psychopharmacology 2, Part 1: Preclinical Psychopharmacology* (eds Grahame-Smith, D.G. & Cowen, P.J.), pp. 364–391. Elsevier, Amsterdam.
Pertwee, R.G. (1988) The central neuropharmacology of psychotropic cannabinoids. *Pharmacol. Therapeutics* **36**, 189–261.
Pertwee, R. (1993) The evidence for the existence of cannabinoid receptors. *Gen. Pharmac.* **24**, 811–824.
Pertwee, R.G., Stevenson, L.A., Elrick, D.B., Mechoulam, R. & Corbett, A.D. (1992) Inhibitory effects of certain enantiomeric cannabinoids in the mouse vas deferens and the myenteric plexus preparation of guinea-pig small intestine. *Brit. J. Pharmacol.* **105**, 980–984.
Puder, M., Weidenfeld, J., Chowers, I., Nir, I., Conforti, N. & Siegel, R.A. (1982) Corticotrophin and corticosterone secretion following Δ^1-tetrahydrocannabinol, in intact and in hypothalamic deafferentated male rats. *Exp. Brain Res.* **46**, 85–88.
Rabin, R.A. & Molinoff, P.B. (1981) Activation of adenylate cyclase by ethanol in mouse striatal tissue. *J. Pharmacol. Exp. Ther.* **216**, 129–134.
Rabin, R.A., Bode, D.C. & Molinoff, P.B. (1986) Relationship between ethanol-induced alterations in fluorescence anisotropy and adenylate cyclase activity. *Biochem. Pharmacol.* **35**, 2331–2335.
Rawitch, A.B., Schultz, G.S., Ebner, K.E. & Vardaris, R.M. (1977) Competition of Δ^9-tetrahydrocannabinol with estrogen in rat uterine estrogen receptor binding. *Science* **197**, 1189–1191.
Reichman, M., Nen, W. & Hokin, L.E. (1987) Effects of Δ^9-tetrahydrocannabinol on prostaglandin formation in brain. *Mol. Pharmacol.* **32**, 686–690.
Reichman, M., Nen, W. & Hokin, L.E. (1988). Δ^9-tetrahydrocannabinol increases arachidonic acid levels in guinea pig cerebral cortex slices. *Mol. Pharmacol.* **34**, 823–828.
Reichman, M., Nen, W. & Hokin, L.E. (1991). Δ^9-tetrahydrocannabinol inhibits arachidonic acid acylation of phospholipids and triacylglycerols in guinea pig cerebral cortex slices. *Mol. Pharmacol.* **40**, 547–555.
Richardson, S., Mirasadeghi, S., Lynn, A., Johnson, M.R., Melvin, L.S., Herkenham, M. & Rice, K.C. (1991) Synthetic and biological studies of potential affinity ligands for the cannabinoid receptor based on CP-55, 244. *NIDA Res. Monograph* **105**, 140–146.
Rinaldi-Carmona, M., Barth, F., Héaulme, M., Shire, D., Calandra, B., Congy, C., Martinez, S., Maruani, J., Néliat, G., Caput, D., Ferrara, P., Soubrié, P., Brelière, J.-C. & Le Fur, G. (1994) SR141716A, a potent and selective antagonist of the brain cannabinoid receptor. *FEBS Letts.* **350**, 240–244.
Rosell, S., Agurell, S. & Martin, B. (1976) Effects of cannabinoids on isolated smooth muscle preparations. In *Marihuana: Chemistry, Biochemistry and Cellular Effects* (ed. Nahas, G.G.), pp. 397–406. Springer-Verlag, New York.
Roth, S.H. (1978) Stereospecific presynaptic inhibitory effect of Δ^9-tetrahydrocannabinol on cholinergic transmission in the myenteric plexus of the guinea pig. *Can. J. Physiol. Pharmacol.* **56**, 968–975.
Roth, S.H. & Williams, P.J. (1979) The non-specific membrane binding properties of Δ^9-tetrahydrocannabinol and the effects of various solubilizers. *J. Pharm. Pharmacol.* **31**, 224–230.
Salesse, R. & Garnier, J. (1984) Adenylate cyclase and membrane fluidity. The repressor hypothesis. *Mol. Cell. Biochem.* **60**, 17–31.
Sauer, M.A., Rifka, S.M., Hawks, R.L., Cutler, G.B., Jr. & Loriaux, D.L. (1983) Marijuana: interaction with the estrogen receptor. *J. Pharmacol. Exp. Therap.* **224**, 404–407.
Savarese, T.M. & Fraser, C.M. (1992) *In vitro* mutagenesis and the search for structure-function relationships among G protein-coupled receptors. *Biochem. J.* **283**, 1–19.
Schatz, A.R., Kessler, F.K. & Kaminski, N.E. (1992) Inhibition of adenylate cyclase by Δ^9-tetrahydrocannabinol in mouse spleen cells: a potential mechanism for cannabinoid-mediated immunosuppression. *Life Sci.* **51**, 25–30.
Seeman, P., Chau-Wong, M. & Moyyen, S. (1972). The membrane binding of morphine,

diphenylhydantoin, and tetrahydrocannabinol. *Can. J. Physiol. Pharmacol.* **50**, 1193–1200.
Simon, M.I., Strathmann, M.P. & Gautam, N. (1991) Diversity of G proteins in signal transduction. *Science* **252**, 802–808.
Sternweis, P.C. & Pang, I.H. (1990) The G protein-channel connection. *Trends Neurosci.* **13**, 122–126.
Strader, C.D., Sigal, I.S. & Dixon, R.A. (1989) Structural basis of β-adrenergic receptor function. *FASEB J.* **3**, 1825–1832.
Vaysse, P.J., Gardner, E.L. & Zukin, R.S. (1987) Modulation of rat brain opioid receptors by cannabinoids. *J. Pharmacol. Exp. Therap.* **241**, 534–539.
White, H.L. & Tansik, R.L. (1980) Effects of Δ^9-tetrahydrocannabinol and cannabidiol on phospholipase and other enzymes regulating arachidonate metabolism. *Prost. Med.* **4**, 409–417.
Wrenn, J.M. & Friedman, M.A. (1978) Evidence for highly specific interactions between Δ^8- and Δ^9-tetrahydrocannabinol and hydrocortisone in the rodent tyrosine aminotransferase system. *Toxicol. Appl. Pharmacol.* **43**, 569–576.
Yang, D.P., Banijamali, A., Charalambous, A., Marciniak, G. & Makriyannis, A. (1991) Solid state ^{2}H-NMR as a method for determining the orientation of cannabinoid analogs in membranes. *Pharmacol. Biochem. Behav.* **40**, 553–557.

CHAPTER 7

FUNCTIONAL SIGNIFICANCE OF CANNABINOID RECEPTORS IN BRAIN

Sam A. Deadwyler, Robert E. Hampson and Steven R. Childers

Center for the Neurobiological Investigation of Drug Abuse, Department of Physiology and Pharmacology, Bowman Gray School of Medicine, Wake Forest University, Winston, Salem, NC 27157, USA

Table of Contents

Cannabinoid Receptors
ISBN 0-12-551460-3

7.1 Introduction

The now confirmed existence of cannabinoid receptors in brain (Howlett *et al.*, 1988; Matsuda *et al.*, 1990) and the important subsequent observation that there are possibly endogenous cannabinoid-like substances also in brain tissue (Devane *et al.*, 1992) provokes the question of the functional significance of cannabinoid receptors in neuronal systems. The task at hand is therefore to identify those processes which are mediated by cannabinoid receptors and result in a change in sensory, motor, perceptual or cognitive functions. There are at present, however, several gaps in our knowledge which prevent extrapolation of existing information on receptor and cellular actions of cannabinoids to such functional endpoints. In this chapter, information which is rapidly accumulating on the cellular actions of cannabinoids will be reviewed within the context of existing data on known physiological, behavioural and cognitive effects of cannabinoids. An integration of recent findings in the hippocampus and cerebellum will also be provided in an attempt to gain insight into specific roles cannabinoids play in the regulation of brain activity.

7.2 Cellular actions of cannabinoids

7.2.1 Receptor-mediated effects of cannabinoids on membrane conductances

The binding of high-affinity ligands such as CP 55,940 (Pfizer Co.) and WIN 55,212–2 (Sterling Co.) to the cannabinoid receptor places these substances among the potent activators of cannabinoid receptor-mediated physiological processes. In several cases this potency has been confirmed with respect to a number of different *in vitro* and *in vivo* test systems (Howlett *et al.*, 1990; Abood and Martin, 1992, see also Chapter 2). However, perhaps the most important aspect of this relationship is the eventual effector mechanisms which cannabinoid receptors influence within the cell. It is well known that cannabinoid receptors belong to the general family of $G_{I/O}$ protein coupled receptors. Thus cannabinoid agonists inhibit adenylate cyclase through classic G protein-mediated pathways (see Chapter 6). Until recently there was little information on the eventual biological signals mediated by the cannabinoid inhibited cyclic AMP system, or even on whether other effectors might be activated by direct actions of G protein-coupled cannabinoid receptors. One of the most immediate and crucial actions of a G protein-coupled cannabinoid receptor is the potential to change nerve cell excitability or responsiveness. This has been tested recently using two well-characterized neuronal membrane conductance mechanisms, calcium and potassium channels.

7.2.2 Effects on calcium currents

The effects of cannabinoid ligands on calcium channel conductance were first reported by Mackie and Hille (1992) and confirmed by Caulfield and Brown (1992). Tests in NG108–15 neuroblastoma cells showed that both CP 55,940 and WIN 55,212–2, as well as Δ^9-tetrahydrocannabinol (Δ^9-THC) in higher concentrations, decrease calcium conductance through a putative N-type (ω-conotoxin-sensitive) calcium channel. It was demonstrated that these effects were sensitive to pertussis toxin (PTX) but were not mediated by changes in cyclic AMP. The degree of reduction in calcium conductance was shown to be 300–400 pA, producing approximately 40% inhibition of total current. These changes in Ca^{2+} conductance involved both low- and high-voltage sensitive channels, and did not totally eliminate the noradrenaline voltage-sensitive component of calcium conductance in these same cells. In comparison with other physiological effects of cannabinoids, the sensitivity of NG108–15 neuroblastoma cells is impressive (10 nM), only an order of magnitude different from receptor binding in isolated membranes, and certainly within the range shared by other biochemical cascades coupled to the receptor, such as inhibition of adenylate cyclase (Howlett, 1985). However, the rather slow onset of the effect (2–15 min) on Ca^{2+} currents is unusual for direct G protein-coupled ion channel responses. Since the decreased Ca^{2+} conductance is PTX sensitive but is not cyclic AMP dependent, it is possible that cannabinoid-induced changes in Ca^{2+} conductance and adenylate cyclase activity are mediated by separate G proteins. This dissociation of Ca^{2+} conductance changes from changes in cyclic AMP levels may be an important factor with respect to understanding other cannabinoid receptor-mediated changes in cellular function (Howlett *et al.*, 1990).

7.2.3 Effects on voltage-sensitive potassium A-current

The voltage-sensitive potassium channel producing A-current (I_A) in hippocampal neurons has been characterized extensively (cf. Rudy 1988; Brown *et al.*, 1990; Storm, 1990; Adams & Nonner, 1992; Hille 1992). I_A is activated to varying degrees during depolarization of the membrane (Segal *et al.*, 1984; Zbicz and Weight, 1985). I_A in hippocampal cells recorded *in vitro*, shows steady-state inactivation at membrane voltages ranging between −100 and −60 mV, exhibits a transient inactivation time-constant (τ) of 40–50 ms, is 4-aminopyridine (4-AP) sensitive and is refractory for 200 ms (Segal and Barker, 1984; Numann *et al.*, 1987; Zona *et al.*, 1988; Wu and Barish, 1992).

Whole-cell patch clamp recording techniques were employed to investigate shifts in voltage-dependent steady-state inactivation of I_A produced by cannabinoid analogues. All drugs were applied through pressure application from a second pipette whose tip was located immediately adjacent to the recorded cell (Deadwyler *et al.*, 1993). Changes in steady-state inactivation of I_A involved varying amplitude prepulses (−96 to −42 mV) of 100 ms duration which were followed by a constant depolarizing step to +80 mV for 300 ms (Figure 1). A second protocol assessed the

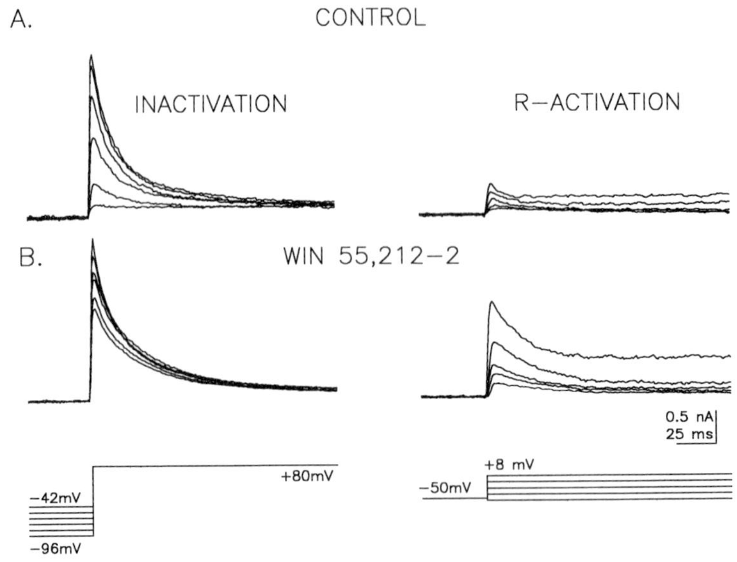

Figure 1 Two protocols used to test steady-state inactivation and relative activation (R-activation) of I_A. A-Left: Inactivation command voltage protocol. Potassium currents elicited via a depolarizing step to +80 mV from different prepulse voltages (–96 to –42 mV). Rapid onset and rapid decay from peak current level characterizes I_A. Control (top trace) shows systematic reduction in peak I_A as a function of prepulse voltage. B-Left: The effect of WIN 55,212–2 (2 μM) on I_A was to decrease steady-state inactivation at all prepulse levels below maximum. A-Right: R-activation protocol designed to test net effect of changes in inactivation and activation on I_A with depolarizing steps from resting (holding) potential (–50 mV) through +8 mV. B-Right: Note increased I_A (middle trace) at the same voltage steps after exposure to WIN 55,212–2 (2 μM). Command voltage protocols used to test inactivation and R-activation of I_A are shown at the bottom.

relative-activation (R-activation) of I_A and employed a series of depolarizing (–40 mV, –24 mV, –8 mV, +8 mV) step pulses (400 ms duration) from a holding potential of –50 mV, near the normal resting potential for these cells. Formalization of the observed changes in parameters of A-current was accomplished by fitting Boltzmann functions to the current generated by each protocol (Figure 1). Relative I_A (I_A/I_{max}) was converted to relative conductance (G_A/G_{max}) and plotted (see Figure 2) for comparison of changes in steady-state I_A voltage dependence across different drug conditions and different cells (Deadwyler *et al.*, 1993).

Figure 1 shows the effect of the potent cannabinoid receptor ligand WIN 55,212–2 on I_A produced by the two voltage-step protocols. Prepulses between –96 and –42 mV produced progressive increases in steady-state inactivation of I_A in control recordings. However, following exposure to WIN 55,212–2 (2 μM), inactivation at these same prepulses was significantly decreased, as evidenced by a larger I_A (Figure 1B). At the same time there was no appreciable effect on transient

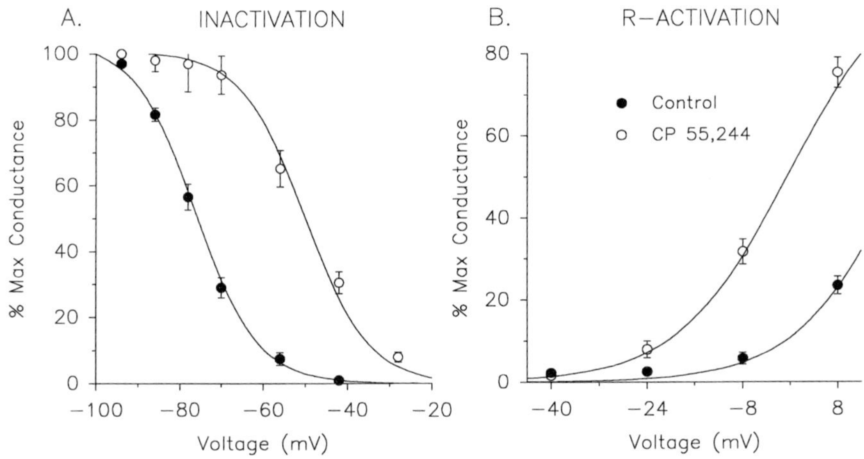

Figure 2 Boltzmann curves for I_A inactivation and R-activation derived from the protocols illustrated in Figure 1. Boltzmann functions fitted to mean $\%G_{max}$ data points. Peak potassium current (I_A) was converted to conductance (G_A) and the Boltzmann equations were fitted to % maximum peak conductance (G/G_{max} or $\%G_{max}$) data points at each voltage step before and after exposure to CP 55,244 (2 μM). (A) plot of inactivation of peak I_A with Boltzmann curves fitted to mean $\%G_{max}$ data points (mean ± SEM $\%G_{max}$ across all cells tested). Exposure to CP 55,244 resulted in a positive (right) shift in voltage dependence resulting from a decreased steady-state inactivation of I_A. (B) Boltzmann curves fitted to I_A data from the relative activation (R-activation) protocol before and after exposure to CP 55,244. The apparent negative shift in voltage dependence of R-activation is the result of the decreased inactivation of I_A assessed at the holding potential (−50 mV) used in the R-activation protocol.

inactivation, as measured by the area under the time decay curve of the activated I_A, in these cells as a result of exposure to cannabinoid receptor ligands. The R-activation protocol in Figure 1A showed very little I_A in control conditions, but was markedly increased (Figure 1B) following application of WIN 55,212–2 (2 μM). These effects dissipated within 2–4 min of continuous wash after the 30 s drug application period.

Figure 2 plots Boltzmann curves fitted to the mean I_A data points. The change in $V^{1/2}$ values derived from the Boltzmann equations (control $V^{1/2}$=−76.2 mV versus CP 55,244 $V^{1/2}$=−50.1 mV) produced by the inactivation protocol reflected the positive shift in the voltage dependence of I_A produced by cannabinoids (Figure 2A). A corresponding negative shift in voltage dependence of I_A activation was revealed in the R-activation plot (Figure 2B) and in corresponding changes in $V^{1/2}$s between control (18.8 mV) and CP 55,244 (3.9 mV) exposure (Figure 2B). The calculated slopes (k) of the Boltzmann equations fitted to mean data points ($\%G_{max}$) did not differ significantly for drug versus control conditions, indicating that changes in A-channel kinetics were not responsible for the cannabinoid induced shifts in I_A voltage dependence. Comparing Figures 2A and 2B, it can be seen that the major increase in $\%G_{max}$ in the R-activation protocol (Figure 2B) resulted primarily from decreased inactivation of I_A (Figure 2A). Thus, cannabinoids increased I_A in cultured hippocampal neurons through a mechanism that did not affect the time course of activation or inactivation but altered the voltage dependence of both processes.

The concentration dependence of the effect of cannabinoid analogues on I_A was assessed across six different concentrations (0.1 to 10 μM) of CP 55,244. The EC_{50} for the increase in $\%G_{max}$ was 0.8 μM, with a threshold at 0.1–0.5 μM and maximal effect between 5–10 μM (Figure 3). The decrease in steady-state inactivation of I_A was significant for five of the six concentrations of CP 55,244 tested (i.e. all but 0.1 μM). For R-activation of I_A the EC_{50} was 1.2 μM (Figure 3), which was increased slightly from the EC_{50} for the inactivation protocol. Threshold and maximal effective concentrations were the same, however, for both activation and inactivation of I_A in these cells. The concentration-dependence of cannabinoid actions on I_A was consistent and systematic across a relatively narrow range (0.5 to 5.0 μM). This dose range is similar to other reports of cannabinoid actions *in vitro* (Howlett *et al.*, 1990, see this volume) as well as the effective range in several *in vivo* tests (see below).

Other cannabinoid receptor agonists have been characterized and tested in several different systems. These include CP 55,940, levonantradol and aminoalkylindoles in the WIN series in addition to WIN 55,212–2 (Johnson *et al.*, 1988; Howlett *et al.*, 1990; Abood and Martin 1992; Ward *et al.*, 1989, 1991). The rank order potencies of cannabinoid compounds tested to date on I_A in the inactivation protocol is CP 55,244=WIN 55,212–2>CP 55,940>levonantradol. This rank order is similar to the relative potency for binding to the cannabinoid receptor (Howlett *et al.*, 1989; Herkenham *et al.*, 1990; Pacheco *et al.*, 1991), for reducing adenylate cyclase activity (Howlett *et al.*, 1990; Pacheco *et al.*, 1993), for inhibiting mouse vas deferens contractions (Ward *et al.*, 1991; Pertwee *et al.*, 1992) and for effecting changes in a battery of behavioural tests (Compton *et al.*, 1992).

7.2.4 Involvement of G proteins

Cannabinoid inhibition of adenylate cyclase is mediated through receptor-coupled PTX-sensitive G proteins (Howlett *et al.*, 1986; Bidaut-Russell *et al.*, 1990; Pacheco *et al.*, 1991). Figure 4 shows that modulation of I_A by WIN 55,212–2 (2 μM) was blocked by pretreatment of cells with PTX. This was shown by comparison of the Boltzmann equations fitted to the data points for cells treated with PTX versus non-treated cells. There was no significant change in mean $\%G_{max}$ for control versus PTX-treated cells in either the inactivation or R-activation protocols. These effects did not result from non-specific actions of the toxin since the voltage dependence of I_A in PTX-treated cells was no different than in control cells prior to exposure to cannabinoid analogues (compare Figures 2 and 4). Thus, pretreatment with PTX selectively eliminated the modulatory effect of cannabinoids on I_A. Further evidence that the effect of cannabinoid analogues on I_A involved G-protein activation was provided by experiments in which cells were dialyzed through the recording pipette

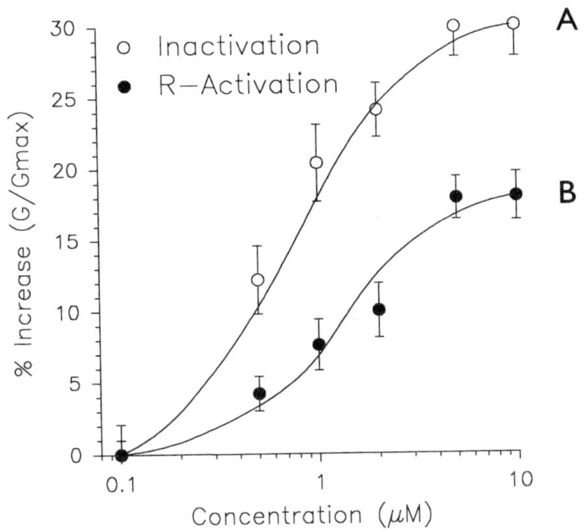

Figure 3 Concentration effect curves for CP 55,244 (range=0.1 to 10 μM). A, the effect on I_A was plotted as a function of per cent increase in mean $\%G_{max}$ from (no drug) control at each concentration. The EC_{50} for CP 55,244 in the inactivation protocol was 0.8 μM. B, Change in mean $\%G_{max}$ of I_A measured in R-activation protocol as a function of concentration of CP 55,244. EC_{50} for R-activation was 1.2 μM. All values are mean ± SEM. Protocols for inactivation and R-activation were as shown in Figure 1.

with a non-hydrolysable GTP analogue, GTP-γ-S (200 μM). The Boltzmann equations fitted to the data showed a positive voltage shift in steady-state inactivation in GTP-γ-S exposed cells (Deadwyler *et al.*, 1993). Application of WIN 55,212–2 to cells dialysed with GTP-γ-S produced no further increase in I_A. Thus, cells dialysed with 200 μM GTP-γ-S exhibited effects that mimicked those exposed to cannabinoid analogues, i.e. decreased steady-state inactivation and increased R-activation of I_A (see also Section 6.2.3.2 of Chapter 6).

7.2.5 Cannabinoid stimulation of low K_m GTPase

Receptor-stimulated low K_m GTPase represents a quantitative measure of G protein-coupled receptor function (Cassel and Selinger, 1976; Koski and Klee, 1981). Cannabinoid-stimulated low K_m GTPase activity was measured in membrane preparations of adult rat hippocampus in which the molecular layer of the dentate gyrus, which contains the highest levels of hippocampal cannabinoid receptors (Herkenham *et al.*, 1991), was isolated and assayed. WIN 55,212–2 and CP 55,940 stimulated low K_m GTPase activity by 61% and 41%, respectively (Deadwyler *et al.*, 1993). The EC_{50} values of both CP 55,940 and WIN 55,212–2 stimulating low K_m GTPase activity were approximately 0.5 μM. These values were similar to the EC_{50} values for cannabinoid-altered voltage dependence of I_A in cultured hippocampal

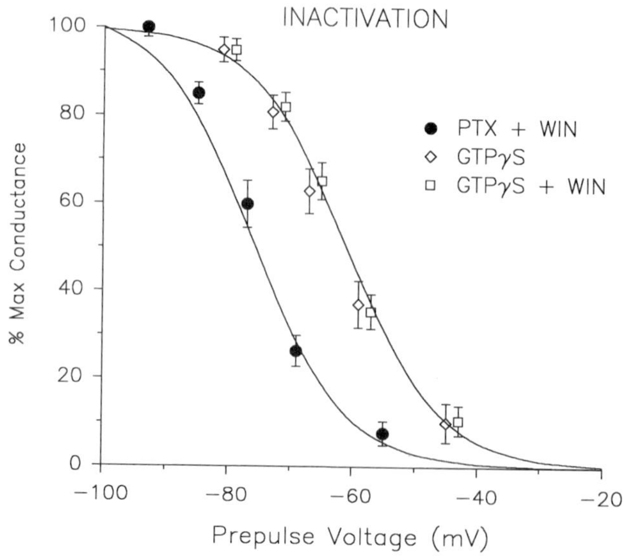

Figure 4 G protein involvement in cannabinoid modulation of I_A. Cannabinoid-induced positive shift in voltage dependence of I_A was absent for PTX-treated cells exposed to WIN 55,212–2 (circles). Cells dialysed with 200 μM GTP-γ-S in the recording pipette (diamonds) showed a positive shift in I_A voltage dependence, similar to that produced by cannabinoids. Application of 2 μM WIN 55,212–2 (squares) produced no additional change in I_A in GTP-γ-S-treated cells.

cells (Deadwyler *et al.*, 1993) and the reported IC_{50} values for these same agonists in inhibiting adenylate cyclase in cerebellar membranes (Pacheco *et al.*, 1991, 1993) suggesting that the physiological effects of the cannabinoids are mediated via activation of a $G_{I/O}$ protein involved in the cyclic AMP cascade within the cell.

7.2.6 Effects of adenylate cyclase and cyclic AMP modulation on I_A

As mentioned above, the decrease in Ca^{2+} conductance produced by cannabinoids was shown not to be mediated by changes in cyclic AMP (Mackie and Hille, 1992). In contrast, a recent set of experiments in our laboratory showed I_A in cultured hippocampal cells to be sensitive to changes in levels of cyclic AMP, via stimulation of adenylate cyclase with forskolin and via application of 8-bromo cyclic AMP. In each case I_A was significantly reduced from control levels (Mu *et al.*, 1993). The forskolin (but not the 8-bromo-cyclic AMP) generated reduction in I_A was reversed by subsequent application of WIN 55,212–2 (Mu *et al.*, 1993). The results suggest that the effect of cannabinoids on I_A are, in fact, mediated by a decrease in cyclic AMP produced by the well-documented cannabinoid receptor inhibition of adenylate cyclase

(Pacheco *et al.*, 1991, Howlett *et al.*, 1990). This was confirmed in recent experiments examining G-protein dependent changes in cyclic AMP levels in cultured hippocampal neurons exposed to similar concentrations of cannabinoid receptor agonists (Deadwyler *et al.*, 1994). Thus, I_A in hippocampal neurons in culture may be modulated by levels of cyclic AMP within the cell which can be altered by cannabinoid receptor-mediated inhibition of adenylate cyclase (Mu *et al.*, 1993, Deadwyler *et al.*, 1994). This is the first functional endpoint described for the cannabinoid receptor-mediated cyclic AMP system. At present, however, analyses of the process by which cyclic AMP alters A-channel operation are not complete but several candidate mechanisms exist for such a linkage (cf. Pfaffinger and Sieglebaum, 1992).

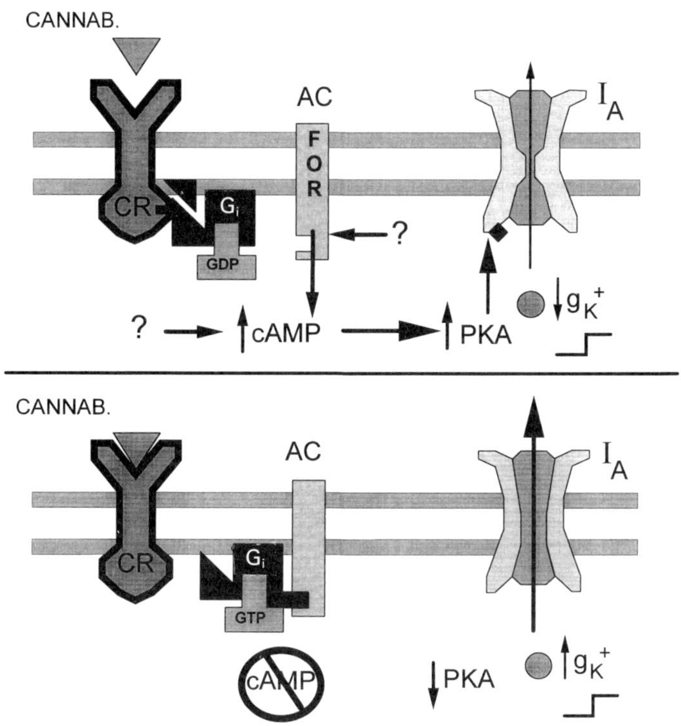

The above schematic shows the proposed interaction of the cannabinoid receptor (CR) with the voltage-sensitive potassium A-current channel (I_A). In the top panel the cannabinoid receptor is depicted in the unoccupied state with the inhibitory G-protein (GDP) inactivated. Under these conditions moderate levels of adenylate cyclase (AC) stimulated cyclic AMP presumably activate a protein kinase (↑PKA), which leads to phosphorylation of A-channel proteins producing an increase in steady-state inactivation and a decreased potassium conductance (↓g_K^+). This can also be produced by activation of AC directly via forskolin (FOR), as in our most recent studies (Mu *et al.*, 1993, Deadwyler *et al.*, 1994). It can be seen in the bottom panel that cannabinoid receptor occupancy by cannabinoid ligands (CANNAB.) inhibits adenylate cyclase (AC) through the PTX sensitive G protein (GTP) complex, reducing cyclic AMP levels in the cell. The reduction in cyclic AMP leads to an increase in A-channel conductance ($\uparrow g_K^+$), presumably through a reduction in the activity of a protein kinase A ($\downarrow$PKA). The PTX-sensitive decrease in steady-state inactivation of I_A (Deadwyler *et al.*, 1993) and the opposite action produced by 8-bromo-cyclic AMP (Mu *et al.*, 1993; Deadwyler *et al.*, 1994) supports the above hypothesized action of cannabinoid ligands in hippocampal neurons.

7.3 Cannabinoid effects on synaptic mechanisms

There have been several attempts to study the effects of cannabinoids, primarily Δ^9-THC, on synaptic mechanisms in various CNS preparations (Weisz *et al.*, 1982; Kujtan *et al.*, 1983; Deadwyler *et al.*, 1985). Turkanis and Karler (1983, 1986) have examined the effects of Δ^9-THC on synaptic processes in different preparations. Intracellular recordings of excitatory and inhibitory postsynaptic potentials were studied in cat motoneurons and in the lobster and frog neuromuscular junctions. In the motoneurons EPSPs were increased and IPSPs were decreased following exposure to Δ^9-THC. In the neuromuscular preparations, low concentrations of Δ^9-THC increased, and high concentrations decreased, mean quantal content of the endplate potential, suggesting dose-dependent differential effects of Δ^9-THC on transmitter release. It has been suggested by these investigators that Δ^9-THC has biphasic dose-dependent actions in several different test systems, including measures of seizures, sedation and motoric convulsions, etc. (Turkanis *et al.*, 1977; Turkanis and Karler, 1981). The newer more potent cannabinoid analogues have yet to be tested in the above systems. In addition, these effects on synaptic mechanisms need to be assessed with respect to G protein involvement. Nevertheless, it appears likely that cannabinoid receptor activation may affect synaptic transmission via modulation of transmitter release (see below).

7.4 Endogenous cannabinoid substances in brain

A major facet of cannabinoid function is the possible existence of endogenous cannabinoid ligands in brain. Is it necessary to postulate the existence of an endogenous ligand for a cannabinoid receptor? After all, although opioid peptides were isolated from brain extracts, the searches for other receptor ligands, including the benzodiazepine and PCP/sigma, have not been nearly as successful. Nevertheless, all G protein-linked receptors should have endogenous ligands because of the special nature of this class of receptor. Unlike other receptors like the GABA/benzodiazepine sites, G protein-coupled receptors are single-subunit molecules with no allosteric binding sites (Gilman, 1984). Thus, the binding site measured by binding assays of G protein-coupled receptors is the true ligand recognition site. This is different from the GABA/benzodiazepine receptor, where the benzodiazepine-binding site is an allosteric regulatory site for GABA. For G protein-coupled receptors, the binding site is straightforward and every such receptor has an endogenous ligand associated with its ligand-binding site.

There are at least three groups of investigators who have heavily invested time and effort in the identification of endogenous cannabinoid ligands. Howlett and colleagues reported the Ca^{2+}-dependent release from brain slices of substances which inhibited [^{3}H]CP 55,940 binding (Evans *et al.*, 1992). This suggested that brain contained endogenous cannabinoid-like substances that are released like neurotransmitters. Devane *et al.* (1992) have isolated an ethanolamine derivative of arachidonic acid, anandamide, from porcine brain (see Chapter 8). Reports suggest that this compound is cannabinoid-like in receptor specificity (Felder *et al.*, 1993; Fride and Mechoulam, 1993; Vogel *et al.*, 1993; Childers *et al.*, 1994). Moreover, further chemical analysis has suggested that different fatty acid derivatives of anandamide exist in brain; thus, anandamide may indeed represent a family of such compounds (R. Mechoulam, personal communication; see also Chapters 1, 2 and 8). Our laboratory (S. Childers) has also isolated an endogenous cannabinoid-like ligand from brain; however, unlike anandamide, this compound is relatively polar. Although the structure of this compound has not yet been elucidated, its chromatographic behaviour and metabolic stability show that it is clearly different from anandamide (Figure 5). The cannabinoid specificity of this compound was demonstrated by the finding that it inhibited specific binding of two different cannabinoid radioligands ([^{3}H]WIN 55,212–2 and [^{3}H]CP 55,940), but did not inhibit binding of ligands to four other G protein-coupled receptors, including δ, μ and κ opioid, and α_2-adrenergic.

Is it possible that different cannabinoid ligands exist in brain? Anandamide has only moderate affinity in binding to cannabinoid receptors, with a K_i of 90 nM. Moreover, as a non-polar lipid which is rapidly metabolized by brain enzymes (Childers *et al.*, 1994), it is not clear how such a compound would be packaged and released as a neurotransmitter. Another more polar compound may play a more traditional neurotransmitter-like role than anandamide. Furthermore, the existence of high levels of cannabinoid receptors throughout different brain regions leaves open the possibility that multiple endogenous cannabinoid substances may mediate many of these actions.

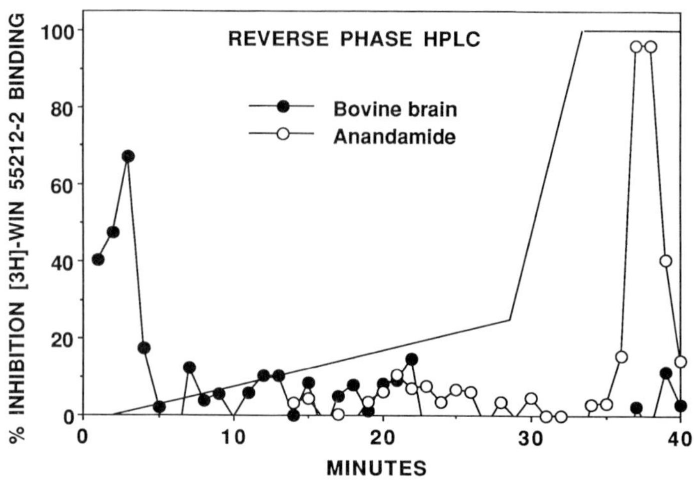

Figure 5 Comparison of reverse-phase high-performance liquid chromatography (HPLC) elution of synthetic anandamide with purified bovine brain cannabinoid substance. The substance from bovine brain was isolated from boiling acetic acid extracts of bovine brain then purified through a series of low-pressure and high-pressure chromatographic steps. These results represent two separate column runs on an ODP-50 C-18 HPLC column where synthetic anandamide and the bovine substance were eluted from the column in separate runs, and aliquots from 1-ml fractions were assayed for inhibition of [^{3}H]WIN 55,212–2 binding to rat cerebellar membranes; data are expressed as per cent inhibition of binding. The gradient line represents percentage methanol in the mobile phase; anandamide elutes during the 100% methanol wash phase.

7.5 Cannabinoid actions in cerebellum

One of the brain regions with highest density of cannabinoid receptors is the cerebellum (Herkenham *et al.*, 1990; Jansen *et al.*, 1992), which is also the region which contains the highest levels of cannabinoid-inhibited adenylate cyclase (Pacheco *et al.*, 1991). Additivity experiments, in which maximally effective concentrations of agonists for several inhibitory receptor systems (including $GABA_B$) were incubated in adenylate cyclase assays, together with cannabinoid agonists, suggested that $GABA_B$ and cannabinoid receptors exist on the same cells (Pacheco *et al.*, 1993; see also Section 7.6). Since $GABA_B$ receptors have been identified on cerebellar granule cells (Wojcik and Neff, 1984; Wojcik *et al.*, 1985), these data suggest that cannabinoid receptors are also present on cerebellar granule cells. To examine this possibility, inhibition of adenylate cyclase activity was measured in cerebellar membranes of neurologically mutant mouse strains exhibiting a loss of either granule cells or Purkinje cells (Pacheco *et al.*, 1993). These strains included Weaver, Staggerer and Nervous mice as well as their heterozygous wild-type littermate controls. Since these are recessive mutations, heterozygous littermates do not demonstrate significant

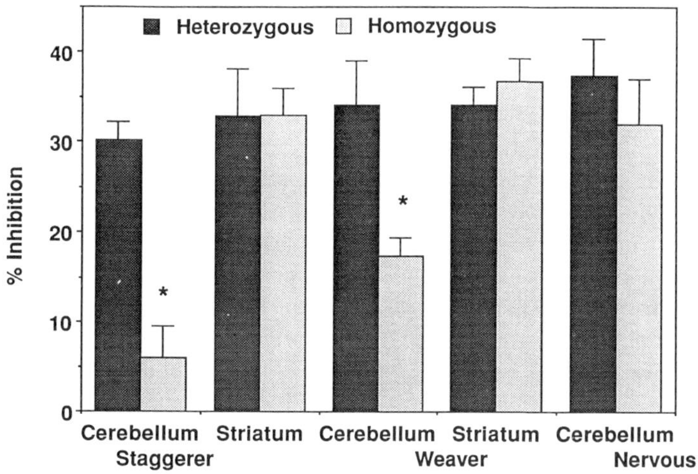

Figure 6 Inhibition of adenylate cyclase by CP 55,940 in striatal and cerebellar membranes from homozygous and heterozygous littermates of Staggerer, Weaver, and Nervous mice. Membranes were isolated from each set of animals and assays of adenylate cyclase were conducted using maximally effective concentration (1 μM) of CP 55,940. Data are expressed as per cent of basal adenylate cyclase activity. *$P<0.02$, homozygous versus heterozygous (Student's *t*-test) (From Pacheco *et al.*, 1993, with permission).

cerebellar deficits. Results showed significant loss of receptor-mediated inhibition of adenylate cyclase in cerebellar membranes from the homozygous Weaver and Staggerer mutants compared with their heterozygous littermates (Figure 6). Conversely, $GABA_B$ and cannabinoid inhibition of adenylate cyclase was not significantly reduced in cerebellar membranes from homozygous Nervous mice – animals in which the cerebellar defect does not involve granule cells. Unlike the cerebellum, maximal inhibition by the cannabinoids and baclofen was not different in the striatal membranes from both heterozygous and homozygous Weaver and Staggerer mice. These data supported the hypothesis that loss of receptor activity was present in the cerebellum but not the striatum, and that the cannabinoid receptors may be localized on cerebellar granule cells.

Cultured cerebellar granule cells have now been established as an effective model system for the study of neuronal actions of cannabinoids (Pacheco *et al.*, 1993). Although these cultures do not have the homogeneity of transformed cell lines such as NG108–15 cells which also contain significant levels of cannabinoid receptors (Howlett, 1985), the cerebellar granule cells have the advantage of being a non-transformed neuronal system that forms synaptic connections with other neurons in culture. In addition, these cells synthesize and release glutamate (Gallo *et al.*, 1982); since glutamate release is inhibited in these cells by agonists acting at $G_{I/O}$-coupled receptors (Travagli *et al.*, 1991), they represent an excellent model system to study the effects of cannabinoids on modulation of neurotransmitter release.

One property examined in these cells has been the ability of cannabinoid agonists to inhibit intracellar cyclic AMP levels (Pacheco *et al.*, 1993). When basal cyclic AMP levels in cultures of cerebellar granule cells were stimulated with 1 μM forskolin, all cannabinoid agonists inhibited the forskolin-stimulated cyclic AMP levels by 50%–65%. The IC_{50} values for these agonists ranged from 0.13 μM to 2.4 μM, with CP 55,940 and levonantradol being the most potent and Δ^9-THC being the least potent. Similar IC_{50} values for these agonists were observed in cerebellar membranes from adult rats. This inhibition of intracellular cyclic AMP levels by cannabinoid agonists was mediated by $G_{I/O}$ proteins, since treatment of cultures with 100 ng/ml of pertussis toxin for 24 h completely abolished the inhibitory effects of several cannabinoid agonists.

A major unanswered question is the role of cannabinoid-inhibited adenylate cyclase in cerebellar granule cells. As discussed above, cannabinoid receptor-mediated inhibition of adenylate cyclase increases potassium conductance through the potassium A-channel in cultured hippocampal cells. Although this electrophysiological action of cannabinoids has not yet been examined in cerebellar granule cells, in the hippocampus these channels may be presynaptic in location and are well situated to modulate neurotransmitter release. Preliminary data in cerebellar granule cells have demonstrated that the potent cannabinoid agonist, WIN 55,212–2, inhibited glutamate release from these cultures by approximately 50%. Studies to explore the signal transduction mechanisms responsible for this effect are currently being performed.

7.6 Receptor convergence in cerebellar granule cells

Several $G_{I/O}$-linked receptors (including $GABA_B$ and adenosine A_1 receptors) coexist on cerebellar granule cells and share common effector systems (Wojcik *et al.*, 1985). This coexistence is demonstrated by a lack of additivity in inhibiting adenylate cyclase when agonists of different receptor systems are added together. To examine whether cannabinoid receptors exhibit the same type of effect, additivity experiments were performed in cultured cerebellar granule cells with maximally effective concentrations for levonantradol, WIN 55,212–2, and baclofen (Figure 7, top). Maximal inhibition of forskolin-stimulated levels of cyclic AMP by the different agonists alone was 40–60% (Figure 7, top). Baclofen, in the presence of WIN 55,212–2 or levonantradol, inhibited forskolin-stimulated cyclic AMP levels no more than 60–65%. The combination of the two cannabinoid agonists inhibited cyclic AMP 50–60%, no more than either agonist alone (Figure 7, top). The lack of additivity seen with any combination of two agonists (cannabinoid or $GABA_B$) suggests that the two receptors are associated with the same cell.

Other experiments explored whether the lack of additivity between cannabinoid and $GABA_B$ receptors was mediated at the effector or the G protein level by assaying receptor-stimulated low K_m GTPase activity. Such experiments would determine

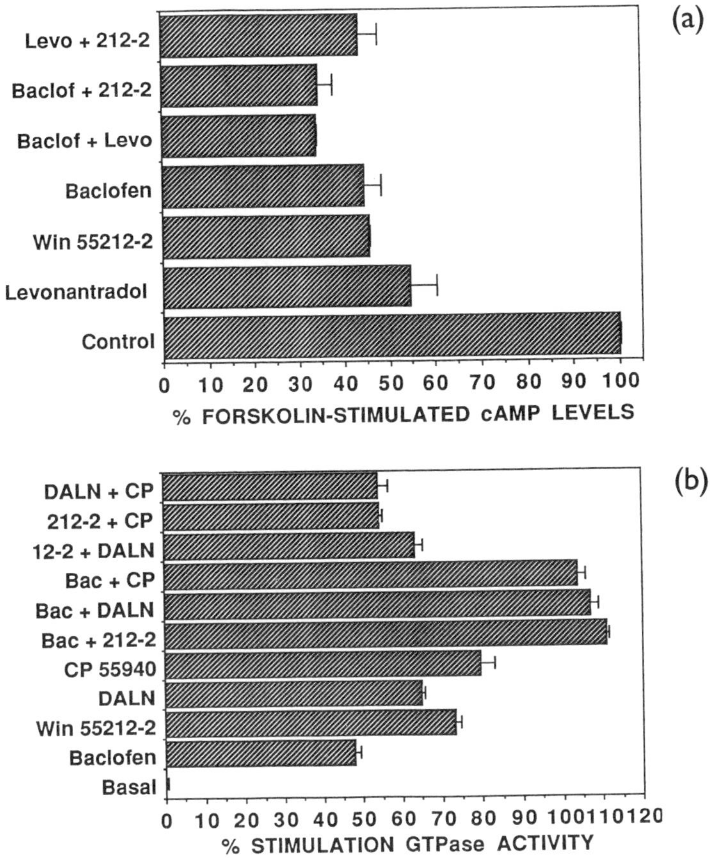

Figure 7 Additivity of agonist responses in inhibiting cyclic AMP levels in rat cerebellar granule cells (a) and in stimulating low K_m GTPase activity in rat cerebellar membranes (b). For the cyclic AMP assays, cells were incubated at 37°C for 10 min with levonantradol (5 μM), WIN 55,212–2 (5 μM), and baclofen (300 μM) in the presence of 1 μM forskolin. Data are expressed as percentage of forskolin-stimulated cyclic AMP levels. For the GTPase assays, low K_m GTPase activity was determined with 0.5 μM [γ-^{32}P]GTP in the presence of maximally effective concentrations for CP 55,940 (10 μM), WIN 55,212-2 (10 μM), baclofen (500 μM) and desacetyllevonantradol (10 μM). Activity is expressed as percentage of basal low K_m GTPase activity (from Pacheco *et al.*, 1993, with permission).

whether the cannabinoid receptor was coupled to a common G protein pool with the $GABA_B$ receptor. Maximally effective concentrations for three cannabinoid agonists (WIN 55,212–2, CP 55,940, and desacetyllevonantradol, DALN), stimulated GTPase activity by 65–75% while baclofen stimulated activity by 50% in cerebellar P_2 membranes (Figure 7, bottom). The combination of any two cannabinoid agonists together stimulated GTPase no more than either agonist alone (50–60%), consistent with their actions being on the same receptor. However, when the

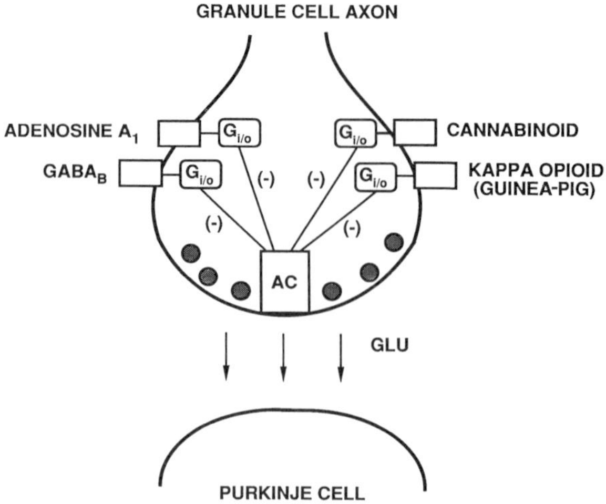

Figure 8 Schematic diagram illustrating receptor convergence of inhibitory G protein receptors, including $GABA_B$, adenosine A_1, cannabinoid and kappa opioid (in guinea-pig), on cerebellar granule cell axons. These receptors are pharmacologically distinct and may couple to their own inhibitory G proteins, but share common adenylate cyclase catalytic units.

cannabinoid agonists were added together with baclofen, the combination stimulated GTPase activity by 100% or more. Therefore, in contrast to the adenylate cyclase system, the $GABA_B$ and cannabinoid receptor responses for GTPase activity were additive. These data are similar to those reported for $GABA_B$ and adenosine A_1 agonists in stimulating GTPase in cerebellar membranes (Wojcik *et al.*, 1985).

These results are consistent with a model of receptor convergence (Figure 8) in which pharmacologically distinct receptors share common effector systems (see also Section 6.2.2.2 of Chapter 6). In the case of the cerebellar granule cells, several $G_{I/O}$-coupled receptors share common adenylate cyclase catalytic units, but not common G proteins. This model explains how agonists acting at different receptor sites can produce common biological signals.

7.7 Functional consequences of cannabinoid receptor activation

7.7.1 Movement

The most prominent clustering of cannabinoid receptors in the brain is in areas that have been closely linked to the control and modulation of motor systems (i.e., striatum, substantia nigra and cerebellum, see Chapter 5). There is every reason to believe that cannabinoid receptors in these areas, because of their density and location, would play a major role in the control and initiation of movements. Historically, there is a precedent for this assumption since one of the most sensitive tests of marijuana and its psychoactive derivatives is the dog static-ataxia test (Dewey, 1986), a test involving cerebellar control of posture and balance. Moss *et al.* (1981) showed a similar acute effect of high doses of Δ^9-THC in rats which became extremely cataleptic. More recently, there have been extensive studies of the effects of various cannabinoid analogues on behavioural test batteries examining various types of performance variables (see Chapter 2). The effects on behaviours such as locomotion, immobility, catalepsy and hypothermia indicate that cannabinoid receptors may mediate several types of motor responses (Thomas *et al.*, 1991) and that the effects correlate with the potency of the cannabinoid substances to bind to the cannabinoid receptor (Compton *et al.*, 1992, 1993). There is also a general consensus in the human literature showing that marijuana and/or Δ^9-THC produce direct effects on posture and hand-steadiness in simple hand–eye coordination tasks (Consroe and Sandyk, 1992). While yet to be tested exclusively in relation to other effects of cannabinoids, the information available indicates that effects on motor system function should be one of the most sensitive reflections of cannabinoid action.

7.7.2 Analgesia

The development of new cannabinoid analogues with high analgesic properties as well as high affinity for the cannabinoid receptor (Johnson *et al.*, 1988) suggests a strong link of cannabinoid receptor action to the pain system (Abood and Martin, 1992). These effects are not mediated by opiate receptors since naloxone does not block the analgesic effects of cannabinoids. However, in several brain and spinal cord regions, cannabinoid and opioid receptors have been shown to exist in close proximity (Herkenham *et al.*, 1991; Herkenham, 1992). The possible involvement of opiate receptors in spinal cord analgesia produced by intrathecal cannabinoids has been reported recently (Welch and Stevens, 1992; Welch, 1993). In addition, the antinociceptive properties of cannabinoid analogues are detectable in the same dose range that produces marked changes in motor activity and hypothermia (Compton *et al.*, 1992). The similarity in dose–dependency of these compounds clearly implicates a common type of cannabinoid receptor mediating several different functions.

7.7.3 Perception and attention

It can be concluded from both the anecdotal and experimental literature concerning cannabinoid actions in humans that only minor distortions in sensory awareness occur during marijuana intoxication. In fact, human studies report several instances of heightened sensory awareness during marijuana intoxication (Hollister, 1986). Studies in monkeys indicate no serious effects of acute or chronic exposure to marijuana smoke on simple visual discrimination tasks (Paule *et al.*, 1988; Schulze *et al.*, 1988). There are, however, significant effects of cannabinoids on attentional mechanisms in animals and humans. Studies conducted in rats indicate dose-dependent effects on accuracy and latency to respond in a differential tone discrimination task (Campbell *et al.*, 1986a,b). Signal detection performance is also severely disrupted in a dose-dependent fashion by Δ^9-THC (Heyser *et al.*, 1989; Deadwyler *et al.*, 1990). In monkeys trained to respond in a choice reaction time task, performance was disrupted by acute exposure to marijuana smoke (Slikker *et al.*, 1992). Humans tested on both attention span and divided attention tasks show deficits during marijuana intoxication (Chait and Pierri, 1992). Thus, in cases where focused attention and persistent perception to detail are not required, cannabinoid receptor occupancy does not appear to affect performance. However, these same discriminatory functions can be impaired under the influence of cannabinoids if they require sustained or divided attention or are made more difficult by increasing the similarity between stimuli.

7.7.4 Memory and cognition

The main effects of cannabinoids on human cognition relate to impairment of short-term memory (Miller and Branconnier, 1983; Chait and Pierri, 1992). Cannabinoids have been consistently implicated in memory loss and distortion of recently acquired information. While long-term memory appears least affected, there is no definitive evidence that established memories are 'immune' from disruption. Other data available suggest that there are no lasting or residual effects of cannabinoid exposure on memory processes and that the effects on memory do not persist beyond the time the drug is in the system (Stiglick and Kalant, 1985; Slikker *et al.*, 1992). However, cannabinoid induced impairment of short-term memory and recall can be quite severe and these effects can be dissociated from cannabinoid influences on other cognitive processes (Chait and Pierri, 1992). In animal models, cannabinoids have been reported to produce deficits that resemble those produced by damage to the hippocampus (Deadwyler *et al.*, 1990; Heyser *et al.*, 1993; see below).

7.8 Promising leads – potential role of cannabinoid receptors in hippocampus

The prescribed actions of cannabinoids in brain will rapidly become constrained to criteria dependent on mediation by a G protein-coupled receptor. Such requirements will rule out non-specific actions of cannabinoids caused by their high lipophilicity and membrane perturbation properties at higher concentrations. Actions on neuronal processes will necessarily reflect G protein toxin sensitivity. When evaluating the role of cannabinoids in brain it will therefore be necessary to demonstrate a link to G protein-coupled cellular processes.

The finding of significant cannabinoid-stimulated GTPase activity in membranes from the molecular layer of the dentate gyrus in the hippocampus together with the high receptor density in that region suggests that cannabinoid receptors perform a particular function in this region (Herkenham *et al.*, 1991). The level of cannabinoid stimulated GTPase activity in the dentate gyrus is large compared with other types of receptor-stimulated GTPase activities in brain membranes (Devane *et al.*, 1988; Konkoy and Childers, 1993 unpublished). In addition, the concentration dependence for this effect is similar to that for cannabinoid modulation of I_A (Figure 3).

The above evidence suggests localized cannabinoid receptor actions in the region of the perforant path terminals (molecular layer) which innervate the granule cells of the dentate gyrus. This finding is consistent with a presynaptic localization of cannabinoid receptors in the dentate gyrus of the adult hippocampus; as demonstrated recently for strio-nigral projections (Herkenham *et al.*, 1992). Since it has been shown that cannabinoids inhibit the electrically evoked release of acetylcholine in guinea-pig ileum (Roth, 1978; Nye *et al.*, 1985) and noradrenaline in the mouse vas deferens (Ward *et al.*, 1991), the demonstration of presynaptic-located cannabinoid receptors in the dentate gyrus would suggest an important role for the cannabinoid produced alterations in I_A (Figure 2).

A shift in the voltage sensitivity of I_A in presynaptic afferent terminals in the dentate gyrus would conceivably reduce transmitter release by increasing potassium channel conductances, thus short-circuiting the effects of action potential depolarization in the terminal. The net effect of the positive shift in voltage dependence would be to increase the interval between successive action potential depolarizations, as well as to decrease action potential duration (Connor and Stevens, 1971; Segal and Barker, 1984; Storm, 1988; Hille, 1992). Thus in the presence of cannabinoid ligands, terminal depolarization would be reduced and consequently result in a decrease in transmitter release. Since cannabinoids also affect N-type calcium currents directly (Caulfield and Brown, 1992; Mackie and Hille, 1992), this, in combination with the altered voltage dependence of I_A, would further decrease the likelihood of transmitter release from presynaptic terminals in the presence of cannabinoid-receptor ligands. A similar function has recently been proposed for $GABA_B$ receptors on presynaptic terminals by Gage and co-workers (Saint *et al.*, 1990; Gage, 1992).

The above hypothesized mechanism would account for the actions of cannabinoids on hippocampus-dependent behaviours and correlated cell activity (Landfield *et al.*, 1988; Deadwyler *et al.*, 1990; Heyser *et al.*, 1993). The perforant path is the main cortical afferent to hippocampal pyramidal and granule cells, originating in cells in the superficial layers of the entorhinal cortex (Witter and Amaral, 1991). Given the above mechanism of action, a key strategic role for cannabinoid receptors could be the regulation of sensory input to the hippocampus via decreased synaptic transmission in perforant path axons (Figure 9). By decreasing transmission in cortical afferents, the hippocampus would have reduced access to sensory information in the presence of cannabinoid ligands and hence reduced capacity to 'register' or 'store' that information for later recognition or recall.

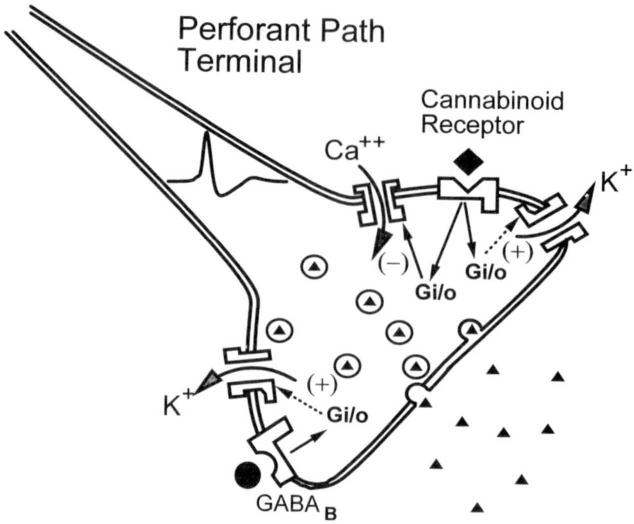

Figure 9 Cannabinoid receptor mediation of transmitter release in perforant path terminals in the dentate gyrus. The drawing shows a perforant path terminal with voltage-dependent Ca^{2+} and A-channels (K^+) coupled to the cannabinoid receptor. Another A-channel is shown coupled to a $GABA_B$ receptor. Triangles inside circles represent transmitter in vesicles subsequently released during Ca^{2+} influx. Cannabinoid receptor occupancy results in (1) a decrease (−) in Ca^{2+} conductance mediated through a G_I protein and (2) an increase in K^+ conductance through the A-channel, also mediated by a G_I protein. The dashed line indicates that the coupling to the A-channel is not direct but likely mediated by cyclic AMP (Mu *et al.*, 1993, with permission).

Evidence for this function of cannabinoids has been provided in concurrent behavioural and electrophysiological studies, as shown in Figure 10 (cf. Heyser *et al.*, 1993). In those studies, Δ^9-THC decreased performance of a delayed matching to sample (DMTS) task, in a delay and dose-dependent manner (Heyser *et al.*, 1993) consistent with effects produced by hippocampal lesions in the same task (Heyser *et al.*, 1989). The effect on DMTS performance was related directly to a reduction or elimination of neural activity associated with the sample, or information storage, phase of the task (Figure 10). The selective action of cannabinoids on this aspect

---

Figure 10 Effects of Δ^9-THC (i.p.) on performance of a delayed-match-to-sample (DMTS) task adapted for rats. (A) Delay-dependent behavioural performance in the DMTS task. Mean percentage of correct (match) responses are plotted as a function of delay interval ranging from 1 to 30 s in 5-s blocks. Performance is shown for vehicle control sessions (unfilled circles) and sessions after 2 mg/kg Δ^9-THC (filled circles). Error bars: min. and max. SEM for all means (points) plotted (n = 12 animals). (B) Hippocampal unit activity recorded during DMTS task. Top: composite (mean) trial-based histogram constructed from nine different isolated and identified hippocampal pyramidal cells over all trials recorded in vehicle control sessions. Trials with different delays were summed together and resynchronized after the line break at LNP, which constitutes the end of the delay period. Firing

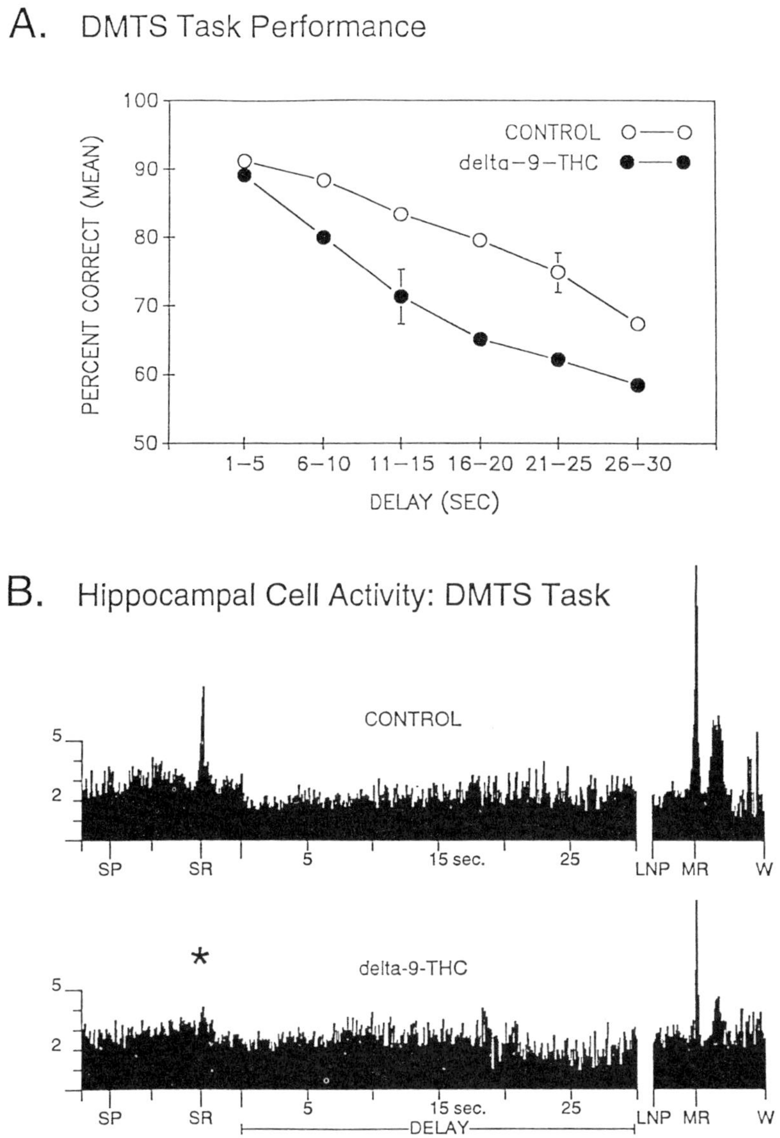

increases were observed following the sample lever response (SR), match response (MR) and during the reinforcement phase preceding consumption of the water reward (W). Bottom: composite histogram of hippocampal cell activity recorded in DMTS sessions following administration of 2 mg/kg Δ^9-THC. Firing peaks were present during the match response and preceding consumption of reinforcement (W), but not during the sample response (asterisk). SP, sample lever presentation; Delay, reflects all delay intervals 1–30 s; LNP, last nose-poke in the delay interval signifying the end of the delay. Numbers reflect time in seconds between events within the trial. Histogram bin width = 50 ms. All firing rates were normalized with respect to number of trials (see Heyser *et al.*, 1993).

of the task suggests that the 'sample' information may be transmitted via axons of the perforant path and that this pathway is compromised in the presence of cannabinoids by the altered voltage dependence of I_A in perforant path terminals (Figure 9). The corresponding enhanced I_A would decrease voltage-dependent transmitter release at the perforant path–dentate granule cell synapse during the sample phase, thereby decreasing the likelihood that such information could be successfully retrieved later during the match phase of the DMTS task (Hampson *et al.*, 1993; Heyser *et al.*, 1993).

7.9 Summary

Cannabinoid receptors in brain probably mediate a variety of different functions. Since they are ubiquitously located and exist in a number of different brain systems their functional significance may be quite different within different systems. The cellular actions of cannabinoids are also varied, suggesting coupling of cannabinoid receptors to many different and potentially divergent cellular mechanisms. At present, G protein-mediated cellular actions appear to be one requirement of this receptor population; however, other elements within these cascades, such as alterations in adenylate cyclase or cyclic AMP levels, need not be involved to produce receptor-coupled changes in effector mechanisms. A third aspect of cannabinoid receptor function relates to co-activation of similar G protein-mediated systems with other types of receptors. This possibility is most likely in the cerebellar granule cells where adenosine A_1, $GABA_B$ and cannabinoid receptors mediate reductions in adenylate cyclase through different G protein substrates. Given this convergence of receptors in the control of cellular and effector processes in other cannabinoid receptor locations, and the moderate to marked range of change in physiological reactivity effected by receptor occupancy, it is extremely unlikely that future research will fail to identify major functional roles for cannabinoid receptors in the mammalian central nervous system.

Acknowledgements

This research was supported by NIDA grants to SAD (DA03502, DA07625 and DA00119) and to SRC (DA 06784). We appreciate the contributions of Gina King, Barbara Bennett and Mary Pacheco to the research reported here. Our appreciation also to Dr Larry Melvin of Pfizer, Inc., who supplied the CP compounds.

References

Abood, M.E. & Martin, B.R. (1992) Neurobiology of marijuana abuse. *Trends Pharmacol. Sci.* **13**, 201–206.

Adams, D.J. & Nonner, W. (1992) Voltage-dependent potassium channels gating ion permeation and block. In *Potassium Channels: Structure, Classification, Function and Therapeutic Potential* (ed. Cook, N.), pp. 41–69. Wiley, NY.

Bidaut-Russell, M., Devane, W.A. & Howlett, A.C. (1990) Cannabinoid receptors and modulation of cyclic AMP accumulation in the rat brain. *J. Neurochem.* **55**, 21–26.

Brown, D.A., Gähwiler, B.H., Griffith, W.H. & Halliwell, J.V. (1990) Membrane currents in hippocampal neurons. *Progr. Brain Res.* **83**, 141–160.

Campbell, K.A., Foster, T.C., Hampson, R.E. & Deadwyler, S.A. (1986a) Effects of delta-9-tetrahydrocannabinol on sensory-evoked discharges of granule cells in the dentate gyrus of behaving rats. *J. Pharmacol. Exp. Ther.* **239**, 941–945.

Campbell, K.A., Foster, T.C., Hampson, R.E. & Deadwyler, S.A. (1986b) Delta-9-tetrahydrocannabinol differentially affects sensory-evoked potentials in the rat dentate gyrus. *J. Pharmacol. Exp. Therap.* **239**, 936–940.

Cassel, D. & Selinger, Z. (1976) Catecholamine-stimulated GTPase activity in turkey erythrocyte membranes. *Biochim. Biophys. Acta.* **452**, 538–551.

Caulfield, M.P. & Brown, D.A. (1992) Cannabinoid receptor agonists inhibit Ca current in NG108–15 neuroblastoma cells via a pertussis toxin-sensitive mechanism. *Br. J. Pharmacol.* **106**, 231–232.

Chait, L.D. & Pierri, J. (1992) Effects of smoked marijuana on human performance: A critical review. In *Marijuana/cannabinoids: Neurobiology and Neurophysiology* (eds Murphy, L. & Bartke, A.) pp. 387–423. CRC Press, Boca Raton, FL.

Childers, S.R., Sexton, T. & Roy, M.B. (1994) Effects of anandamide on cannabinoid receptors in rat brain membranes. *Biochem. Pharmacol..* **47**, 711–715.

Compton, D.R., Gold, L.H., Ward, S.J., Balster, R.L. & Martin, B.R. (1992) Aminoalkylindole analogs: cannabimimetic activity of a class of compounds structurally distinct from delta 9-tetrahydrocannabinol. *J. Pharmacol. Exp. Therap..* **263**, 1118–1126.

Compton, D.R., Rice, K.C., de Costa, B.R., Razdan, R.K., Melvin, L.S., Johnson, M.R. & Martin, B.R. (1993) Cannabinoid structure-activity relationships: correlation of receptor binding and *in vivo* activities. *J. Pharmacol. Exp. Therap.* **265**, 218–226.

Connor, J.A. & Stevens, C.F. (1971) Inward and delayed outward membrane currents in isolated neural somata under voltage clamp. *J. Physiol*, **213**, 1–19.

Consroe, P. & Sandyk, R. (1992) Potential role of cannabinoids for therapy of neurological disorders. In *Marijuana/Cannabinoids: Neurobiology and Neurophysiology* (eds. Murphy, L., & Bartke, A.) pp. 459–524. CRC Press, Boca Raton.

Deadwyler, S.A., Hampson, R.E. & Marlow, G.A. (1985) Excitatory effects of delta-9-tetrahydrocannabinol on hippocampal neurons recorded in vitro. *Soc. Neurosci. Abstr.* **11**, 852.

Deadwyler, S.A., Hampson, R.E., Bennett, B.A., Edwards, T.A., Mu, J., Pacheco, M.A., Ward, S.J. & Childers, S.R. (1993) Cannabinoids modulate potassium current in cultured hippocampal neurons. *Receptors and Channels.* **1**, 121–134.

Deadwyler, S.A., Heyser, C.J., Michaelis, R.C. & Hampson, R.E. (1990) The effects of delta-9-THC on mechanisms of learning and memory. In *Neurobiology of Drug Abuse: Learning and Memory* (ed. Erinoff, L.). National Institute on Drug Abuse, Rockville, MD pp. 79–93.

Deadwyler, S.A., Mu, J., King, V.C., Hampson, R.E. & Childers, S.R. (1994) Inhibition of cAMP dependent protein kinase blocks cannabinoid receptor modulation of potassium A-current in cultured rat hippocampal neurons. *Soc. Neurosci. Abstr.* **20**, 724.

Devane, W.A., Dysarz, F.A., Johnson, M.R., Melvin, L.S. & Howlett, A.C. (1988) Determination and characterization of a cannabinoid receptor in rat brain. *Mol Pharmacol.* **34**, 605–613.

Devane, W.A., Hanuš, L., Breuer, A., Pertwee, R.G., Stevenson, L.A., Griffin, G., Gibson, D., Mandelbaum. A., Etinger, A. & Mechoulam, R. (1992) Isolation and structure of a brain constituent that binds to the cannabinoid receptor. *Science* **258**, 1946–1949.

Dewey, W.L. (1986) Cannabinoid pharmacology. *Pharmacol. Rev.* **38**, 151–178.

Evans, D.M., Johnson, M.R. & Howlett, A.C. (1992) Calcium-dependent release from rat brain of a cannabinoid binding activity. *J. Neurochem.* **58**, 780–782.

Felder, C.C., Briley, E.M., Axelrod, J., Simpson, J.T., Mackie, K. & Devane, W.A. (1993) Anandamide, an endogenous cannabimimetic eicosanoid, binds to the cloned human cannabinoid receptor and stimulates receptor-mediated signal transduction. *Proc. Natl Acad. Sci. USA* **90**, 7656–7660.

Fride, E. & Mechoulam, R. (1993) Pharmacological activity of the cannabinoid receptor agonist, anandamide, a brain constituent. *Eur J. Pharmacol.* **231**, 313–314.

Gage, P.W. (1992) Activation and modulation of neuronal K^+ channels by GABA. *Trends Neurosci.* **15**, 46–51.

Gallo, V., Ciotti, M.T., Coletti, A., Aloisi, F. & Levi, G. (1982) Selective release of glutamate from cerebellar granule cells differentiating in culture. *Proc. Natl Acad. Sci. USA* **79**, 7919–7923.

Gilman, A.G. (1984) G proteins and dual control of adenylyl cyclase. *Cell* **36**, 577–579.

Hampson, R.E., Heyser, C.J. & Deadwyler, S.A. (1993) Hippocampal cell firing correlates of delayed match to sample performance in the rat. *Behav. Neurobiol.* **107**, 715–739.

Herkenham, M. (1991) Characterization and localization of cannabinoid receptors in brain: an *in vitro* technique using slide-mounted tissue sections. *NIDA Res Monogr.* **112**, 129–145.

Herkenham, M. (1992) Cannabinoid receptor localization in brain: relationship to motor and reward systems. *Ann. NY Acad. Sci.* **654**, 19–32.

Herkenham, M., Lynn, A.B., Little, M.D., Johnson, M.R., Melvin, L.S., de Costa, B.R. & Rice, K.C. (1990) Cannabinoid receptor localization in brain. *Proc. Natl Acad. Sci. USA* **87**, 1932–1936.

Herkenham, M., Lynn, A.B., Johnson, M.R., Melvin, L.S., de Costa, B.R. & Rice, K.C. (1991) Characterization and localization of cannabinoid receptors in rat brain: a quantitative *in vitro* autoradiographic study. *J. Neurosci.* **11**, 563–583.

Heyser, C.J., Hampson, R.E. & Deadwyler, S.A. (1989) Characteristics of hippocampal complex spike neurons during performance of a delayed match to sample task. Selective influences of delta-9-THC. *Soc. Neurosci. Abst.* **15**, 1170.

Heyser, C.J., Hampson, R.E. & Deadwyler, S.A. (1993) Effects of delta-9-tetrahydrocannabinol on delayed match to sample performance in rats: alterations in short-term memory associated with changes in task specific firing of hippocampal cells. *J. Pharmacol. Exp. Therap.* **264**, 294–307.

Hille, B. (1992) *Ionic Channels of Excitable Membranes*. Sinauer Associates, New York.

Hollister, L.E. (1986) Health aspects of Cannabis. *Pharmacol. Rev.* **38**, 1–20.

Howlett, A.C. (1985) Cannabinoid inhibition of adenylate cyclase. Biochemistry of the response in neuroblastoma cell membranes. *Mol. Pharmacol.* **27**, 429–436.

Howlett, A.C., Qualy, J.M. & Khachatrian, L.L. (1986) Involvement of Gi in the inhibition of adenylate cyclase by cannabimimetic drugs. *Mol. Pharmacol.* **29**, 307–313.

Howlett, A.C., Johnson, M.R., Melvin, L.S. & Milne, G.M. (1988) Nonclassical cannabinoid analgetics inhibit adenylate cyclase: development of a cannabinoid receptor model. *Mol. Pharmacol.* **33**, 297–302.

Howlett, A.C., Scott, D.K. & Wilken, G.H. (1989) Regulation of adenylate cyclase by cannabinoid drugs. Insights based on thermodynamic studies. *Biochem. Pharmacol.* **38**, 3297–3304.

Howlett, A.C., Bidaut-Russell, M., Devane, W.A., Melvin, L.S., Johnson, M.R. & Herkenham, M. (1990) The cannabinoid receptor: biochemical, anatomical and behavioural characterization. *Trends Neurosci.* **13**, 420–423.

Jansen, E.M., Haycock, D.A., Ward, S.J. & Seybold, V.S. (1992) Distribution of cannabinoid receptors in rat brain determined with aminoalkylindoles. *Brain Res.* **575**, 93–102.
Johnson, M., Devane, W., Howlett, A., Melvin, L. & Milne, G. (1988) Structural studies leading to the discovery of a cannabinoid binding site. *NIDA Res. Monogr. Ser.* **90**, 129–135.
Koski, G. & Klee, W.A. (1981) Opiates inhibit adenylate cyclase by stimulating GTP hydrolysis. *Proc. Natl Acad. Sci. USA* **78**, 4185–4189.
Kujtan, P.W., Carlen, P.L. & Kapur, B.M. (1983) Delta-9-tetrahydrocannabinol and cannabidiol: dose-dependent effects on evoked potentials in the hippocampal slice. *Can. J. Physiol. Pharmacol.* **61**, 420–426.
Landfield, P.W., Cadwallader, L.B. & Vinsant, S. (1988) Quantitative changes in hippocampal structure following long-term exposure to delta-9-tetrahydrocannabinol: possible mediation by glucocorticoid systems. *Brain Res.* **443**, 47–62.
Mackie, K. & Hille, B. (1992) Cannabinoids inhibit N-type calcium channels in neuroblastoma-glioma cells. *Proc. Natl. Acad. Sci. USA* **89**, 3825–3829.
Matsuda, L.A., Lolait, S.J., Brownstein, M.J., Young, A.C. & Bonner, T.I. (1990) Structure of a cannabinoid receptor and functional expression of the cloned cDNA. *Nature* **346**, 561–564.
Miller, L.L. & Branconnier, R.J. (1983) Cannabis: effects on memory and the cholinergic limbic system. *Psychol. Bull.* **93**, 441–456.
Moss, D.E., McMaster, S.B. & Rogers, J. (1981) Tetrahydrocannabinol potentiates reserpine induced hypokinesia. *Pharmacol. Biochem. Behav.* **15**, 779–783.
Mu, J., Kirby, M.T., King, V.C., Bennett, B.A., Hampson, R.E., Childers, S.R. & Deadwyler, S.A. (1993) Cannabinoids increase potassium A-current and decrease synaptic transmission via cAMP-dependent processes in hippocampus. *Soc. Neurosci. Abstr.* **19**, 280.
Numann, R.E., Wadman, W.J. & Wong, R.K.S. (1987) Outward currents of single hippocampal cells obtained from the adult guinea-pig. *J. Physiol.* **393**, 331–353.
Nye, J.S., Seltzman, H.H., Pitt, C.G. & Snyder, S.H. (1985) High-affinity cannabinoid binding sites in brain membranes labeled with [^{3}H]-5′-trimethylammonium delta-8-tetrahydrocannabinol. *J. Pharmacol. Exp. Therap.* **234**, 784–791.
Pacheco, M., Childers, S.R., Arnold, R., Casiano, F. & Ward, S.J. (1991) Aminoalkylindoles: Actions on specific G-protein-linked receptors. *J. Pharmacol. Exp. Therap.* **257**, 170–183.
Pacheco, M.A., Ward, S.J. & Childers, S.R. (1993) Identification of cannabinoid receptors in cultures of rat cerebellar granule cells. *Brain Res.* **603**, 102–110.
Paule, M.G., Schulze, G.E. & Slikker, W. Jr. (1988) Complex brain function in monkeys as a baseline for studying the effects of exogenous compounds. *Neurotoxicology.* **9**, 463–470.
Pertwee, R.G., Stevenson, L.A., Elrick, D.B., Mechoulam, R. & Corbett, A.D. (1992) Inhibitory effects of certain enantiomeric cannabinoids in the mouse vas deferens and the myenteric plexus preparation of guinea-pig small intestine. *Br J. Pharmacol.* **105**, 980–984.
Pfaffinger, P.J. & Sieglebaum, S.A. (1992) K^+ channel modulation by G-proteins and second messengers. In *Potassium Channels: Structure, Classification, Function and Therapeutic Potential* (ed. Cook, N.) pp. 118–153. Wiley, NY.
Roth, S.H. (1978) Stereospecific presynaptic inhibitory effect of delta-9-tetrahydrocannabinol on cholinergic transmission in the myenteric plexus of the guinea pig. *Can. J. Physiol. Pharmacol.* **56**, 968–975.
Rudy, B. (1988) Diversity and ubiquity of K channels. *Neuroscience* **25**, 729–749.
Saint, D.A., Thomas, T. & Gage, P.W. (1990) GABA-B agonists modulate a transient potassium current in cultured mammalian hippocampal neurons. *Neurosci. Lett.* **118**, 9–13.
Schulze, G.E., McMillan, D.E., Bailey, J.R., Scallet, A., Ali, S.F., Slikker, W. Jr. & Paule, M.G. (1988) Acute effects of delta-9-tetrahydrocannabinol in rhesus monkeys as measured by performance in a battery of complex operant tests. *J. Pharmacol. Exp. Therap.* **245**, 178–186.

Segal, M. & Barker, J.L. (1984) Rat hippocampal neurons in culture: potassium conductances. *J. Neurophysiol.* **51**, 1409–1433.

Segal, M., Rogawski, M.A. & Barker, J.L. (1984) A transient potassium conductance regulates the excitability of cultured hippocampal and spinal neurons. *J. Neurosci.* **4**, 604–609.

Slikker, W. Jr., Paule, M.G., Ali, S.F., Scallet, A.C. & Bailey, J.R. (1992) Behavioural, neurochemical, and neurohistological effects of chronic marijuana smoke exposure in the nonhuman primate. In *Marijuana/Cannabinoids. Neurobiology and Neurophysiology* (eds. Murphy, L. & Bartke, A.). CRC Press, Boca Raton. pp. 219–273.

Stiglick, A. & Kalant, H. (1985) Residual effects of chronic cannabis treatment on behaviour in mature rats. *Psychopharmacol.* **85**, 436–439.

Storm, J.F. (1988) Temporal integration by a slowly inactivating K^+ current in hippocampal neurons. *Nature* **336**, 379–381.

Storm, J.F. (1990) Potassium currents in hippocampal pyramidal cells. *Prog. Brain Res.* **83**, 161–187.

Thomas, B.F., Compton, D.R., Martin, B.R. & Semus, S.F. (1991) Modeling the cannabinoid receptor: a three-dimensional quantitative structure-activity analysis. *Mol. Pharm.* **40**, 656–665.

Travagli, R.A., Ulivi, M. & Wojcik, W.J. (1991) γ-Aminobutyric acid-B receptors inhibit glutamate release from cerebellar granule cells: consequences of inhibiting cyclic AMP formation and calcium in flux. *J. Pharmacol. Exp. Therap.* **258**, 903–909.

Turkanis, S.A., Chiu, P., Borys, H.K. & Karler, R. (1977) Influence of delta-9-tetrahydrocannabinol and cannabidiol on photically evoked after-discharge potentials. *Psychopharmacol.* **52**, 207–212.

Turkanis, S.A. & Karler, R. (1981) Excitatory and depressant effects of delta-9-tetrahydrocannabinol and cannabidiol on cortical evoked responses in the conscious rat. *Psychopharmacol.* **75**, 294–298.

Turkanis, S.A. & Karler, R. (1983) Effects of delta-9-tetrahydrocannabinol on cat spinal motoneurons. *Brain Res.* **288**, 283–287.

Turkanis, S.A. & Karler, R. (1986) Cannabidiol-caused depression of spinal motoneuron responses in cats. *Pharmacol. Biochem. Behav.* **25**, 89–94.

Vogel, Z., Barg, J., Levy, R., Saya, D., Heldman, E. & Mechoulam, R. (1993) Anandamide, a brain endogenous compound, interacts specifically with cannabinoid receptors and inhibits adenylate cyclase. *J. Neurochem.* **61**, 352-355.

Ward, S.J., Baizman, E., Bell, M., Childers, S., D'Ambra, T., Eissenstat, M., Estep, K., Haycock, D., Howlett, A., Luttinger, D., Miller, M. & Pacheco, M. (1991) Aminoalkylindoles (AAIs): a new route to the cannabinoid receptor? *NIDA Res Monogr.* **105**, 425–426.

Ward, S.J., Childers, S.R. & Pacheco, M. (1989) Pravadoline and aminoalkylindole (AAI) analogues: actions which suggest a receptor interaction. *Br. J. Pharmacol.* **98** (Suppl), 831P.

Weisz, D.J., Gunnell, D.L., Teyler, T.J. & Vardaris, R.M. (1982) Changes in hippocampal CA1 population spikes following administration of delta-9-THC. *Brain Res. Bull.* **8**, 155–162.

Welch, S.P. (1993) Blockade of cannabinoid-induced antinociception by norbinaltorphimine, but not *N,N*-diallyl-tyrosine-Aib-phenylalanine-leucine, ICI 174,864, or naloxone in mice. *J. Pharmacol. Exp. Therap.* **265**, 633–640.

Welch, S.P. & Stevens, D.L. (1992) Antinociceptive activity of intrathecally administered cannabinoids alone, and in combination with morphine, in mice. *J. Pharmacol. Exp. Therap.* **262**, 10–18.

Witter, M.P. & Amaral, D.G. (1991) Entorhinal cortex of the monkey: V. Projections to the dentate gyrus, hippocampus, and subicular complex. *J. Comp. Neurol.* **307**, 437–459.

Wojcik, W.J., Cavalla, D. & Neff, N.H. (1985) Co-localized adenosine A1 and γ-aminobutyric acid-B (GABA-B) receptors of cerebellum may share a common adenylate cyclase catalytic unit. *J. Pharmacol. Exp. Therap.* **232**, 62–66.

Wojcik, W.J. & Neff, N.H. (1984) γ-Aminobutyric acid B receptors are negatively coupled to adenylate cyclase in brain, and in the cerebellum these receptors may be associated with granule cells. *Mol. Pharmacol.* **25**, 24–28.

Wu, R.L. & Barish, M.E. (1992) Two pharmacologically and kinetically distinct transient potassium currents in cultured embryonic mouse hippocampal neurons. *J. Neurosci.* **12**, 2235–2246.

Zbicz, K.L. & Weight, F.F. (1985) Transient voltage and calcium-dependent outward currents in hippocampal CA3 pyramidal neurons. *J. Neurophysiol.* **53**, 1038–1058.

Zona, C., Pirrone, G., Avoli, M. & Dichter, M. (1988) Delayed and fast transient potassium currents in rat neocortical neurons in cell culture. *Neurosci. Lett.* **94**, 285–290.

CHAPTER 8

THE UNPAVED ROAD TO THE ENDOGENOUS BRAIN CANNABINOID LIGANDS, THE ANANDAMIDES

R. Mechoulam and E. Fride

Department of Natural Products, Hebrew University, Medical Faculty, Jerusalem 91120, Israel

Table of Contents

8.1 Introduction

Opium and cannabis, their preparations and derivatives, are the most widely used illicit drugs in most regions of the world, yet our understanding of cannabis chemistry, pharmacology and biochemistry has always lagged behind that of the opiates: morphine was isolated in 1803 (Casy and Parfitt, 1986) – Δ^9-tetrahydrocannabinol (Δ^9-THC) in 1964 (Gaoni and Mechoulam, 1964). The very complicated structure of morphine was elucidated by Robinson in 1923 (Gulland and Robinson, 1923) and its first synthesis was achieved in 1952 (Gates and Tschudi, 1952). Chemists had to wait till the mid-1960s to reach the same level of structural understanding with the considerably simpler Δ^9-THC (Mechoulam *et al.*, 1967). The receptors and

Cannabinoid Receptors
ISBN 0-12-551460-3

transmitters of the opiate system were discovered in the mid-1970s (Casy and Parfitt, 1986); a receptor of THC was found only in 1988 (Devane *et al.*, 1988) and the first endogenous cannabis ligand, anandamide, was described in 1992 (Devane *et al.*, 1992b). Why the discrepancy?

8.2 Technical and conceptual problems

The basic reasons for the time gap in research progress between opium and cannabis were both technical and conceptual.

On the technical side one can find the simple crystallization of morphine – an alkaloid which forms numerous easily isolable crystalline salts – while THC, an oily substance present in a complex mixture of closely related cannabinoids, can only be separated by careful chromatography. Reliable chromatographic techniques were not developed until the middle of this century. The crude methods of structural elucidation by molecular degradation available to Robinson were sufficient owing to the purity of the large amounts of morphine salts available. In contrast, the structure of THC could not be determined as it was not available in pure form until 1964, and synthetic routes could not be investigated until its structure was well established.

Two main conceptual problems, which are specific for cannabinoids but not for opiates, hampered work aimed at the identification of a cannabinoid receptor and a mediator. One of these was the presumed lack of stereoselectivity (Dewey *et al.*, 1984). It is assumed that compounds acting through a receptor generally show a very high degree of stereoselectivity. As all known receptors are asymmetric, it is conceivable that their interactions with asymmetric ligands will be limited to one enantiomer only. For many years it was assumed that cannabinoid activity exhibits a relatively low degree of stereoselectivity (Martin, 1986). Synthetic (+)-Δ^9-THC showed cannabimimetic activity of 5–10% compared with that of (–)-Δ^9-THC (Dewey *et al.*, 1984). This observation casts a doubt over the existence of a specific cannabinoid receptor and hence of a cannabinoid mediator. During the last decade it was established that cannabinoid activity is highly stereoselective and that the previous observations resulted from separation problems (see below). However, this presumed low degree of stereoselectivity delayed research aimed at the identification of a receptor-mediator cannabinoid system. The structures of the (+)- and (–)-enantiomers of Δ^9-THC are shown in Figure 1.

The second problem was summarized by Paton in his classic review in 1975. Paton pointed out that 'underlying much of the pharmacology of cannabis is the high lipophilicity of its active principles which is responsible for the slowness of its kinetics, its cumulation, [and] its persistence'. Cannabis pharmacology parallels in many ways that of the general anaesthetics, yet it does not produce the expected surgical anaesthesia. Paton suggested that 'considering the strongly hydrophobic character of Δ^9-THC, it is possible that there is too great a physicochemical disparity between it

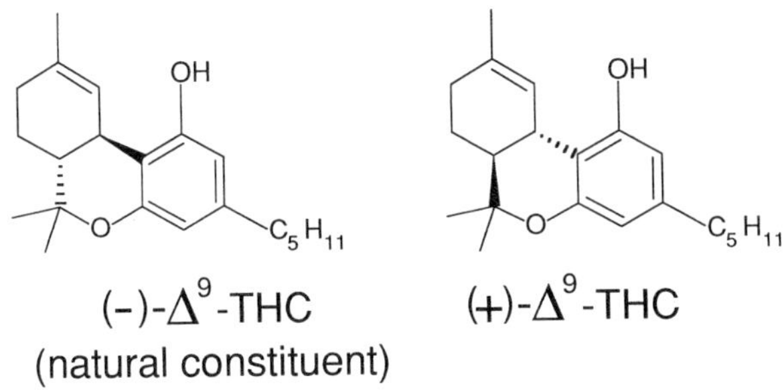

Figure 1 Structures of (–)-Δ^9-THC and (+)-Δ^9-THC.

and biological membranes into which it is inserted for a volume fraction to be achieved sufficient to produce the phenomenon of full anesthesia'.

Δ^9-THC should hence be considered, according to Paton, to belong to the group of biologically active lipophiles and its effects should be compared with 'the chronic effects of anesthetics and solvents at lower dose levels'.

Following this line of thought it is possible to explain the action of cannabinoids without postulating the existence of a specific cannabinoid receptor and of an endogenous mediator of cannabinoid action.

Lawrence and Gill (1975) and Gill and Lawrence (1976) expanded the above ideas, proposed an attractive hypothesis and found experimental evidence that accounts for cannabinoid stereoselectivity without invoking the existence of a classical cannabinoid receptor.

First, they compared the activity of cannabinoids and of steroid anaesthetics on the order parameter of spin-labelled liposomes. They reported that the steroid alphaxolone increased the fluidity of a liposome bilayer in a manner comparable to that produced by the volatile anaesthetic halothane. The same type of effect was observed with (–)-Δ^9-THC. The psychotropically inactive enantiomer (+)-Δ^9-THC (of unrecorded optical purity) produced the same effect which, however, at high membrane loading was significantly smaller than that of the (–)-enantiomer. The non-cannabimimetic cannabinol and cannabidiol actually decreased the molecular disorder of the lipid bilayer. The absence of anaesthesia on THC administration could be explained by the considerably lower degree of fluidization produced by it in comparison with clinically useful anaesthetics. From these and related experiments Gill and Lawrence (1976) concluded that: 'The molecular pertubation produced by the psychoactive cannabinoids is qualitatively the same as that produced by the general anesthetics, namely, an increased fluidization and disordering of the lipid phase of the cell membrane. The effectiveness of the various cannabinoids in producing this increased fluidization correlates well with their potency as psychoactive agents, which encourages us in the belief that this effect is relevant to the *in vivo* actions of the cannabinoids. Furthermore, as the liposomal membrane is apparently able to discriminate between the various cannabinoids in a way similar to the nerve cell, it is unnecessary to postulate the existence of a more complex macromolecular receptor substance to account for the observed structure–activity relationships.'

The above technical and conceptual problems deterred investigators in initiating a search for a specific receptor and a specific ligand.

8.3 Myelin basic protein – a potential cannabinoid ligand

The technical and conceptual problems described above did not prevent at least one group making a serious effort to find a receptor–neurotransmitter system. Nye *et al.* (1985) identified a high-affinity binding site using a positively charged, water-soluble derivative of Δ^9-THC, [^{3}H]5′-trimethylammonium-Δ^8-THC ([^{3}H]TMA). [^{3}H]TMA binds saturably and reversibly to brain membranes (K_D=89 nM). Δ^9-THC competitively inhibits [^{3}H]TMA binding. The site discriminates cannabinoids from numerous other compounds and drugs; however, binding does not parallel the well-established structure–activity relationships (SAR) formulated for the cannabinoids. For example it binds several psychotropically weak or inactive cannabinoids with high affinity. It was suggested that this site does not mediate the psychoactive effects but may mediate other pharmacological actions of the cannabinoids.

The same group also reported the purification and chemical identification of an endogenous inhibitor of [^{3}H]TMA binding (Nye *et al.*, 1988). It was isolated from rat brain by acid extraction followed by reverse-phase matrix and gel filtration chromatography. The inhibitor was shown to have a molecular weight of 14 500 daltons by polyacrylamide gel electrophoresis. It comigrates with the small form of rat myelin basic protein (MBP) on gel filtration, reacts with MBP antibodies and has an amino acid composition related to that of MBP. Inhibition of [^{3}H]TMA binding by the endogenous inhibitor was non-competitive, with a K_i=4 nM.

The physiological significance of this inhibitor is far from clear. There is no evidence that MBP is an endogenous mediator of cannabinoid activity. As there is no correlation between binding of cannabinoids to the TMA-site and their psychoactivity (see above) it is indeed difficult to perceive a role for MBP in this area. Furthermore, several large synthetic polycations, such as polylysine and polyarginine were also found to bind to the TMA site (Nye *et al.*, 1988). Such polycations have not been associated with cannabinoid-like activity. Nye *et al.* (1988) have suggested a possible cannabinoid-related role for MBP in the nervous system: by binding to the TMA site on myelin, MBP may protect myelin from the effects of the lipo-soluble cannabinoids (synthetic or endogenous) which bind to the same site. It would be of interest to establish such an interaction.

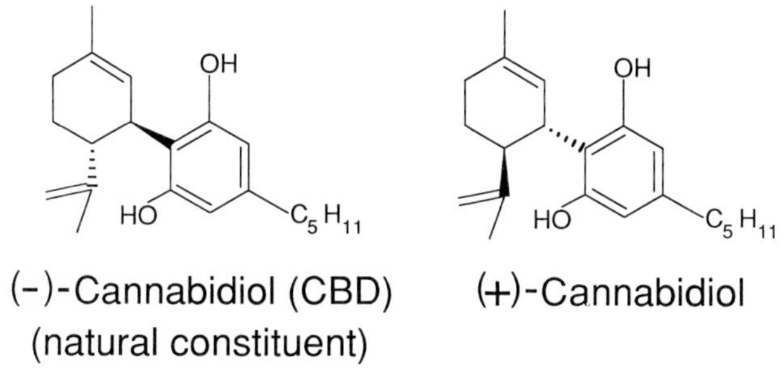

Figure 2 Structures of (−)cannabidiol and (+)cannabidiol.

8.4 Stereoselectivity of cannabimimetic action

Gradually, during the early 1980s, it became clear that while some of the interactions of cannabinoids with membranes are indeed non-specific, others are highly specific (Martin, 1986). One can presume that effects on membrane-associated enzymes or other biological systems can also be both specific and non-specific. For example, while the psychotropic effects of THCs are highly stereoselective (see below), the anticonvulsant effects of CBD are not: both (+) cannabidiol (CBD) and (−) CBD (Figure 2) have been found to show the same level of activity (Leite *et al.*, 1982). More recently Felder *et al.* (1992) have shown stereoselectivity for binding and functional inhibition of cyclic AMP but not for release of arachidonic acid and intracellular calcium.

The presumed lack of stereoselectivity of the psychotropic effects was gradually shown to result from technical difficulties. While several syntheses of THC-type cannabinoids with the natural (−)-stereochemistry are available and have been widely used, cannabinoids with the enantiomeric (+)-stereochemistry are exclusively prepared via a route published by our group over 25 years ago (Mechoulam *et al.*, 1967). The starting material for this synthesis is (+)-α-pinene. However, commercial (+)-α-pinene has only about 90–95% optical purity and, if the intermediates or the final products are not further purified, THCs with a corresponding level of enantiomeric purity will be obtained. We have recently repeated the above synthesis (unpublished data). The intermediate crystalline *cis* verbenol was recrystallized to nearly 100% optical purity and was further employed in the above-mentioned synthesis (Figure 3). (+)-Δ^8-THC and (+)-Δ^9-THC thus obtained were tested in the mouse ring test. While the natural (−)-THCs showed significant activity at 1 mg/kg, no activity was observed with the (+)-enantiomers at doses of up to 50 mg/kg.

An example of essentially absolute stereoselectivity is the case of the enantiomers of the dimethylheptyl homologues of 11-hydroxy-Δ^8-THC (11-OH-Δ^8-THC-DMH). The synthesis of these enantiomers is described in Figure 4 (Mechoulam *et al.*, 1990). It should be pointed out that in this synthesis, both the intermediate ketone and the 11-OH-Δ^8-THC-DMHs are crystalline compounds which, by further crystallization, can be obtained with absolute enantiomeric purity.

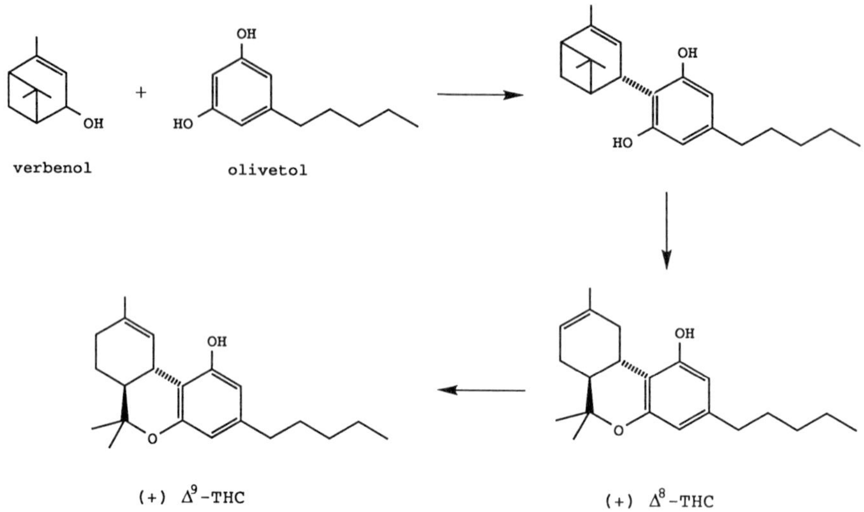

Figure 3 Synthesis of (+)-Δ^9-THC.

The effects of the enantiomers of 11-OH-Δ^8-THC-DMH on spontaneous activity, rectal temperature, tail-flick latency, and catalepsy were studied in mice and in the dog static-ataxia model (Little *et al.*, 1989; see also Chapter 2). The (−)-enantiomer (HU-210) was active in all tests at doses between 3 and 100 μg/kg, while the (+)-enantiomer (HU-211) was inactive at 30 mg/kg in the mouse and 1 mg/kg in the dog. Thus, the (−)-enantiomer (HU-210) was 100 to 800 times more potent than Δ^9-THC in the mouse. Related results were obtained in the mouse ring test and in the rotarod tests in rats (Mechoulam *et al.*, 1988). In drug stimulus generalization tests it was found that, in the pigeon, HU-210 was about 87 times more active than Δ^9-THC (Järbe *et al.*, 1989). The (+)-enantiomer (HU-211) was inactive in pigeon and rat at doses that were, respectively, at least 4500 and 1000 times higher than the corresponding ED_{50}s of the (−)-enantiomer (Järbe *et al.*, 1989). Pertwee *et al.* (1992) have examined the inhibitory effect of (−)HU-210 and (+)HU-211 in mouse vas deferens (MVD) and in myenteric plexus (GP-MP) preparations of guinea-pig small intestine. HU-210 was about 40 times more potent than Δ^9-THC in both preparations; HU-210 decreased twitch heights by 50% at concentrations of 0.15 nM and 1.4 nM in the MVD and GP-MP preparations respectively. (+)HU-211 was at least 1000 times less potent in the MVD preparation and at least 500 times less potent in the GP-MP preparation.

Other enantiomeric pairs have presented the same overall picture (Figure 5). With (−)-CP 55,940 – (+)-CP 56,667 and with (−)-CP 55,244 – (+)-CP 55,243 it was found that the (−) enantiomers were highly cannabimimetic in a battery of *in vivo* tests. The (+)-enantiomers were essentially inactive (Little *et al.*, 1988; Thomas *et al.*, 1990).

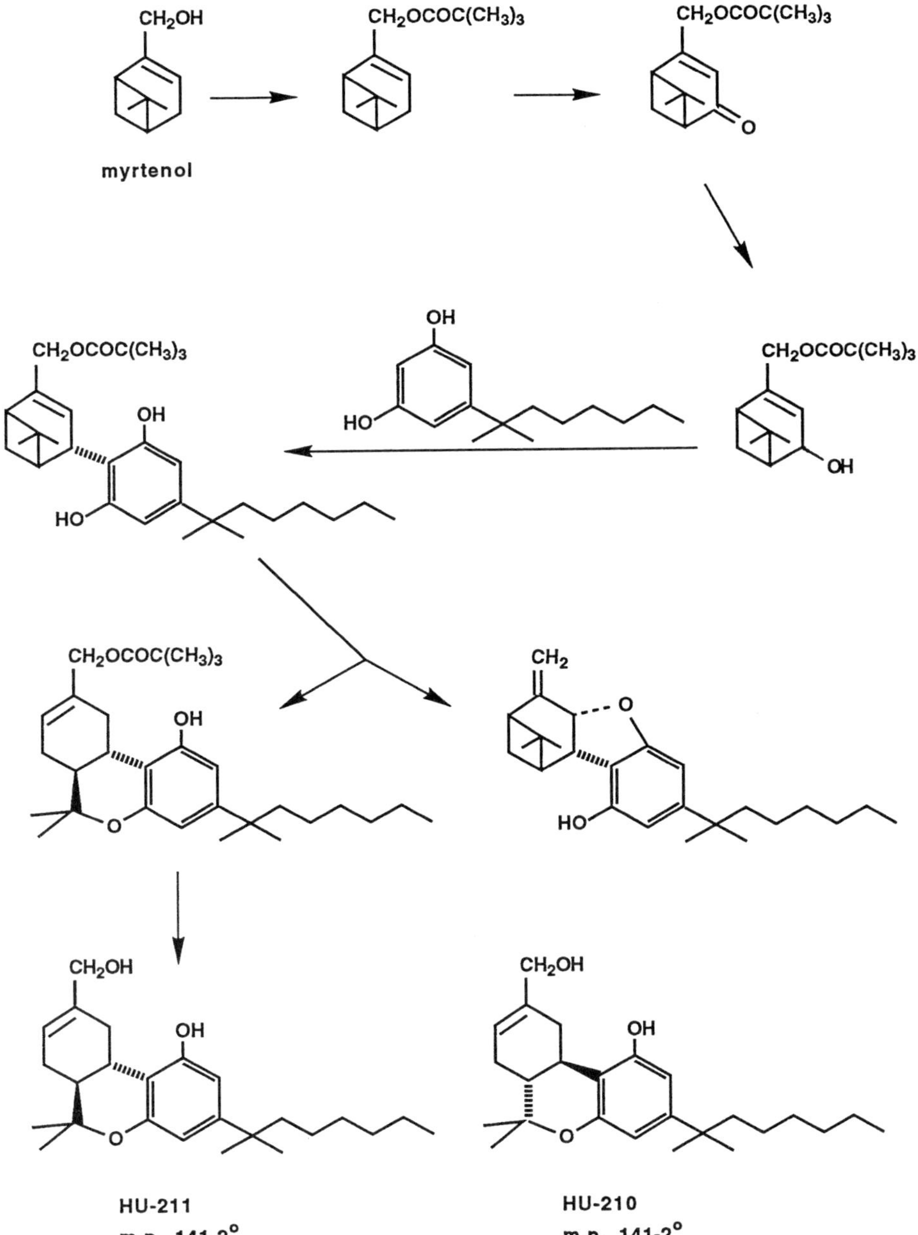

Figure 4 Synthesis of HU-210 and HU-211.

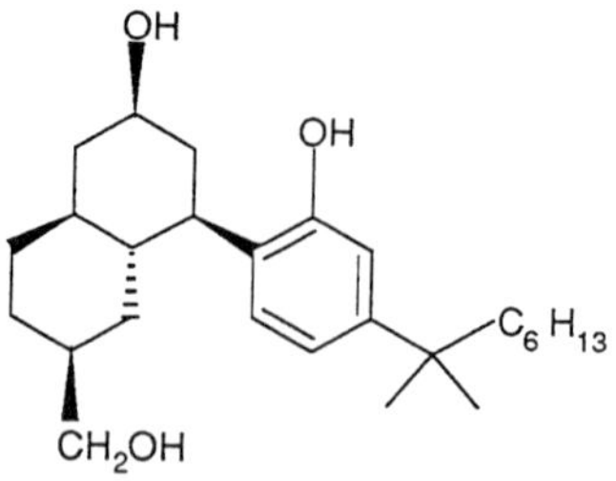

(−)-CP 55,244

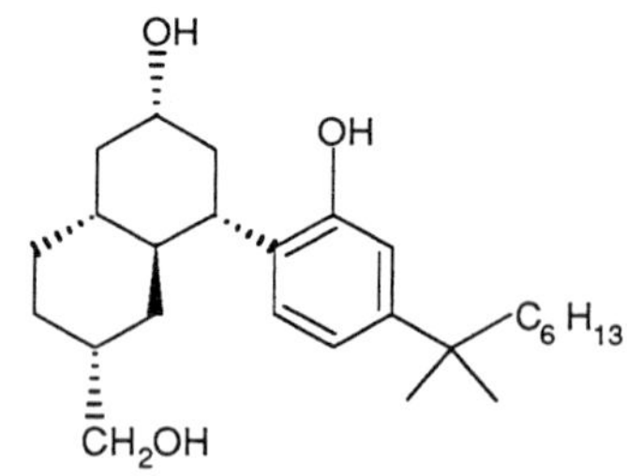

(+)-CP 55,243

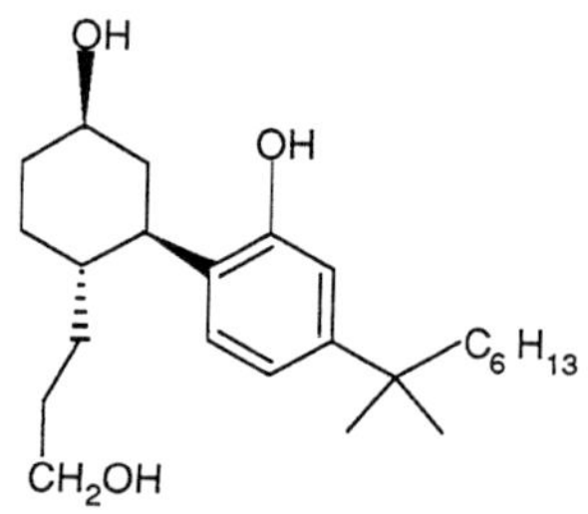

(−)-CP 55,940

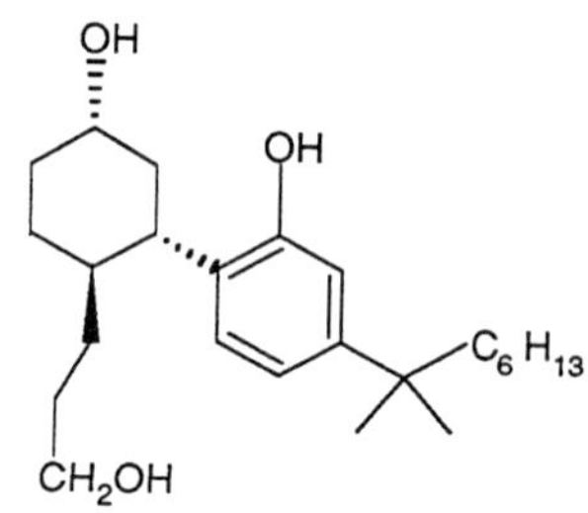

(+)-CP 56,667

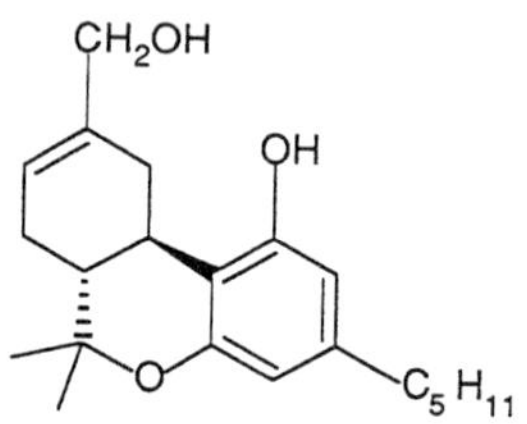

(−)-11-OH-Δ^8-THC

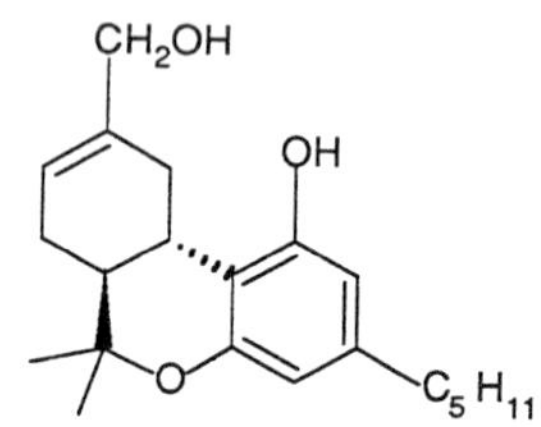

(+)-11-OH-Δ^8-THC

Figure 5 Several cannabinoid enantiomeric pairs.

Recently, the primary metabolite of (−)-Δ^8-THC, namely (−)-11-hydroxy-Δ^8-THC and its enantiomer, (+)-11-hydroxy-Δ^8-THC, were prepared with very high enantiomeric purity (>99%) following the synthesis described in Figure 4 (the difference being a C_5H_{11} side-chain on the aromatic ring rather than a dimethylheptyl one). They were tested in discrimination tests in Mongolian gerbils and in pigeons (Järbe *et al.*, 1994). (−)-11-OH-Δ^8-THC exhibited cannabimimetic activity with a potency similar to that of Δ^9-THC. No cannabimimetic activity was seen after administration of (+)-11-OH-Δ^8-THC. For a review of the stereoselectivity of cannabinoid activity see Mechoulam *et al.* (1992).

The establishment of the very high stereoselectivity of cannabinoid action removed a further obstacle in the development of the field and it was now open for a renewed effort towards the identification of a specific brain cannabinoid receptor. This was achieved in 1988 in Allyn Howlett's laboratory in St. Louis (Devane *et al.*, 1988). A peripheral receptor (CB_2) was identified in the spleen by Munro *et al.* (1993). The CB_2 receptor has only 44% homology with the receptor (CB_1) that is found in the brain (see Chapter 4).

8.5 Various approaches towards the identification of endogenous cannabinoids

The identification of a specific cannabinoid receptor (see Chapter 1) inevitably posed the question of the existence of cannabinoid mediator(s) in the brain. Several groups approached the problem using various techniques (see also Chapter 7). Childers and Pacheco (1992) extracted brain tissue in hot acid. The mixture obviously contained water-soluble constituents (sugars, low molecular weight peptides) as well as various amines, in the form of salts. The mixture displaced [^{3}H]WIN 55,212–2 binding but did not compete for μ opioid binding. To date, this approach has not led to defined chemical entities. Howlett *et al.* (1992) have looked into the binding of a long list of known brain constituents, neurotransmitter agonists and antagonists, and representatives of numerous types of compounds active on various biological systems in the body. In this screening-type approach 136 compounds were tested: seven adrenergic ligands, five dopaminergic ligands, seven opioid ligands, nine cholinergic ligands, four sigma ligands, eight GABA receptor ligands, seven serotonergic ligands, nine steroid receptor ligands as well as neuroleptics, hormones, neuromodulators, amino acids, peptides, arachidonic acid derivatives, prostaglandins, leukotrienes and miscellaneous agents. At concentrations between 1–10 μM none of these compounds caused significant inhibition of [^{3}H]CP 55,940 binding.

A neurotransmitter should be released from cells on stimulation mostly in a Ca^{2+}-dependent fashion. Following this line of reasoning Evans *et al.* (1992) examined the ability of A-23187 (a calcium ionophore) to release compounds that can interact with the cannabinoid receptor. Indeed, they found that A-23187 can release an unidentified material with cannabinoid receptor-binding activity in the presence, but not in the absence of Ca^{2+}. Release of this material was dependent on the time of exposure to the ionophore as well as on its concentration. The significance of these observations is not yet clear.

8.6 Anandamide[1]

8.6.1 HU-243 – a novel cannabinoid receptor probe

Our approach to the identification of an endogenous mediator, which we planned in 1990, was to prepare initially a novel probe for the cannabinoid receptor. The probe, which had been used for the identification of the cannabinoid receptor, namely labelled CP 55,940, was not available to us at that time and, in any case its structure is only marginally related to natural Δ^9-THC. We believed that a novel probe with a typical THC-type structure, a K_D in the picomolar range and high stereoselectivity with respect to both pharmacological activity and binding could serve as an excellent tool in such a project. We had previously reported the synthesis of HU-210 (see Figure 4; Mechoulam *et al.*, 1990) which at that time was possibly the most potent cannabimimetic known. In numerous pharmacological tests it was shown to be 100–800 times more potent than Δ^9-THC (see above). HU-210 competes fully for the specific binding of [^{3}H]CP 55,940 to membranes from rat brain in heterologous displacement studies (K_i=234 pM; Howlett *et al.*, 1990). The (+)-enantiomer of HU-210, namely HU-211, was generally 1500–5000 times less active than HU-210 in the above tests. The classical cannabinoid structure of HU-210, its very high potency and the stereoselective profile of its activity made this cannabinoid an excellent candidate for labelling. The obvious labelling route seemed to be reduction of the double bond with tritium. This route was dependent on the retention of activity by the reduced product. We had already shown that hydrogenation of Δ^9-THC or Δ^8-THC led to two epimeric hexahydrocannabinols, both of which were cannabimimetic, the C-9 equatorial epimer being more active than the axial one (Gaoni and Mechoulam, 1966; Edery *et al.*, 1971). 11-Hydroxyhexahydrocannabinols have also been prepared (Mechoulam *et al.*, 1980). Again, the equatorial epimer was considerably more active. In discrimination tests the equatorial epimer was about 3–17 times more active than the axial one (Järbe *et al.*, 1986). On the basis of this background we believed that hydrogenation of HU-210 (or tritiation in the labelling experiments) would produce highly active products which could be used as cannabinoid probes. Indeed, hydrogenation over platinum oxide led to an active mixture of 11-hydroxyhexahydrocannabinol dimethylheptyl homologues. As receptor probes should preferably be well-defined single moieties, and not mixtures, we sought and found routes to both epimers at C-9. The more active equatorial epimer (HU-243) was prepared with high stereoselectivity (93%) by reduction of HU-210 with a mixture of Wilkinson's catalyst and Kagan's catalyst. It was also prepared in labelled form by reduction of HU-210 with tritium (Figure 6) and served as a most useful tool in our search for the endogenous cannabinoid ligand. The specific activity of [^{3}H]HU-243 prepared and used by us was 54 Ci/mmol (Devane *et al.*, 1992a).

[1] The first person-plural (we) used in this portion of the review refers to the group of dedicated students and collaborators who participated in the cannabinoid research project, including work on anandamide, in the Department of Natural Products.

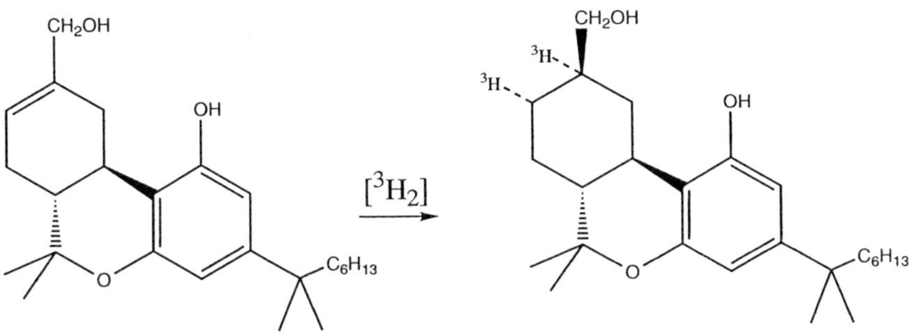

HU-210 **[^{3}H]HU-243**

Figure 6 Synthesis of [^{3}H]HU-243.

8.6.2 Isolation of anandamide

Our approach towards the isolation of an endogenous cannabinoid in the brain was based on the assumption that it is a lipid-soluble compound. The SARs in the cannabinoid series indicate that increase in liposolubility, at least up to a certain point, leads to increase in cannabimimetic activity. Thus, the dimethylheptyl homologue of Δ^8-THC is considerably more active than Δ^8-THC; Δ^9-THC itself is more active than the natural propyl homologue (Edery *et al.*, 1971; Razdan, 1986). Other examples are also known. On this somewhat tenuous basis we assumed that the endogenous cannabinoid ligands are lipids. However, in view of the identification of peptides as modulators of a long list of receptors (including the opiate receptor) the decision to focus most of our attention on the lipid-soluble fractions was essentially intuitive.

We extracted animal brains with solvents appropriate for lipid constituents, namely chloroform–methanol (Devane *et al.*, 1992b). Initially, we extracted cow brains and dog brains but ultimately settled on porcine brains for practical purposes – availability and inexpensiveness. Several separation routes were investigated. One of these was direct silica gel chromatography of the extract. Numerous problems were encountered or were identified later. The lipid content of the brain is very high; our extracts presumably contained only miniscule amounts of active material in a complex mixture of various types of lipids which, on chromatography, possibly served as coelutants and the active compound, therefore, was dispersed throughout the chromatographic fractions. By repeated chromatography on silica gel (normal-phase and later reverse-phase), we were able gradually to concentrate the active fractions. A problem that became obvious to us only towards the end of the endless purifications was the lability of the product: we were seldom able to elute from the column the amount of active material deposited on it. Although concentration and purification of the active compounds was achieved during the chromatographic process, the amount of activity gradually decreased.

The isolation of active constituents from mixtures, such as plant or animal extracts, requires monitoring of activity by a chemical or a biological assay. Extracts from natural sources may contain hundreds of chemicals, with the required active material being present in minute concentrations. It is extremely difficult, ineffective, or even impossible, to isolate in pure form all constituents and then to test them for the required biological activity. The well-established route to the identification of natural active constituents is to fractionate a mixture (usually by chromatography), test the crude fractions for the desired activity, disregard the inactive fractions and then repeat the fractionation using the same or a different separation method. It is expected that gradually the activity will increase with the elimination of inactive fractions. In our case, although inactive fractions were indeed discarded, the activity actually decreased, owing to the lability of the active material.

The activity was monitored by the ability of fractions to displace the above-described radiolabelled probe, [^{3}H]HU-243, in a centrifugation-based ligand-binding assay (Devane *et al.*, 1992a,b). As discussed above, many fractions inhibited the binding of [^{3}H]HU-243 to the cannabinoid receptor. A further problem was that HU-243, like many cannabinoids, binds to the siliconized polypropylene microfuge tubes in which the assay was conducted. This three-way equilibrium of the probe among the synaptosomal receptors, the solution and the microfuge tube was a major obstacle in the interpretion of the binding values. For practical reasons, we chose to continue work with the fractions which had least effect on the adherence of the radioligand to the microfuge tube. These fractions were further purified by low- and medium-pressure column chromatography and by thin-layer chromatography (TLC) in both normal-phase and reversed-phase systems. Ultimately, we isolated a fraction which gave one spot on TLC.[1] It eluted mainly as one main peak on gas chromatography (GC) when a mass spectrometer was used as a detector. This compound represented the first example of a purified brain constituent which exhibited some of the properties (see below) of Δ^9-THC. We assumed that it is a natural ligand for the cannabinoid receptor.

We named it anandamide – based on the Sanskrit word ananda meaning bliss and on its chemical nature (see below). This constituent inhibited the specific binding of [^{3}H]HU-243 in a manner typical of competitive ligands with a K_i value of 52 ± 1.8 nM. Surprisingly, this value is almost identical to that of Δ^9-THC in this system (K_i = 46 ± 3 nM). Figure 7 represents the competitive inhibition of [^{3}H]HU-243 by natural anandamide.

In addition to the specific binding to the cannabinoid receptor it seemed to us of considerable importance to determine the activity of natural anandamide in an additional bioassay. Pertwee *et al.* (1992) had reported that cannabinoids inhibit the twitch response of murine isolated vasa deferentia. This assay is not specific for cannabinoids; it had previously been extensively used in opioid research. With

[1] TLC: Eluents were petroleum ether:chloroform:methanol (6:40:4). The plate is sprayed with sulfuric acid:methanol (1:1) and heated on an electric plate (150°C) until a spot appears.

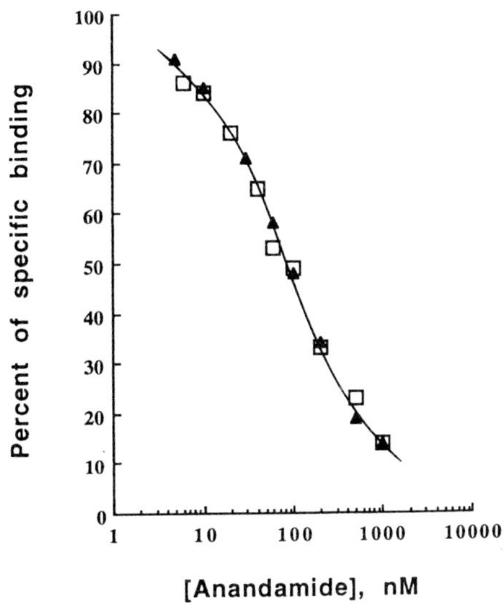

Figure 7 Competitive inhibition of [^{3}H]HU-243 binding by natural anandamide. The data were normalized to 100% of specific binding which was determined with unlabelled HU-243. Specific binding accounted for 77–82% of the total radioactivity bound to the membranes (Devane *et al.*, 1992b). Courtesy of AAAS, publishers of *Science*.

cannabinoids it had been shown to be stereospecific and sensitive. Thus, a concentration of (–)HU-210 that decreased the twitch height in the mouse vas deferens by 50% was 0.15 nM, while the enantiomeric (+)HU-211 had no inhibitory effect up to 30 nM. Δ^9-THC decreased the twitch height by 50% at 6.3 nM. The inhibition of the twitch response by opioids is blocked by the opioid antagonist naloxone. The cannabinoid effect is not attenuated by naloxone. Hence, this effect is not mediated by the opioid receptor system. We assumed that this assay was suitable for anandamide in particular in view of its sensitivity and of the minute amounts of natural material available. Indeed, anandamide elicited a concentration-dependent inhibition of the twitch response, decreasing the twitch height by 50% at a concentration of 90 nM (Figure 8). The inhibition was not attenuated by addition of naloxone at a concentration of 300 nM. It should be noted that in this test also, the potency of anandamide was comparable to that observed in the receptor binding assay and was relatively close to that of Δ^9-THC.

As pointed out above anandamide is a labile compound. This instability is not surprising since it is a derivative of arachidonic acid, which is also labile. We have found that anandamide in ethanol solution is stable over several months when kept at –20°C. Such a solution is not stable at room temperature and numerous additional spots are observed on TLC after several days. Anandamide as an oil (not in a solution) can also be kept at –20°C, preferably under nitrogen, for at least several weeks.

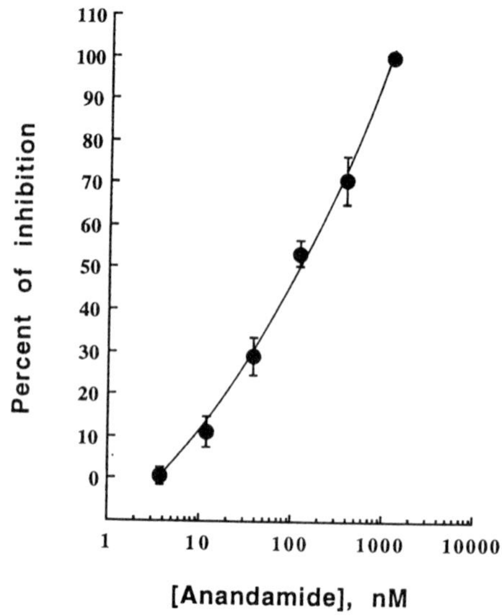

Figure 8 Inhibition of the mouse vas deferens twitch response by natural anandamide (Devane *et al.*, 1992b). Courtesy of AAAS, publishers of *Science.*

8.6.3 Elucidation of the structure of anandamide

The laborious isolation procedures described above led to minute amounts of purified material. We never had more than several-hundred micrograms of natural anandamide. The isolation of very small amounts of a natural constituent from a complicated mixture inevitably poses problems for the elucidation of structure, owing to the presence of minor impurities associated with the isolation process. Anandamide, as ultimately isolated, contained traces of materials that originated from the plastic labware as well as traces of solvents, tightly bound to the viscous anandamide.

The structure of anandamide was deduced from mass spectrometric and nuclear magnetic resonance (NMR) measurements (Devane *et al.*, 1992b). The initial measurements were by high-resolution mass spectrometry (MS) which suggested the elemental composition $C_{22}H_{37}NO_2$ (m/z 347.2762), indicating the presence of five double bonds (or rings). No ultraviolet (u.v.) absorption above 220 nm, consistent with a conjugated system, was noted. The first indication of the structure was the observation that the NMR peaks at δ 5.30–5.45, presumably owing to double-bond protons, were coupled with peaks at δ 2.75–2.90 which we assumed to be signals of doubly allylic protons. Such protons, their couplings and the 4:3 ratio of vinylic to doubly allylic protons are typically observed in all-*cis*, non-conjugated polyunsaturated fatty acids such as linoleic and arachidonic acids (Christie, 1982). At this point we assumed that we had a *N*-derivative of such a

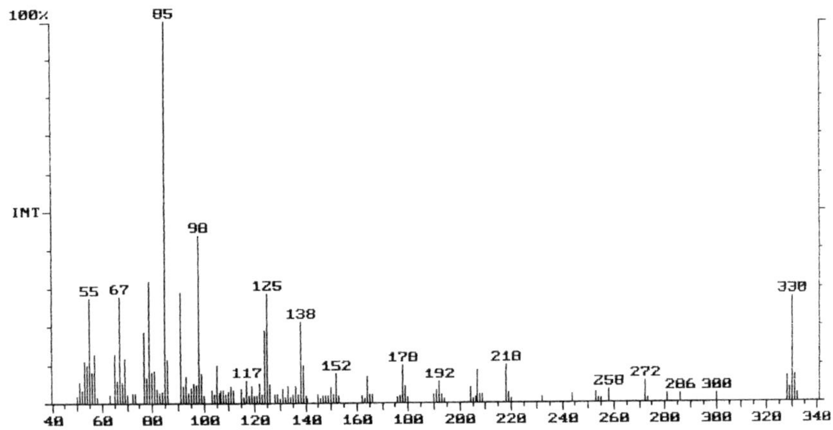

Figure 9 GC-MS spectrum (electron impact) of anandamide in an ion-trap instrument. Anandamide undergoes thermal dehydration under these conditions; hence the spectrum shown is that of the MH^+ ion of the corresponding 2-oxazoline (Devane *et al.*, 1992b). Courtesy of AAAS, publishers of *Science.*

fatty acid. MS spectra supplied additional data which clarified the structure. Collision-induced dissociation (CID) measurements of the MH^+ ion (m/z 348), obtained from direct exposure chemical ionization, gave several major significant ions at m/z 387, 62 and 44. The m/z 62 ion had an elemental composition C_2H_8NO, which best fits a protonated ethanolamine ion, $HOCH_2CH_2NH_3^+$; the m/z 44 ion could represent dehydrated ethanolamine (protonated form); n/z 287 ion corresponds to MH^+ less ethanolamine. The existence of an ethanolamine moiety was supported by the MS of a trimethylsilyl derivative (TMS) of anandamide. An m/z 419 ion indicated the formation of a mono TMS derivative and hence the existence of a single hydroxyl group. The CID of a major m/z 404 ion ($M\text{-}CH_3^+$) gave mainly two ions which corresponded to $Me_2Si^+OCH_2CH_2NH_2$ (m/z 118) and $Me_2Si^+OCH_2CH_2NCOCH{=}CH_2$ (m/z 172).

Gas chromatography-MS measurements (GC–MS) of anandamide supplied supporting evidence. Thermal dehydration gave rise to an m/z 330 MH^+ ion under chemical ionization (CI). Both m/z 330 MH^+ and m/z 329 were formed on electron impact (EI) in an ion-trap instrument (Figure 9). The fragmentation pattern in the low mass range included m/z 85 (McLaferty rearrangement product) and m/z 98 (product of α-cleavage). Additional MS and NMR data are available in the original publication (Devane *et al.*, 1992b).

The NMR and MS data led to the assumption that anandamide is the ethanolamide of a C-20 fatty acid with four unconjugated double bonds, presumably arachidonic acid. This assumption was proved by a simple synthesis (Devane *et al.*, 1992b). Arachidonic acid was converted into its acyl chloride with oxalyl chloride and in a second step the arachidonyl chloride was reacted with ethanolamine, leading to synthetic arachidonylethanolamide (see Figure 10) identical to

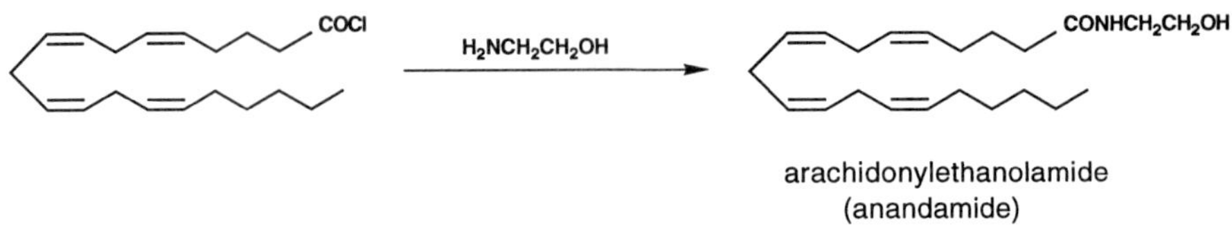

Figure 10 Synthesis of anandamide (anandamide, 20:4, *n*-6).

natural anandamide in its infrared, NMR and MS spectra and in its ability to inhibit the twitch response of isolated vasa deferentia (Devane *et al.*, 1992b; Pertwee *et al.*, 1993). Synthetic anandamide binds to the cannabinoid receptor with a K_i of 39 ± 5 nM, that is slightly lower than that of natural anandamide (K_i of 52 ± 1.8 nM), presumably owing to minor impurities in the natural product.

In addition to anandamide, two other unsaturated fatty acid ethanolamides have been isolated from pig brain: homo-γ-linolenylethanolamide and docosatetraenylethanolamide (Figure 11, see also Chapters 1 and 2). They bind to the brain cannabinoid receptor with K_is essentially identical to that of arachidonylethanolamide (Hanuš *et al.*, 1993). It has been suggested that the acylethanolamides that bind to the brain cannabinoid receptor be named 'anandamides' with each individual member identified with the parent fatty acid (indicated in parenthesis) following the generally accepted fatty acid shorthand designation (Hanuš *et al.*, 1993): hence the anandamide derived from arachidonic acid is anandamide (20:4, *n*-6), that from homo-γ-linolenic acid is anandamide (20:3, *n*-6), and that from docosatetraenoic acid is anandamide (22:4, *n*-6). When just anandamide is mentioned, it is generally understood to mean the anandamide (20:4, *n*-6).

8.6.4 Activity of anandamide

The identification of a brain constituent which binds to the cannabinoid receptor and acts on the vas deferens in the same manner as THC does not sufficiently define this ligand as a new neuromodulator. Numerous criteria have yet to be fulfilled. For example, a neuromodulator has to produce a response at the target cells, or organs, and show a pharmacological profile related to that of THC. Inhibition of adenylate cyclase, parallel to that of THC must be observed. A neuromodulator must be released on stimulation and a mechanism must exist for its removal by uptake or metabolism. Some of the above criteria have been fulfilled.

Shortly after the initial publication on anandamide we published the first report on its pharmacological activity (Fride and Mechoulam, 1993).

The cannabinoids produce a unique syndrome of behavioural effects in animals: at low doses a mixture of depressant and stimulatory effects is observed, while at higher doses central depression predominates. We examined anandamide for behavioural effects using tests frequently employed in this area: (1) the ring immobility test, which measures per cent of time mice remain motionless on a ring (Pertwee, 1972); (2) the open field test, which measures locomotor and rearing activity; (3)

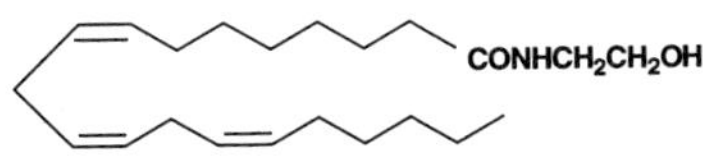

homo-γ-linolenylethanolamide

(anandamide, 20:3, n-6)

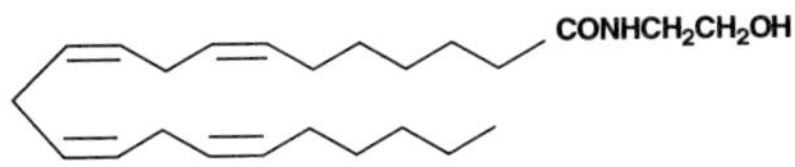

7,10,13,16-docosatetraenylethanolamide

(anandamide, 22:4, n-6)

Figure 11 Structures of homo-γ-linolenylethanolamide (anandamide, 20:3, *n*-6) and of docosatetraenylethanolamide (anandamide, 22:4, *n*-6).

hypothermia and (4) antinociception (hot plate latency). Martin *et al.* (1991) have shown that, when taken individually, these assays are not selective for any particular class of compounds. However, this tetrad of tests, evaluated together 'has proven to be highly predictive of cannabinoids' (see Chapter 2). The results with anandamide are presented in Figures 12–16 and are compared with those obtained with Δ^8-THC (single high dose). Peak activity on i.p. administration is observed for anandamide after around 10 min (10 mg/kg) and for Δ^8-THC after around 30–60 min. No activity was found in rats on i.p. administration (Fride, 1993, unpublished).

At a low dose (0.01 mg/kg) anandamide caused significant stimulation in the open field, but no significant changes were observed in the other tests. At higher doses anandamide produced significant depression in the open field, increase in the time spent on the ring, hypothermia and antinociception. By comparison with the effects produced at a dose level of 100 mg/kg, we determined that at 20 mg/kg anandamide had reached its maximal effect, except in the ring test. In this test the per cent catalepsy reached at 100 mg/kg was 79.2 ± 3.0.

Oral administration of anandamide to Sabra mice (so far examined at a single high dose) causes ring immobility (Fride, 1993, unpublished observations). The results were: 6.0 ± 6.3% immobility for the controls versus 73 ± 5.5 for anandamide-treated mice at a dose of 50 mg/kg ($P<0.001$; $n=3$ per group).

The results presented in the Figures are for Sabra strain female mice. We have obtained similar, significant results with Sabra male mice and with C57Bl (male and female) mice (10 mg/kg) (Table 1) (Fride, 1993, unpublished). Smith *et al.* (1994) have found that i.v. administration to mice produces effects similar to those reported above.

The action of cannabinoids is predominently central (Pertwee, 1988). Therefore the observations reported now indicate that anandamide rapidly crosses the blood–brain barrier.

Howlett (1987) has shown that cannabimimetic compounds decrease the cyclic AMP content of neuronal cells. This action is due to inhibition of adenylate cyclase in plasma membranes. As discussed in Chapter 6, considerable evidence suggests that a cannabinoid receptor mediates these effects on adenylate cyclase.

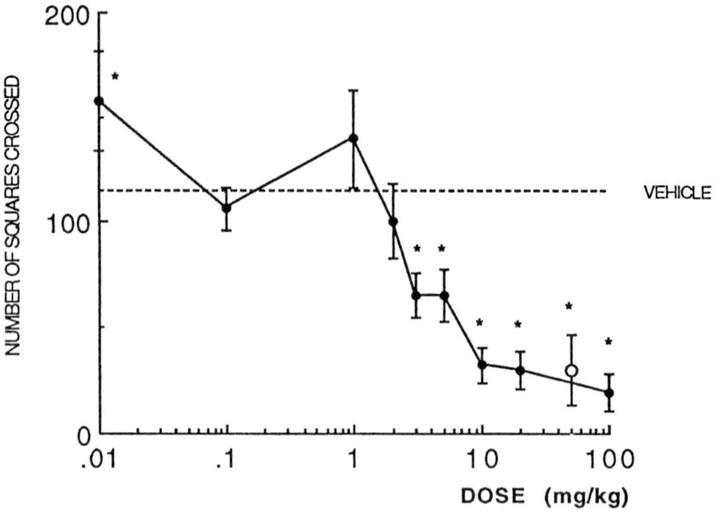

Figure 12 Effect of anandamide on ambulation in an open field. Female Sabra mice were injected i.p. with anandamide (0.01–100 mg/kg, presented on log scale) or Δ^8-THC (50 mg/kg). Ten minutes later, the number of squares crossed during an 8-min exposure in an open field (20 × 30 cm, divided into 12 equal parts) was recorded. Solid circles: anandamide. Open circle: Δ^8-THC. Results from vehicle treated animals are presented by the dotted line. Data were analysed by analysis-of-variance. (*) Indicates significant difference from vehicle ($P<0.05$, Fisher's protected least significant difference test).

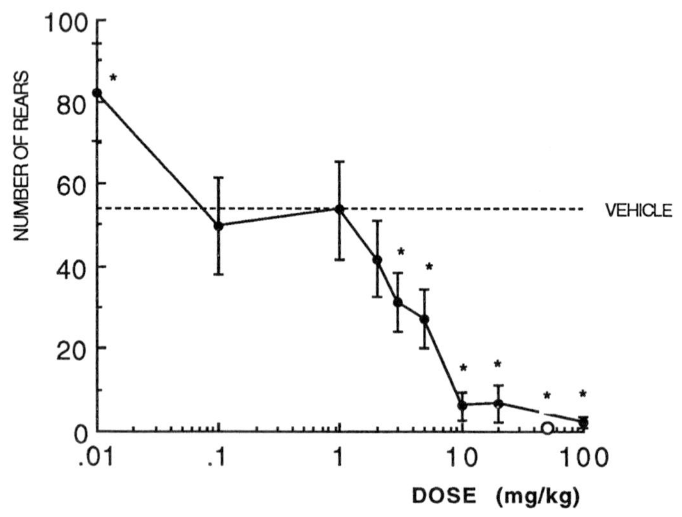

Figure 13 Effect of anandamide on rearing in an open field. Female Sabra mice were injected i.p. with anandamide (0.01–100 mg/kg, presented on log scale) or Δ^8-THC (50 mg/kg). Ten minutes after injection, the number of rears was recorded during an 8-min exposure in the open field. For further details and symbols see Figure 12.

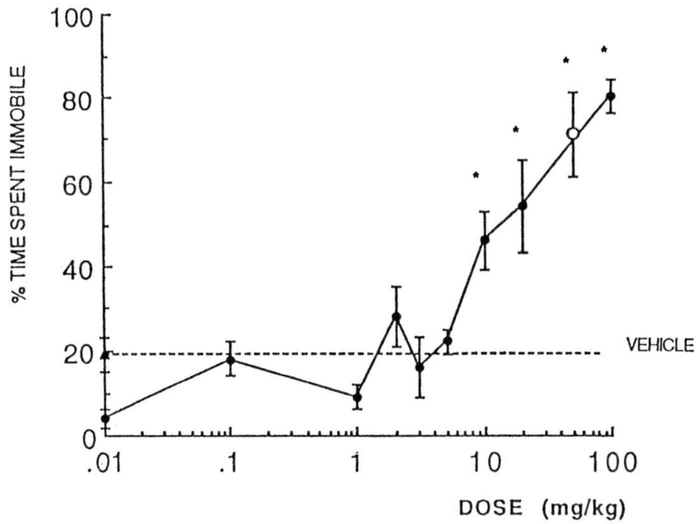

Figure 14 Effect of anandamide on immobility in the ring test. Catalepsy after i.p. injection of anandamide (0.01±100 mg/kg, presented on log scale) or Δ^8-THC (50 mg/kg) was assessed by percentage immobility (Pertwee, 1972) during a 4-min exposure to a 5.5 cm diameter ring. The ring test was initiated immediately after completion of the open field exposure. Symbols are the same as for Figure 12.

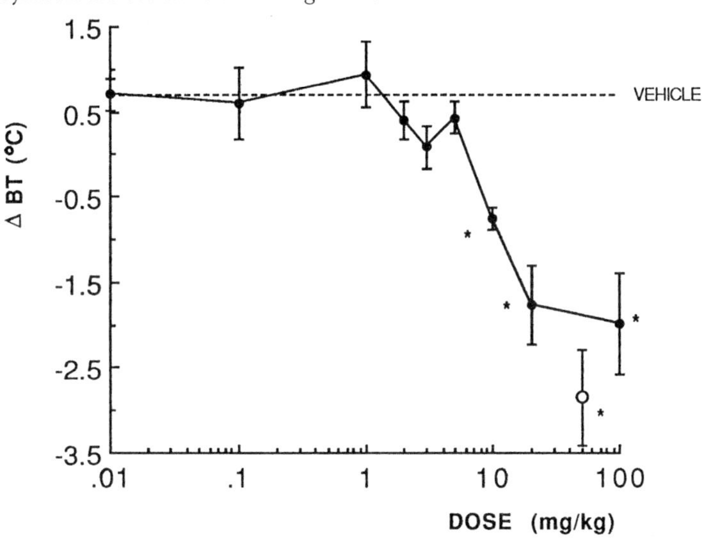

Figure 15 Effect of anandamide on body temperature (BT). Body temperature was measured with a Yellow Springs Instrument Co. telethermometer immediately before injection, and after completion of the ring test (i.e., 22 min after injection). The difference (°C) between the two measurements is presented for anandamide (0.01–100 mg/kg, presented on log scale) and Δ^8-THC (50 mg/kg). Symbols are the same as for Figure 12.

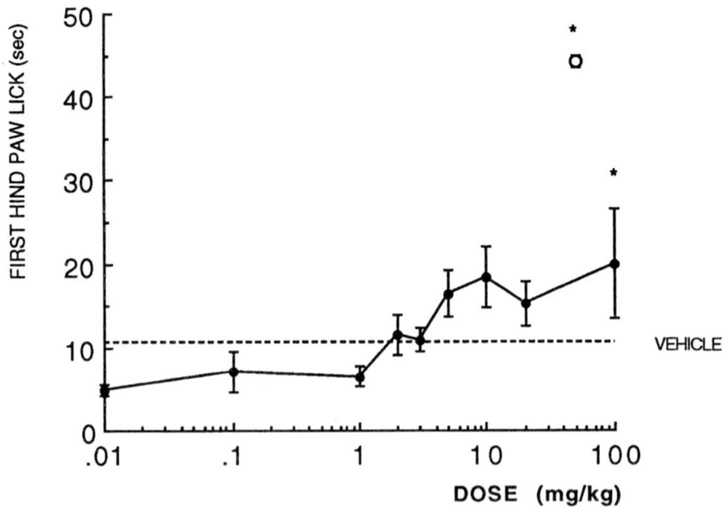

Figure 16 Effect of anandamide on antinociception. Antinociception was assessed after i.p. injections of anandamide (0.01–100 mg/kg, presented on log scale) or Δ^8-THC (50 mg/kg). Animals were placed on a hot plate kept at 54°C (Eddy *et al.*, 1953). They were allowed to adapt to the apparatus for 1 min after placing cardboard over the hot plate in order to insulate the mice from the heat. After removal of the insulation, latency to the first hind paw lick was recorded. A cut-off time of 45 s was used. (No tissue damage was observed by Ankier (1974) in mice remaining on a hot plate kept at 59°C for up to 60 s). Animals were placed on the hot plate after the second measurement of body temperature. Symbols are the same as for Figure 12.

Of particular interest is the observation that the regulation of adenylate cyclase by cannabinoids requires a functional G_i protein: it was shown that treatment of intact cells or membranes with pertussis toxin blocked the cannabinoid response. An endogenous cannabinoid ligand would therefore be expected to inhibit adenylate cyclase. Indeed, Vogel *et al.* (1993) have fully validated this prediction. First, they showed that specific binding of [^{3}H]HU-243 was obtained with membranes from rat cerebellum, $N_{18}TG_2$ mouse neuroblastoma cells and cells transfected with the rat cannabinoid receptor DNA. No binding was observed with membranes from non-transfected cells. Anandamide was shown to compete efficiently with the above binding in a concentration-dependent manner. The amount of binding of [^{3}H]HU-243 displaced by 10 μM anandamide was equivalent to 1.30 pmol/mg protein for cerebellar membrane.

Vogel *et al.* (1993) then showed that HU-210 and anandamide inhibited forskolin-stimulated adenylate cyclase in $N_{18}TG_2$ cells. The same effect was observed with transfected cells but not with non-transfected cells. The activity recorded for Δ^9-THC, HU-210 and anandamide was IC_{50} 9.1, 0.9 and 201 nM respectively. The (+)-enantiomers, (+)-Δ^9-THC and (+)HU-211 which exhibit no cannabimimetic activity had essentially no effect on adenylate cyclase. The inhibi-

Table 1 Effects of anandamide in Sabra and C57BL/6 male and female mice

Strain/sex	Open field (number of squares)	Ring (% immobility)	Time on hot plate (s)
Sabra Male			
Vehicle	154±19	21±10.0	10±2.9
Anandamide	50±10*	34±7.4	14±5.5
Sabra Female			
Vehicle	95±14	26±4.7	8±1.2
Anandamide	32±8*	46±7.8*	18±3.6*
C57BL/6HsD Male			
Vehicle	170±24	18±5.6	8±1.2
Anandamide	86±21*	46±9.5*	16±7.3
C57BL/6HsD Female			
Vehicle	181±20	22±6.4	7±0.5
Anandamide	63±11*	63±7.3*	12±1.4

A dose of 10 mg/kg of anandamide was administered i.p. Each group consisted of n=5. Two-way ANOVA yielded significant drug effects ($P<0.0001$) for all parameters studied ±S.E. (not shown: rearing and number of faecal pellets in open field and change in body temperature).
* $P<0.05$.

tion of adenylate cyclase was mediated by pertussis toxin-sensitive GTP-binding protein, as this toxin completely blocked the above effects of both the cannabinoids and anandamide. The high IC_{50} value recorded for anandamide is probably a result of enzymatic breakdown. Felder *et al.* (1993) have recorded similar findings. They also found inhibition of N-type calcium channels in N-18 neuroblastoma cells. This inhibition is voltage-dependent (Mackie *et al.*, 1993). It is of some interest that anandamide differs from WIN 55,212–2 and CP 55,940 as an inhibitor of N-type calcium channels in that it behaves as a partial agonist. Further evidence that anandamide inhibits calcium channels comes from the experiments of Johnson *et al.* (1993) who found that brain extracts contain a substance that inhibits binding of the calcium channel antagonist, 1,4-dihydropyridine, and were able to identify anandamide as the active constituent.

Pertwee *et al.* (1993) have shown that cross-tolerance between Δ^9-THC and anandamide occurs in the mouse isolated vas deferens, and Wickens and Pertwee (1993) found that both Δ^9-THC and anandamide enhance the ability of muscimol to induce catalepsy in the globus pallidus of rats.

Weidenfeld *et al.* (1994) have found that intracerebroventricular injection of anandamide or THC (50 or 150 μg per rat) increased significantly the serum levels of ACTH and corticosterone, while depleting CRF in the median eminence.

These data suggest that anandamide, like THC, is capable of activating the hypothalamo–pituitary–adrenal axis.

Snowman and Snyder (1993) explored the effects of anandamide on several non-cannabinoid receptors and obtained no significant inhibition at 10 μM concentration. The receptors (with the labelled probe indicated in parenthesis) were: α_1-adrenoceptors (^{3}H-prazosin), β-adrenoceptors ([^{3}H]dihydroalprenolol), muscarinic cholinoceptors ([^{3}H]quinuclidinyl benzilate), $GABA_A$ receptors ([^{3}H]muscimol), benzodiazepine receptors ([^{3}H]flunitrazepam and [^{3}H]PK11195) and glutamate NMDA receptors ([^{3}H]MK801). Anandamide also failed to displace [^{3}H]WIN 35–428 from dopamine transporter binding sites.

The above results give support to the assumption that anandamide is indeed a specific endogenous mediator of cannabinoid action.

8.7 Conclusion

The identification of a brain constituent which acts as a ligand for the cannabinoid receptor will undoubtedly make possible a deeper understanding of cannabinoid action and may open a path to the establishment of a new neurochemical system. Numerous basic problems remain to be solved. The first ones to come to mind are: (1) analytical procedures for the quantitative determination of anandamide in body fluids; (2) the presence of additional anandamide-type compounds; (3) the levels of anandamide under different physiological conditions – in stress, at rest, etc.; (4) the conditions for release from and uptake of anandamide into cells; (5) the relation, if any, of anandamide to memory, coordination of movements and emotions; (6) the possibility that lack of, or excess of, anandamide may cause a disease, presumably neurological or mental character; (7) the elucidation of the biosynthesis of and the determination of sensitivity to anandamide from the foetal stage through to adulthood.

Little is known about the metabolism of anandamide. Deutsch and Chin (1993) have found that anandamide is enzymatically hydrolysed by an amidase to arachidonic acid, and presumably ethanolamine. This reaction can be blocked by phenylmethylsulfonyl fluoride (PMSF). The rapid degradation of anandamide by amidase may explain the considerable diversity of binding activity recorded under experimental conditions in which variable amounts of protein, and hence of the enzyme, are presumably present. The metabolic pathways of anandamide may not be confined to hydrolysis. One can visualize the existence of routes that proceed via leukotriene or prostaglandin-type compounds, thus following the arachidonic acid example, or via oxidation of the primary alcohol to a carboxyl moiety, thus forming the amide of glycine with arachidonic acid – a compound with potentially a considerable attraction to the neurochemist.

The above problems are related to anandamide-type compounds alone. Are there additional ligands which are not ethanolamides of unsaturated fatty acids? Do the non-anandamide type natural products investigated by Nye *et al.* (1988) and Childers and Pacheco (see Martin *et al.*, 1993) represent specific ligands for the cannabinoid receptor? Is the unidentified material with cannabinoid-receptor binding activity reported by Evans *et al.* (1992) identical or related to anandamide? These problems will, presumably, be solved within a few years. Our intuitive assumption is that the specific ligands of the brain cannabinoid receptor will all be ethanolamides, or amides of amino acids. Such amides may be formed not only with fatty acids but also with leukotrienes and prostaglandins. This line of reasoning follows the logic of the standard biological routine of having only one type of natural mediator for an individual receptor. However, one can expect endogenous ligands of other types to bind to the peripheral (CB_2) receptor.

Thousands of publications on the endogenous opiates have been published since their discovery. We may expect a similar avalanche of research on the endogenous cannabinoid(s).

Acknowledgements

The contributions made by the department of Natural Products are the results of enthusiasm and hard work of many collaborators and students. In the anandamide project, Dr W. Devane, a postdoctoral fellow and Dr L. Hanuš, a visiting scientist from the Czech Republic, should be singled out; they devoted nearly 3 years of hard work on the isolation and purification of anandamide. We collaborated with numerous colleagues outside our department, in this country and abroad. Their knowledge, techniques and, above all, their advice made the anandamide work a pleasure. Any credit should be aportioned equally to all collaborators.

The work in Jerusalem was supported by NIDA, by the Amdursky Fund and the Brettler Center for Medical Research at the Hebrew University, by the Lady Davis Foundation (stipend to Dr W. Devane) and by the Israeli Antidrug Authority.

References

Ankier, S.I. (1974) New hot plate tests to quantify antinociceptive and narcotic antagonist activities. *Eur. J. Pharmacol.* **27**, 1–4.

Casey, A.F. & Parfitt, R.T. (1986) *Opioid Analgesics, Chemistry and Receptors.* Plenum Press, New York.

Christie, W.W. (1982) *Lipid Analysis*, pp. 81–82. Pergamon, Oxford.

Deutsch, D.G. & Chin, S.A. (1993) Enzymatic synthesis and degradation of anandamide, a cannabinoid receptor agonist. *Biochem. Pharmacol.* **46**, 791–796.

Devane, W.A., Dysarz, F.A. III., Johnson, M.R., Melvin, L.S. & Howlett, A.C. (1988) Determination and characterization of a cannabinoid receptor in rat brain. *Mol. Pharmacol.* **34**, 605–613.

Devane, W.A., Breuer, A., Sheskin, T., Järbe, T.U.C., Eisen, M.S. & Mechoulam, R. (1992a) A novel probe for the cannabinoid receptor. *J. Med. Chem.* **35**, 2065–2069.

Devane, W.A., Hanuš, L., Breuer, A., Pertwee, R.G., Stevenson, L.A., Griffin, G., Gibson, D., Mandelbaum, A., Etinger, A. & Mechoulam, R. (1992b) Isolation and structure of a brain constituent that binds to the cannabinoid receptor. *Science* **258**, 1946–1949.

Dewey, W.L., Martin, B.R. & May, E.L. (1984) Cannabinoid stereoisomers: pharmacological effects. In *Handbook of Stereoisomers: Drugs in Psychopharmacology* (ed. Smith, D.F.), pp. 317–326. CRC Press, Boca Raton, FL.

Eddy, N.B., Touchberry, C.F. & Lieberman, J.E. (1953) Synthetic analgesics I. Methadone isomers and derivatives. *J. Pharmacol.* **98**, 121–137.

Edery, H., Grunfeld, Y., Ben-Zvi, Z. & Mechoulam, R. (1971) Structural requirements for cannabinoid activity. *Ann. NY Acad. Sci.* **191**, 40–53.

Evans, D.M., Johnson, M.R. & Howlett, A.C. (1992) Ca^{2+}-dependent release from rat brain of cannabinoid receptor binding activity. *J. Neurochem.* **58**, 780–782.

Felder, C.C., Veluz, J.S., Williams, H.L., Briley, E.M. & Matsuda, L.A. (1992) Cannabinoid agonists stimulate both receptor-mediated and nonreceptor-mediated signal transduction pathways in cells transfected with and expressing cannabinoid receptor clones. *Mol. Pharmacol.* **42**, 838–845.

Felder, C.C., Briley, E.M., Axelrod, J., Simpson, J.T., Mackie, K. & Devane, W.A. (1993) Anandamide, an endogenous cannabimimetic eicosanoid, binds to the cloned human cannabinoid receptor and stimulates receptor-mediated signal transduction. *Proc. Natl Acad. Sci. USA* **90**, 7656–7660.

Fride, E. & Mechoulam, R. (1993) Pharmacological activity of the cannabinoid receptor agonist, anandamide, a brain constituent. *Eur. J. Pharmacol.* **231**, 313–314.

Gaoni, Y. & Mechoulam, R. (1964) Isolation, structure and partial synthesis of an active constituent of hashish. *J. Amer. Chem. Soc.* **86**, 1646–1647.

Gaoni, Y. & Mechoulam, R. (1966) The isomerization of cannabidiol to tetrahydrocannabinols. *Tetrahedron* **22**, 1481–1488.

Gates, M. & Tschudi, G. (1952) The synthesis of morphine. *J. Am. Chem. Soc.* **74**, 1109–1110.

Gill, E.W. & Lawrence, D.K. (1976) The physicochemical mode of action of tetrahydrocannabinol on cell membranes. In: *Pharmacology of Marihuana* (eds. Braude, M.C. & Szara, S.), pp. 147–155. Raven Press, New York.

Gulland, J.M. & Robinson, R. (1923) The morphine group. Part I. A discussion of the constitutional problem. *J. Chem. Soc.* **123**, 980–998.

Hanuš, L., Gopher, A., Almog, S. & Mechoulam, R. (1993) Two new unsaturated fatty acid ethanolamides in brain that bind to the cannabinoid receptor. *J. Med. Chem.* **36**, 3032–3034.

Howlett, A.C. (1987) Cannabinoid inhibition of adenylate cyclase: relative activity of constituents and metabolites of marihuana. *Neuropharmacology*, **26**, 507–512.

Howlett, A.C., Champion, T.M., Wilken, G.H. & Mechoulam, R. (1990) Stereochemical effects of 11-OH-delta-8-tetrahydrocannabinol-dimethylheptyl to inhibit adenylate cyclase and bind to the cannabinoid receptor. *Neuropharmacology* **29**, 161–165.

Howlett, A.C., Evans, D.M. & Houston, D.B. (1992) The cannabinoid receptor. In *Marijuana/Cannabinoids. Neurobiology and Neurophysiology* (eds. Murphy, L. & Bartke, A.), pp. 35–72. CRC Press, Boca Raton, FL.

Järbe, T.U.C., Hiltunen, A.J., Lander, N. & Mechoulam, R. (1986) Cannabimimetic activity (Δ^1-THC cue) of cannabidiol monomethyl ether and two stereoisomeric hexahydrocannabinols in rats and pigeons. *Pharmacol. Biochem. Behav.* **25**, 393–399.

Järbe, T.U.C., Hiltunen, A.J. & Mechoulam, R. (1989) Stereospecificity of the discriminative stimulus functions of the dimethylheptyl homologs of 11-hydroxy-delta-8-tetrahydrocannabinol in rats and pigeons. *J. Pharmacol. Exp. Therap.* **250**, 1000–1005.

Järbe, T.U.C., Mechoulam, R. & Zahalka, J. (1994) Discriminative stimulus and open field effects of the enantiomers of 11-hydroxy-delta-8-tetrahydrocannabinol in pigeons and gerbils. *Pharmacol. Biochem. Behav.* **47**, 113–119.

Johnson, D.E., Heald, S.L., Dally, R.D. & Janis, R.A. (1993) Isolation, identification and synthesis of an endogenous arachidonic amide that inhibits calcium channel antagonist 1,4-dihydropyridine binding. *Prostaglandins Leukotrienes, and Ess. Fatty Acids* **48**, 429–437.

Lawrence, D.K. & Gill, E.W. (1975) The effects of delta-1-tetrahydrocannabinol and other cannabinoids on spin-labeled liposomes and their relationship to mechanisms of general anesthesia. *Mol. Pharmacol.* **11**, 595–602.

Leite, J.R., Carlini, E.A., Lander, N. & Mechoulam, R. (1982) Anticonvulsant effects of (−) and (+) isomers of CBD and their dimethylheptyl homologs. *Pharmacology* **24**, 141–146.

Little, P.J., Compton, D.R., Johnson, M.R., Melvin, L.S. & Martin, B.R. (1988) Pharmacology and stereoselectivity of structurally novel cannabinoids in mice. *J. Pharmacol. Exp. Ther.* **247**, 1046–1051.

Little, P.J., Compton, D.R., Mechoulam, R. & Martin, B.R. (1989) Stereochemical effects of 11-OH-delta-8-THC-dimethylheptyl in mice and dogs. *Pharmacol. Biochem. Behav.* **32**, 661–666.

Mackie, K., Devane, W.A. & Hille, B. (1993) Anandamide, an endogenous cannabinoid, inhibits calcium currents as a partial agonist in N18 neuroblastoma cells. *Mol. Pharmacol.* **44**, 498–503.

Martin, B.R. (1986) Cellular effects of cannabinoids. *Pharmacol. Rev.* **38**, 45–74.

Martin, B.R., Compton, D.R., Thomas, B.F., Prescott, W.R., Little, P.J., Razdan, R.K., Johnson, M.R., Melvin, L.S., Mechoulam, R. & Ward, S.J. (1991) Behavioural, biochemical, and molecular modeling evaluations of cannabinoid analogs. *Pharmacol. Biochem. Behav.* **40**, 471–478.

Martin, B.R., Cabral, G., Childers, S.R., Deadwyler, S., Mechoulam, R. & Reggio, P. (1993) International Cannabis Research Society Meeting Summary, Keystone, CO (June 19–20, 1992). *Drug and Alcohol Dependence* **31**, 219–227.

Mechoulam, R., Braun, P. & Gaoni, Y. (1967) A stereospecific synthesis of (−)-Δ^1 and (−)-Δ^6-tetrahydrocannabinols. *J. Am. Chem. Soc.* **89**, 4552–4554.

Mechoulam, R., Lander, N., Varkony, T.H., Kimmel, I., Becker, O., Ben-Zvi, Z., Edery, H. & Porath, G. (1980) Stereochemical requirements for cannabinoid activity. *J. Med. Chem.* **23**, 1068–1072.

Mechoulam, R., Feigenbaum, J.J., Lander, N., Segal, M., Järbe, T.U.C., Hiltunen, A.J. & Consroe, P. (1988) Enantiomeric cannabinoids: stereospecificity of psychotropic activity. *Experientia* **44**, 762–764.

Mechoulam, R., Lander, N., Breuer, A. & Zahalka, J. (1990) Synthesis of the individual, pharmacologically distinct, enantiomers of a tetrahydrocannabinol derivative. *Tetrahedron (Asymmetry)* **1**, 315–319.

Mechoulam, R., Devane, W.A. & Glaser, R. (1992) Cannabinoid geometry and biological activity. In *Marijuana/Cannabinoids. Neurobiology and Neurophysiology* (eds Murphy, L. & Bartke, A.), pp. 1–33. CRC Press, Boca Raton, FL.

Munro, S., Thomas, K.L. & Abu-Shaar, M. (1993) Molecular characterization of a peripheral receptor for cannabinoids. *Nature* **365**, 61–65.

Nye, J.S., Seltzman, H.H. Pitt, C.G. & Snyder, S.H. (1985) High-affinity cannabinoid binding sites in brain membranes labeled with [^{3}H]-5′-trimethylammonium-Δ^8-tetrahydrocannabinol. *J. Pharmacol. Exp. Ther.* **234**, 784–791.

Nye, J.S., Voglmaier, S., Martenson, R.E. & Snyder, S.H. (1988) Myelin basic protein is an endogenous inhibitor of the high-affinity cannabinoid binding site in brain. *J. Neurochem.* **50**, 1170–1178.

Paton, W.D.M. (1975) Pharmacology of marijuana. *Annu. Rev. Pharmacol.* **15**, 191–220.

Pertwee, R.G. (1972) The ring test: a quantitative method for assessing the 'cataleptic' effect of cannabis in mice. *Br. J. Pharmacol.* **46**, 753–763.

Pertwee, R.G. (1988) The central neuropharmacology of psychotropic cannabinoids. *Pharmacol. Ther.* **36**, 189–261.

Pertwee, R.G., Stevenson, L.A., Elrick, D.B., Mechoulam, R. & Corbett, A.D. (1992) Inhibitory effects of certain enantiomeric cannabinoids in the mouse vas deferens and the myenteric plexus preparation of guinea-pig small intestine. *Brit. J. Pharmacol.* **105**, 980–984.

Pertwee, R.G., Stevenson, L.A. & Griffin, G. (1993) Cross-tolerance between delta-9-tetrahydrocannabinol and the cannabimimetic agents, CP 55,940, WIN 55,212–2 and anandamide. *Br. J. Pharmacol.* **110**, 1483–1490.

Razdan, R.K. (1986) Structure–activity relationships in cannabinoids. *Pharmacol. Rev.* **38**, 75–149.

Smith, P.B., Compton, D.R., Welch, S.P., Razdan, R.K., Mechoulam, R. & Martin, B.R. (1994) The pharmacological activity of anandamide, a putative endogenous cannabinoid, in mice. *J. Pharmacol. Exp. Therap.* **270**, 219–227.

Snowman, A. & Snyder, S.H. (1993) Private communication.

Thomas, B.F., Compton, D.R. & Martin, B.R. (1990) Characterization of the lipophilicity of natural and synthetic analogs of delta-9-tetrahydrocannabinol and its relationship to pharmacological potency. *J. Pharmacol. Exp. Therap.* **255**, 624–630.

Vogel, Z., Barg, J., Levy, R., Saya, D., Heldman, E. & Mechoulam, R. (1993) Anandamide, a brain endogenous compound, interacts specifically with cannabinoid receptors and inhibits adenylate cyclase. *J. Neurochem.* **61**, 352–355.

Weidenfeld, J., Feldman, S. & Mechoulam, R. (1994) The effect of the brain constituent anandamide, a cannabinoid receptor agonist, on the hypotholamo-pituitary-adrenal axis in the rat. *Neuroendrocinol.* **59**, 110–112.

Wickens, A.P. & Pertwee, R.G. (1993) Delta-9-tetrahydrocannabinol and anandamide enhance the ability of muscimol to induce catalepsy in the globus pallidus of rats. *Eur. J. Pharmacol.* **250**, 205–208.

Index